T0252380

A Guide to Sample Size for Animal-based Studies

Penny S. Reynolds

Department of Anesthesiology, College of Medicine
Department of Small Animal Clinical Sciences
College of Veterinary Medicine
University of Florida, Gainesville
Florida, USA

WILEY Blackwell

This edition first published 2024
© 2024 John Wiley & Sons Ltd

All rights reserved. No part of this publication may be reproduced, stored in a retrieval system, or transmitted, in any form or by any means, electronic, mechanical, photocopying, recording or otherwise, except as permitted by law. Advice on how to obtain permission to reuse material from this title is available at http://www.wiley.com/go/permissions.

The right of Penny S. Reynolds to be identified as the author of this work has been asserted in accordance with law.

Registered Offices
John Wiley & Sons, Inc., 111 River Street, Hoboken, NJ 07030, USA
John Wiley & Sons Ltd, The Atrium, Southern Gate, Chichester, West Sussex, PO19 8SQ, UK

For details of our global editorial offices, customer services, and more information about Wiley products visit us at www.wiley.com.

Wiley also publishes its books in a variety of electronic formats and by print-on-demand. Some content that appears in standard print versions of this book may not be available in other formats.

Trademarks: Wiley and the Wiley logo are trademarks or registered trademarks of John Wiley & Sons, Inc. and/or its affiliates in the United States and other countries and may not be used without written permission. All other trademarks are the property of their respective owners. John Wiley & Sons, Inc. is not associated with any product or vendor mentioned in this book.

Limit of Liability/Disclaimer of Warranty
While the publisher and authors have used their best efforts in preparing this work, they make no representations or warranties with respect to the accuracy or completeness of the contents of this work and specifically disclaim all warranties, including without limitation any implied warranties of merchantability or fitness for a particular purpose. No warranty may be created or extended by sales representatives, written sales materials or promotional statements for this work. This work is sold with the understanding that the publisher is not engaged in rendering professional services. The advice and strategies contained herein may not be suitable for your situation. You should consult with a specialist where appropriate. The fact that an organization, website, or product is referred to in this work as a citation and/or potential source of further information does not mean that the publisher and authors endorse the information or services the organization, website, or product may provide or recommendations it may make. Further, readers should be aware that websites listed in this work may have changed or disappeared between when this work was written and when it is read. Neither the publisher nor authors shall be liable for any loss of profit or any other commercial damages, including but not limited to special, incidental, consequential, or other damages.

Library of Congress Cataloging-in-Publication Data applied for

Paperback ISBN: 9781119799979

Cover Design: Wiley
Cover Images: © MOLEKUUL/SCIENCE PHOTO LIBRARY/Getty Images; MOLEKUUL/SCIENCE PHOTO LIBRARY/Getty Images; Verin/Shutterstock; n.tati.m/Shutterstock; Mariia Zotova/Getty Images; RF Pictures/Getty Images

Set in 11.5/13.5pt STIXTwoText by Straive, Pondicherry, India

To Nyx, Mel, Finnegan, and Fat Boy Higgins
Holly, Molly, and Abby
and all their nameless, uncounted kindred
who have done so much to advance science and medicine

Contents

Preface

'How large a sample size do I need for my study'? Although one of the most commonly asked questions in statistics, the importance of proper sample size estimation seems to be overlooked by many preclinical researchers. Over the past two decades, numerous reviews of the published literature indicate many studies are too small to answer the research question and results are too unreliable to be trusted. Few published studies present adequate justification of their chosen sample sizes or even report the total number of animals used. On the other hand, it is not unusual for protocols (usually those involving mouse models) to request preposterous numbers of animals, sometimes in the tens or even hundreds of thousands, 'because this is an exploratory study, so it is unknown how many animals we will require'.

This widespread phenomenon of sample sizes based on nothing more than guesswork or intuition illustrates the pervasiveness of what Amos Tversky and Daniel Kahneman identified in 1971 as the 'belief in the law of small numbers'. Researchers overwhelmingly rely on best judgement in planning experiments, but judgement is almost always misleading. Researchers choose sample sizes based on what 'worked' before or because a particular sample size is a favourite with the research community. Tversky and Kahneman showed that researchers who gamble their research results on small intuitively-based samples consistently have the odds stacked against their findings (even if results are true). They overestimate the stability and precision of their results, and fail to account for sampling variation as a possible reason for observed pattern. The result is research waste on a colossal scale, especially of animals, that is increasingly difficult to justify.

This book was written to assist non-statisticians who use animals in research to 'right-size' experiments, so they are statistically, operationally, and ethically justifiable. A 'right-sized' experiment has a clear plan for sample size justification and transparently reports the numbers of all animals used in the study. For basic and veterinary researchers, appropriate sample sizes are critical to the design and analysis of a study. The best sample sizes optimise study design to align with available resources and ensure the study is adequately powered to detect meaningful, reliable, and generalisable results. Other stakeholders not directly involved in animal experimentation can also benefit from understanding the basic principles involved. Oversight veterinarians and ethical oversight committees are responsible for appraising animal research protocols for compliance with best practice, ethical, and regulatory standards. An appreciation of sample size construction can help assess scientific and ethical justifications for animal use and whether the proposed sample size is fit for purpose. Funding agencies and policymakers use research results to inform decisions related to animal welfare, public health, and future scientific benefit. Understanding the logic behind sample size justification can assist in evaluation of study quality and reliability of research findings, and ultimately promote more informed evidence-based decision-making.

An extensive background in statistics is not required, but readers should have had some basic statistical training. The emphasis throughout is on the upstream components of the research process – statistical process, study planning, and sample size calculations rather than analysis. I have used real data in nearly all examples and provided formulae and code, so sample size approximations can be reproduced by hand or by computer. By training and inclination I prefer SAS, but whenever possible I have provided R code or links to R libraries.

Acknowledgements

Many thanks to Anton Bespalov (PAASP, Heidelberg, Germany); Cori Astrom, Christina Hendricks, and Bryan Penberthy (University of Florida); Cora Mezger, Maria Christodoulou, and Mariagrazia Zottoli (Department of Statistics, University of Oxford); and Megan Lafollette (North American 3Rs Collaborative), who kindly reviewed various chapters of this book whilst it was in preparation and provided much helpful feedback. Thanks to the University of Florida IACUC chairs Dan Brown and Rebecca Kimball, who encouraged researchers to consult the original 10-page handout I had devised for sample size estimation. And last, but certainly not least, special thanks to Tim Morey, Chair of the Department of Anesthesiology, University of Florida, who encouraged me to put that handout into book form.

Thanks are also due to the University of Florida Faculty Endowment Fund for providing me with a Faculty Enhancement Opportunities grant to allow me to devote some concentrated time to writing. A generous honorarium from Scientist Center for Animal Welfare (SCAW) and an award from the UK Animals in Science Education Trust enabled me to upgrade my home computer system, making working on this project immeasurably easier.

The book was nearing completion when I came across the Icelandic word *sprakkar* that means 'extraordinary women'. I have been fortunate to encounter many *sprakkar* whilst writing this book. In addition to the women (and men!) already mentioned, special thanks to researchers Amara Estrada, Francesca Griffin, Autumn Harris, Maggie Hull, Wendy Mandese, and Elizabeth Nunamaker, who generously allowed me to use some of their data as examples. And special thanks to Jane Buck and Julie Laskaris for their wonderful friendship and hospitality over the years. Jane Buck, Professor Emerita of Psychology, Delaware State University, and past president of the American Association of University Professors, continues to amaze and show what is possible for a statistician 'with attitude'. Julie advised me that the only approach to properly edit one's own work on a book-length project was to 'slit its throat', then told me to do as she said, not as she actually did. Cheers.

What is Sample Size?

1 The Sample Size Problem in Animal-Based Research

CHAPTER OUTLINE HEAD

Good Numbers Matter. This is especially true when animals are research subjects. Researchers are responsible for minimising both direct harms to research animals and the indirect harms that result from wasting animals in poor-quality studies (Reynolds 2021). The ethical use of animals in research is framed by the 'Three Rs' principles of Replacement, Reduction, and Refinement. Originating over 60 years ago (Russell and Burch 1959), the 3Rs strategy is framed by the premise that maximal information should be obtained for minimal harms. Harms are minimised by Replacement, methods or technologies that substitute for animals; Reduction, the methods using the fewest animals for the most robust and scientifically valid information; and Refinement, the methods that improve animal welfare through minimising pain, suffering, distress, and other harms (Graham and Prescott 2015).

The focus of this book is on Reduction and methods of 'right-sizing' experiments. A right-sized experiment is an optimal size for a study to achieve its objectives with the least amount of resources, including animals. However, simply minimising the total number of animals is not the same as right-sizing. A right-sized experiment has a sample size that is statistically, operationally, and ethically defensible (Box 1.1). This will mean compromising between the scientific objectives of the study, production of scientifically valid results, availability

---

BOX 1.1
Right-Sizing Checklist

Statistically defensible: Are numbers verifiable? (Calculations)

- Outcome variable identified
- Difference to be detected
- Expected variation in response
- Number of groups
- Anticipated statistical test (if hypothesis tests used)
- All calculations shown

Operationally defensible: Are numbers feasible? (Resources)

- Qualified technical staff
- Time
- Space
- Resources
- Equipment
- Funding

Ethically defensible: Are numbers fit for purpose? (3Rs)

- Appropriate for study objectives?
- Reasonable number of groups?
- Are collateral losses accounted for and minimized?
- Are loss mitigation plans described?
- Are 3Rs strategies described?

Source: Adapted from Reynolds (2021).

A Guide to Sample Size for Animal-based Studies, First Edition. Penny S. Reynolds.
© 2024 John Wiley & Sons Ltd. Published 2024 by John Wiley & Sons Ltd.

of resources, and the ethical requirement to minimise waste and suffering of research animals. Thus, sample size calculations are not a single calculation but a set of calculations, involving iteration through formal estimates, followed by reality checks for feasibility and ethical constraints (Reynolds 2019).

Additional challenges to right-sizing experiments include those imposed by experimental design and biological variability (Box 1.2). In *The Principles of Humane Experimental Technique* (1959), Russell and Burch were very clear that Reduction is achieved by systematic strategies of experimentation rather than trial and error. In particular, they emphasised the role of the statistically based family of experimental designs and design principles proposed by Ronald Fisher, still relatively new at the time. Formal experimental designs customised to address the particular research question increase the experimental signal through the reduction of variation. Design principles that reduce bias, such as randomisation and allocation concealment (blinding) increase validity. These methods increase the amount of usable information that can be obtained from each animal (Parker and Browne 2014).

Although it has now been almost a century since Fisher-type designs were developed many researchers in biomedical sciences still seem

unaware of their existence. Many preclinical studies reported in the literature consist of numerous two-group designs. However, this approach is both inefficient and inflexible and unsuited to exploratory studies with multiple explanatory variables (Reynolds 2022). Statistically based designs are rarely reported in the preclinical literature. In part, this is because the design of experiments is seldom taught in introductory statistics courses directed towards biomedical researchers.

Power calculations are the gold standard for sample size justification. However, they are commonly misapplied, with little or no consideration of study design, type of outcome variable, or the purpose of the study. The most common power calculation is for two-group comparisons of independent samples. However, this is inappropriate when the study is intended to examine multiple independent factors and interactions. Power calculations for continuous variables are not appropriate for correlated observations or count data with high prevalence of zeros. Power calculations cannot be used at all when statistical inference is not the purpose of the study, for example, assessment of operational and ethical feasibility, descriptive or natural history studies, and species inventories.

Evidence of right-sizing is provided by a clear plan for sample size justification and transparent reporting of the number of all animals used in the study. This is why these items are part of best-practice reporting standards for animal research publications (Kilkenny et al. 2010, Percie du Sert et al. 2020 and are essential for the assessment of research reproducibility (Vollert et al. 2020). Unfortunately, there is little evidence that either sample size justification or sample size reporting has improved over the past decade. Most published animal research studies are underpowered and biased (Button et al. 2013, Henderson et al. 2013, Macleod et al. 2015) with poor validity (Würbel 2017, Sena and Currie 2019), severely limiting reproducibility and translation potential (Sena et al. 2010, Silverman et al. 2017). A recent cross-sectional survey of mouse cancer model papers published in high-impact oncology journals found that fewer than 2% reported formal power calculations, and less than one-third reported sample size per group. It was impossible to determine attrition losses, or how many experiments (and therefore animals) were discarded due to failure to achieve statistical

> **BOX 1.2**
> *Challenges for Right-Sizing Animal-Based Studies*
>
> *Ethics and welfare considerations.* The three Rs Replacement, Reduction, and Refinement should be the primary driver of animal numbers.
> *Experimental design.* Animal-based research has no design culture. Clinical trial models are inappropriate for exploratory research. Multifactorial agriculture/industrial design may be more suitable in many cases, and they are unfamiliar to most researchers.
> *Biological variability.* Animals can display significant differences in responses to interventions, making it challenging to estimate an appropriate sample size.
> *Cost and resource constraints.* The financial cost of conducting animal-based research, including the cost of housing, caring for, and monitoring the animals, must be considered in estimates of sample size.

significance (Nunamaker and Reynolds 2022). The most common sample size mistake is not performing any calculations at all (Fosgate 2009). Instead, researchers make vague and unsubstantiated statements such as 'Sample size was chosen because it is what everyone else uses' or 'experience has shown this is the number needed for statistical significance'. Researchers often game, or otherwise adjust, calculations to obtain a preferred sample size (Schultz and Grimes 2005, Fitzpatrick et al. 2018). In effect, these studies were performed without justification of the number of animals used.

Statistical thinking is both a mindset and a set of skills for understanding and making decisions based on data (Tong 2019). Reproducible data can only be obtained by sustained application of statistical thinking to all experimental processes: good laboratory procedure, standardised and comprehensive operating protocols, appropriate design of experiments, and methods of collecting and analysing data. Appropriate strategies of sample size justification are an essential component.

1.1 Organisation of the Book

This book is a guide to methods of approximating sample sizes. There will never be one number or approach, and sample size will be determined for the most part by study objectives and choice of the most appropriate statistically based study design. Although advanced statistical or mathematical skills are not required, readers are expected to have at least a basic course on statistical analysis methods and some familiarity with the basics of power and hypothesis testing. SAS code is provided in appendices at the end of each chapter and references to specific R packages in the text. It is strongly recommended that everyone involved in devising animal-based experiments take at least one course in the design of experiments, a topic not often covered by statistical analysis courses.

This book is organised into four sections (Figure 1.1).

Part I *Sample size basics* discusses definitions of sample size, elements of sample size determination, and strategies for maximising information power without increasing sample size.

Part II *Feasibility*. This section presents strategies for establishing study feasibility with pilot studies. Justification of animal numbers must first address questions of operational feasibility ('*Can it work?*' Is the study possible? suitable? convenient? sustainable?). Once operational logistics are standardised, pilot studies can be performed to establish empirical feasibility ('*Does it work?*' is the output large enough to be measured? consistent enough to be reliable?)

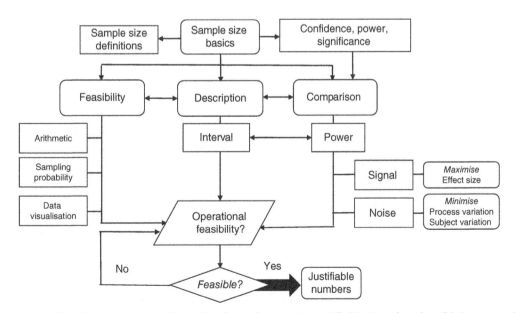

Figure 1.1: Overview of book organisation. For animal numbers to be *justifiable* (Are they feasible? appropriate? ethical? verifiable?), sample size should be determined by formal quantitative calculations (arithmetic, probability-based, precision-based, power-based) and consideration of operational constraints.

and translational feasibility ('*Will it work?*' proof of concept and proof of principle) before proceeding to the main experiments. Power calculations are not appropriate for most pilots. Instead, common-sense feasibility checks include *basic arithmetic* (with structured back-of-the-envelope calculations), simple probability-based calculations, and graphics.

Part III *Description*. This section presents methods for summarising the main features of the sample data and results. Basic descriptive statistics provide a simple and concise summary of the data in terms of central tendency and dispersion or spread. Graphical representations are used to identify patterns and outliers and explore relationships between variables. Intervals computed from the sample data are the range of values estimated to contain the true value of a population parameter with a certain degree of confidence. Four types of intervals are discussed: confidence intervals, prediction, intervals, tolerance intervals, and reference intervals. Intervals shift emphasis away from significance tests and *P*-values to more meaningful interpretation of results.

Part IV *Comparisons*. Power-based calculations for sample size are centred on understanding effect size in the context of specific experimental designs and the choice of outcome variables. Effect size provides information about the practical significance of the results beyond considerations of statistical significance. Specific designs considered are two-group comparisons, ANOVA-type designs, and hierarchical designs.

References

Button, K.S., Ioannidis, J.P.A., Mokrysz, C. et al. (2013). Power failure: why small sample size undermines the reliability of neuroscience. *Nature Reviews Neuroscience* 14: 365–376.

Fitzpatrick, B.G., Koustova, E., and Wang, Y. (2018). Getting personal with the "reproducibility crisis": interviews in the animal research community. *Lab Animal (NY)* 47: 175–177.

Fosgate, G.T. (2009). Practical sample size calculations for surveillance and diagnostic investigations. *Journal of Veterinary Diagnostic Investigation* 21: 3–14. https://doi.org/10.1177/104063870902100102.

Graham, M.L. and Prescott, M.J. (2015). The multifactorial role of the 3Rs in shifting the harm-benefit analysis in animal models of disease. *European Journal of Pharmacology* 759: 19–29. https://doi.org/10.1016/j.ejphar.2015.03.040.

Henderson, V.C., Kimmelman, J., Fergusson, D. et al. (2013). Threats to validity in the design and conduct of preclinical efficacy studies: a systematic review of guidelines for in vivo animal experiments. *PLoS Medicine* 10: e1001489.

Kilkenny, C., Browne, W.J., Cuthill, I.C. et al. (2010). Improving bioscience research reporting: the ARRIVE guidelines for reporting animal research. *PLoS Biology* 8 (6): e1000412. https://doi.org/10.1371/journal.pbio.1000412.

Macleod, M.R., Lawson McLean, A., Kyriakopoulou, A. et al. (2015). Risk of bias in reports of *in vivo* research: a focus for improvement. *PLoS Biology* 13: e1002301. https://doi.org/10.1371/journal.pbio.1002273.

Nunamaker, E.A. and Reynolds, P.S. (2022). "Invisible actors"—how poor methodology reporting compromises mouse models of oncology: a cross-sectional survey. *PLoS ONE* 17 (10): e0274738. https://doi.org/10.1371/journal.pone.0274738.

Parker, R.M.A. and Browne, W.J. (2014). The place of experimental design and statistics in the 3Rs. *ILAR Journal* 55 (3): 477–485.

Percie du Sert, N., Hurst, V., Ahluwalia, A. et al. (2020). The ARRIVE guidelines 2.0: updated guidelines for reporting animal research. *PLoS Biology* 18 (7): e3000410. https://doi.org/10.1371/journal.pbio.3000410.

Reynolds, P.S. (2019). When power calculations won't do: fermi approximation of animal numbers. *Lab Animal (NY)* 48: 249–253.

Reynolds, P.S. (2021). Statistics, statistical thinking, and the IACUC. *Lab Animal (NY)* 50 (10): 266–268. https://doi.org/10.1038/s41684-021-00832-w.

Reynolds, P.S. (2022). Between two stools: preclinical research, reproducibility, and statistical design of experiments. *BMC Research Notes* 15: 73. https://doi.org/10.1186/s13104-022-05965-w.

Russell, W.M.S. and Burch, R.L. (1959). *The Principles of Humane Experimental Technique*. London: Methuen.

Schulz, K.F. and Grimes, D.A. (2005). Sample size calculations in randomised trials: mandatory and mystical. *Lancet* 365 (9467): 1348–1353. https://doi.org/10.1016/S0140-6736(05)61034-3.

Sena, E.S. and Currie, G.L. (2019). How our approaches to assessing benefits and harms can be improved. *Animal Welfare* 28: 107–115.

Sena ES, van der Worp HB, Bath PM, Howells DW, Macleod MR. Publication bias in reports of animal stroke studies leads to major overstatement of efficacy. *PLoS Biology*, 2010 8(3):e1000344. https://doi.org/10.1371/journal.pbio.1000344.

Silverman, J., Macy, J., and Preisig, P. (2017). The role of the IACUC in ensuring research reproducibility. *Lab Animal (NY)* 46: 129–135.

Tong, C. (2019). Statistical inference enables bad science; statistical thinking enables good science. *American Statistician* 73: 246–261.

Vollert, J., Schenker, E., Macleod, M. et al. (2020). Systematic review of guidelines for internal validity in the design, conduct and analysis of preclinical biomedical experiments involving laboratory animals. *BMJ Open Science* 4 (1): e100046. https://doi.org/10.1136/bmjos-2019-100046.

Würbel, H. (2017). More than 3Rs: the importance of scientific validity for harm-benefit analysis of animal research. *Lab Animal* 46: 164–166.

2 Sample Size Basics

CHAPTER OUTLINE HEAD

2.1 Introduction

Investigators frequently assume 'sample size' is the same as 'the number of animals'. This is not necessarily true. Reliable sample size estimates are determined by the correct identification of the experimental units, the true unit of replication (Box 2.1). Replication of experimental units increases both precision of estimates and statistical power for testing the central hypothesis. Replicates on the same subject over time provide an estimate of time dependencies in response. Technical replicates are used to obtain an estimate of measurement error and are essential for quality control of experimental procedures. Pseudo-replication is a serious statistical error that occurs when the number of data points (evaluation units) is confused with the number of independent samples, or experimental units (Hurlbert 2009; Lazic 2010). Incorrect specification of the true sample size results in erroneous estimates of the standard error, inflated type I error rates, and increased number of false positives (Cox and Donnelly 2011). Research results will therefore be biased and misleading.

Definitions of 'replicates' and 'replication' are frequently confused in the literature, and further conflated with study replication. Planning experiments using formal statistical designs can help differentiate between the different types of replicates and sampling units, and determine which is best suited for the intended study.

> **BOX 2.1**
> *What Is Sample Size?*
>
> A *replicate* is one unit in one group.
>
> Sample size is determined by the number of replicates of the *experimental unit*.
>
> *Experimental unit*: Entire entity to which a treatment or control intervention can be independently and individually applied.
>
> *Biological replicate* is a biologically distinct and independent experimental unit.
>
> *Technical replicate* is one of multiple measurements on subsamples of the experimental unit, used to obtain an estimate of measurement error.

2.2 Experimental Unit

The *experimental unit* or *unit of analysis* is the smallest entire entity to which a treatment or control intervention can be independently and randomly

A Guide to Sample Size for Animal-based Studies, First Edition. Penny S. Reynolds.
© 2024 John Wiley & Sons Ltd. Published 2024 by John Wiley & Sons Ltd.

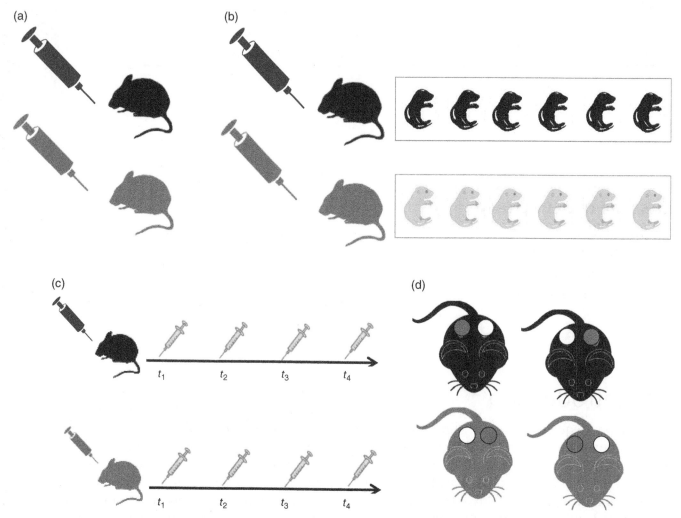

Figure 2.1: Units of replication. (a) Experimental unit = individual animal = biological unit. The entire entity to which an experimental or control intervention can be independently applied. There are two treatment interventions A or B. Here each mouse receives a separate intervention, and the individual mouse is the experimental unit (EU). The individual mouse is also the biological unit. (b). Experimental unit = groups of animals. There are two treatment interventions A or B. Each dam receives either A or B, but measurements are conducted on the pups in each litter. The experimental unit is the dam ($N = 2$), and biological unit is the pup ($n = 8$). For this design, the number of pups cannot contribute to the test of the central hypothesis. (c) Experimental unit with repeated observations. The experimental unit is the individual animal (= biological unit) with four sequential measurements made on each animal. The sample size $N = 2$. (d) Experimental unit = part of each animal. There are two treatment interventions A or B. Treatment A is randomised to either the right or left flank of each mouse, and B is injected into the opposite flank of that mouse. The experimental unit is flank ($N = 8$). The individual mouse is the biological unit. Each mouse can be considered statistically as a block with paired observations within each animal.

applied (Figure 2.1a). Cox and Donnelly (2011) define it as the 'smallest subdivision of the experimental material such that two distinct units might be randomized (randomly allocated) to different treatments.' Whatever happens to one experimental unit will have no bearing on what happens to the others (Hurlbert 2009). If the test intervention is applied to a 'grouping' other than the individual animal (e.g. a litter of mice, a cage or tank of animals, a body part; (Figure 2.1b–d)), then the sample size N will not be the same as the number of animals.

The total sample size N refers to the number of independent experimental units in the sample. The classic meaning of a 'replicate' refers to the number of experimental units within a treatment or intervention group. Therefore, replicating experimental units (and hence increasing N) contributes to statistical power for testing the central statistical hypothesis. Power calculations estimate the number of experimental units required to test the hypothesis. The assignment of treatments and controls to experimental units should be randomised if the intention is to perform statistical hypothesis tests on the data (Cox and Donnelly 2011).

Independence of experimental units is essential for most null hypothesis statistical tests and methods of analysis and is the most important condition for ensuring the validity of statistical inferences (van Belle 2008). Non-independence of experimental units occurs with repeated measures and multi-level designs and must be handled by the appropriate statistically based designs and analyses for hypothesis tests to be valid.

2.3 Biological Unit

The biological unit is the entity about which inferences are to be made. Replicates of the biological unit are the number of unique biological samples or individuals used in an experiment. Replication of biological units captures biological variability between and within these units (Lazic et al. 2018). The biological unit is not necessarily the same as the experimental unit. Depending on how the treatment intervention is randomised, the experimental

unit can be an individual biological unit, a group of biological units, a sequence of observations on a single biological unit or a part of a biological unit (Lazic and Essioux 2013; Lazic et al. 2018). The biological unit of replication may be the whole animal or a single biological sample, such as strains of mice, cell lines or tissue samples (Table 2.1).

2.4 Technical Replicates

Technical replicates or repeats are multiple measurements made on subsamples of an experimental unit (Figure 2.2). Technical replicates are used to obtain an estimate of measurement error, the difference between a measured quantity and its true value. Technical replicates are essential for assessing internal quality control of experimental procedures and processes, and ensuring that results are not an artefact of processing variation (Taylor and Posch 2014). Differences between operators and instruments, instrument drift, subjectivity in determination of measurement landmarks, or faulty calibration can result in measurement error. Cell cultures and protein-based experiments can also show considerable variation from run to run, so *in vitro* experiments are usually repeated several

Table 2.1: Units of Replication in a Hypothetical Single-Cell Gene Expression RNA Sequencing Experiment. Designating a given replicate unit as an experimental unit depends on the central hypothesis to be tested and the study design.

	Replicate 'unit'	Replicate type
Animals	Colonies	Biological
	Strains	Biological
	Cohoused animals in a cage	Biological
	Sex (male, female)	Biological
	Individuals	Biological
Sample preparation	Organs from animals killed for purpose	Biological
	Methods for dissociating cells from tissue	Technical
	Dissociation runs from given tissue sample	Technical
	Individual cells	Biological
	RNA-seq library construction	Technical
Sequencing	Runs from the library of a given cell	Technical
	Readouts from different transcript molecules	Biological or technical
	Readouts with unique molecular identifier (UMI) from a given transcript molecule	Technical

Source: Adapted from Blainey et al. (2014).

times. At least three technical replicates of Western blots, PCR measurements, or cell proliferation assays may be necessary to assess reliability of technique and confirm validity of observed changes in protein levels or gene expression (Taylor and Posch 2014).

The variance calculated from the multiple measurements is an estimate of the precision, and therefore the repeatability, of the measurement. Technical replicates measure the variability between measurements on the same experimental units. Repeating measurements increases the precision only for estimates of the measurement error; they do not measure variability either within or between treatment groups. Therefore, increasing the number of technical replicates does not improve power or contribute to the sample size for testing the central hypothesis. Analysing technical repeats as independent measurements is pseudo-replication.

High-dimensionality studies produce large amounts of output information per subject. Examples include multiple DNA/RNA microarrays; biochemistry assays; biomarker studies; proteomics; metabolomics; inflammasome profiles, etc. These studies may require a number of individual animals, either for operational purposes (for example, to obtain enough tissue for processing) or as part of the study design (for example, to estimate biological variation). Sample size will then be determined by the amount of tissue required for the assay technical replicates, or by design-specific requirements for power. Design features include anticipated

response/expression rates, expected false positive rate, and number of sampling time points (Lee and Whitmore 2002; Lin et al. 2010; Jung and Young 2012).

> ### Example: Experimental Units with Technical Replication
>
> Two treatments A and B are randomly allocated to six individually housed mice, with three mice receiving A and three receiving B. Lysates are obtained from each mouse in three separate aliquots (Figure 2.2).
>
> The individual mouse is the experimental unit because treatments can be independently and randomly allocated to each mouse. There are three subsamples or technical replicates per mouse. The total sample size is $N = 6$, with $k = 2$ treatments, $n = 3$ mice per treatment group, and $j = 3$ technical replicates per mouse. The total sample size N is 6, not 18.

2.5 Repeats, Replicates, and Pseudo-Replication

Confusion of repeats with replicates is a problem of study design, and pseudo-replication is a problem of analysis. Study validity is compromised by incorrect identification of the experimental unit. A replicate is

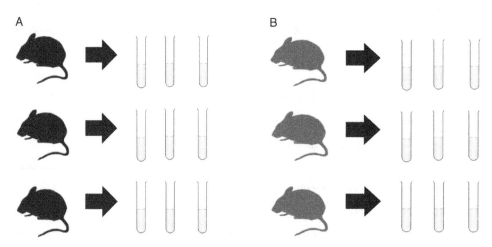

Figure 2.2: Experimental unit versus technical replicates. Two treatments A and B are randomly allocated to six mice. The individual mouse is the *experimental unit*. Three lysate aliquots are obtained from each mouse. These are *technical replicates*. The total sample size *N* is 6, not 18.

a new experimental run on a new experimental unit. Randomisation of interventions to experimental units and randomising the order in which experimental units are measured (sequence allocation randomisation) minimises the effects of systematic error or bias. A repeat is a consecutive run of the same treatment or factor combination. It does not minimise bias and may actually increase bias if there are time-dependencies in the data. Repeats are not valid replicates.

Example: Replication Versus Repeats

In Figure 2.3, the experimental units are eight mice that receive one of two interventions. In the first scenario, both treatment allocated to mouse and the measurement sequence are randomised. Bias is minimised and treatment variance

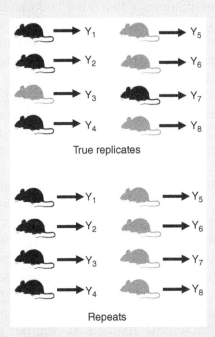

Figure 2.3: Replicates versus repeats. *True replicates* are separate runs of the same treatment on separate experimental units. Both treatment allocation to units and sequence allocation for the processing of individual experimental units are randomised. In this experiment, the eight measurements on eight mice are taken in random order. *Repeat* measurements are taken during the same experimental run or consecutive runs. Unless processing order is randomised, there will be confounding with systematic sources of variability caused by other variables that change over time. In this experiment, eight measurements on eight mice are obtained consecutively with units in the first treatment measured first.

will be appropriately estimated. In the second scenario, treatment intervention may or may not have been randomly allocated to mice, but measurements were obtained for all mice in the first group followed by those in the second group. Bias results from confounding of outcome measurements with potential time-dependencies (for example, increasing skill levels or learning) and difference in assessment, especially if treatment allocation is not concealed (blinded).

2.5.1 Repeats of Entire Experiments

A common practice in mouse studies is to repeat an entire experiment two or three times. It has been argued that this practice provides evidence that results are robust. However, NIH directives are clear that replication is justifiable only for major or key results, and that replications be independent. Repeating an experiment in triplicate by a single laboratory is not independent replication. These repeats can provide only an estimate of the overall measurement error of that experiment for that lab. A major consideration is study quality. If the study is poorly designed and underpowered, replicating it only wastes animals. Unless the purpose of direct internal replications is scientifically justified, experiments are appropriately designed and conducted to maximise internal validity, and experimental, biological, and technical replicates are clearly distinguished, simple direct repeats of experiments on whole animals are rarely ethically justifiable. Chapter 6 provides practical guidelines for experiment replication.

2.5.2 Pseudo-Replication

In a classic paper, Hurlbert (1984) defines *pseudo-replication* as occurring when inferential statistics are used 'to test for treatment effects with data from experiments where either treatments are not replicated (though samples may be) or experimental units are not statistically independent' (Hurlbert 1984, 2009). The extent of pseudo-replication in animal-based research is disturbingly prevalent. Lazic et al. (2018) reported that less than one-quarter of studies they surveyed identified the correct repli-cation unit, and almost half showed pseudo-replication, suggesting that inferences based on hypothesis tests were likely invalid.

Three of the most common types of pseudo-replication are *simple, sacrificial*, and *temporal*. Others are described in Hurlbert and White (1993) and Hurlbert (2009).

Simple pseudo-replication occurs when there is only a single replicate per treatment. There may be multiple observations, but they are not obtained from independent experimental replicates. The artificial inflation of sample size results in estimates of the standard error that are too small, contributing to increased Type I error rate and increased number of false positives.

Example: Mouse Photoperiod Exposure

A study on circadian rhythms was conducted to assess the effect of two different photoperiods on mouse wheel-running. Mice in one environmental chamber were exposed to a long photoperiod with 14 hours of light, and mice in a second chamber to a short photoperiod with 6 hours of light. There were 15 cages in each chamber with four mice per cage. What is the effective sample size?

This is simple pseudo-replication. The experimental unit is the chamber, so the effective sample size is one per treatment. Analysing the data as if there is a sample size of $n = 60$ (or even $n = 15$) per treatment is incorrect. The number of mice and cages in each chamber is irrelevant. This design implicitly assumes that chamber conditions are uniform and chamber effects are zero. However, variation both between chambers and between repeats for the same chamber can be considerable (Potvin and Tardif 1988; Hammer and Hopper 1997). Increasing sample size of mice will not remedy this situation because chamber environment is confounded with photoperiod. It is, therefore, not possible to estimate experimental error, and inferential statistics cannot be applied. Analysis should be restricted to descriptive statistics only. The study should be re-designed either to allow replication across several chambers, or if chambers are limited, as a multi-batch design replicated at two or more time points.

Sacrificial pseudo-replication occurs when there are multiple replicates within each treatment arm, the data are structured as a feature of the design (such as pairing, clustering, or nesting), but design structure is ignored in the analyses. The units are treated as independent, so the degrees of freedom for testing treatment effects are too large. Sacrificial pseudo-replication is especially common in studies with categorical outcomes when the χ^2 test or Fisher's exact test is used for analysis (Hurlbert and White 1993; Hurlbert 2009).

Example: Sunfish Foraging Preferences

Dugatkin and Wilson (1992) studied feeding success and tankmate preferences in 12 individually marked sunfish housed in two tanks. Preference was evaluated for each fish for all possible pairwise combinations of two other tankmates. There were 2 groups × 60 trials, per group × 2 replicate sets of trials, for a total of 240 observations. They concluded that feeding success was weakly but statistically significantly correlated with aggression ($P < 0.001$) based on 209 degrees of freedom, and that fish in each group strongly preferred ($P < 0.001$) the same individual in each of the two replicate preference experiments, based on 60 observations.

The actual number of experimental units is 12, with 6 fish per tank. The correct degrees of freedom for the regression analysis is 4, not 209. Suggested analyses for preference data included one-sample *t*-tests with 5 degrees of freedom or one-tailed Wilcoxon matched-pairs test with $N = 12$. Correct analyses would produce much larger *P*-values, suggesting that interpretation of these data requires substantial revision (Lombardi and Hurlbert 1996).

Temporal (or spatial) pseudo-replication occurs when multiple measurements are obtained sequentially on the same experimental units, but analysed as if they represent an individual experimental unit. Sequential observations (or repeated measures) are correlated within each individual. Repeated measures increase the precision of within-unit estimates, but the number of repeated measures do not increase the power for estimating treatment effects.

Example: Tumour Proliferation in Mouse Models of Cancer

Sequential measurements of solid tumour volume in mice are commonly reported as a measure of disease progression or response to an intervention. Mull et al. (2020) tested the effects of low-dose UCN-01 to promote survival of tumour-bearing mice with lower tumour burden. Mice in four treatment groups were weighed daily for 30 days, then twice weekly to day 75. Differences in tumour volume between groups were assessed by t-tests and one-way ANOVA at five time points.

This is temporal pseudo-replication because the same groups of mice are repeatedly sampled over time, but separate hypothesis tests were performed at different time points. However, successive observations on the same mice are correlated, and sample size is expected to decline as mice die or are humanely euthanised at different times during the study. Traditional ANOVA or repeated-measures ANOVA methods cannot handle missing data or imbalance in the number of repeated responses and do not incorporate the actual correlation structure of the data. Mixed models are much more appropriate, because the true variation in the repeated measurements can be modelled directly by incorporating time dependencies and allowing customisation of the correlation structure; they can also accommodate missing data due to subject loss.

References

Blainey, P., Krzywinski, M., and Altman, N. (2014). Replication. *Nature Methods* 11: 879–880.

Cox, D.R. and Donnelly, C.A. (2011). *Principles of Applied Statistics*. Cambridge: Cambridge University Press.

Dugatkin, L.A. and Wilson, D.S. (1992). The prerequisites for strategic behaviour in bluegill sunfish, *Lepomis macrochirus*. *Animal Behaviour* 44: 223–230.

Hammer, P.A. and Hopper, D.A. (1997). Experimental design. In: *Plant Growth Chamber Handbook* (ed. R. W. Langhans and T.W. Tibbitts), 177–188. Iowa State University NCR-101 Publication No. 340. https://www.controlledenvironments.org/wp-content/uploads/sites/6/2017/06/Ch13.pdf.

Hurlbert, S.H. and White, M.D. (1993). Experiments with freshwater invertebrate zooplanktivores: quality of statistical analysis. *Bulletin of Marine Science* 53: 128–153.

Hurlbert, S. (2009). The ancient black art and transdisciplinary extent of pseudoreplication. *Journal of Comparative Psychology* 123 (4): 434–443.

Hurlbert, S.H. (1984). Pseudoreplication and the design of ecological field experiments. *Ecological Monographs* 54: 187–211.

Jung, S.-H. and Young, S.S. (2012). Power and sample size calculation for microarray studies. *Journal of Biopharmaceutical Statistics* 22: 30–42.

Lazic, S.E. (2010). The problem of pseudoreplication in neuroscientific studies: is it affecting your analysis? *BMC Neuroscience* 11: 5. https://doi.org/10.1186/1471-2202-11-5.

Lazic, S.E. and Essioux, L. (2013). Improving basic and translational science by accounting for litter-to-litter variation in animal models. *BMC Neuroscience* 14: 37.

Lazic, S.E., Clarke-Williams, C.J., and Munafò, M.R. (2018). What exactly is 'N' in cell culture and animal experiments? *PLoS Biology* 16: e2005282.

Lee, M.-L.T. and Whitmore, G.A. (2002). Power and sample size for DNA microarray studies. *Statistics in Medicine* 21: 3543–3570.

Lin, W.-J., Hsueh, H.-M., and Chen, J.J. (2010). Power and sample size estimation in microarray studies. *BMC Bioinformatics* 11: 48. https://doi.org/10.1186/1471-2105-11-48.

Lombardi, C.M. and Hurlbert, S.H. (1996). Sunfish cognition and pseudoreplication. *Animal Behaviour* 52: 419–422.

Millar, R.B. and Anderson, M.J. (2004). Remedies for pseudoreplication. *Fisheries Research* 70: 397–407. https://doi.org/10.1016/j.fishres.2004.08.016.

Mull, B.B., Livingston, J.A., Patel, N. et al. (2020). Specific, reversible G1 arrest by UCN-01 *in vivo* provides cytostatic protection of normal cells against cytotoxic chemotherapy in breast cancer. *British Journal of Cancer* 122 (6): 812–822. https://doi.org/10.1038/s41416-019-0707-z.

Potvin, C. and Tardif, S. (1988). Sources of variability and experimental designs in growth chambers. *Functional Ecology* 2: 123–130.

Taylor, S.C. and Posch, A. (2014). The design of a quantitative Western Blot. *BioMed Research International* https://doi.org/10.1155/2014/361590.

van Belle, G. (2008). *Statistical Rules of Thumb*, 2nd edition. New York: Wiley.

Vaux, D., Fidler, F., and Cumming, G. (2012). Replicates and repeats—what is the difference and is it significant? *EMBO Reports* 13: 291–296.

3

Ten Strategies to Increase Information (and Reduce Sample Size)

CHAPTER OUTLINE HEAD

3.1 Introduction

Reduction of animal numbers is a key tenet of the 3Rs strategy, but at times may seem to conflict with the goal of maximising statistical power. Large power results in part from increasing sample size. However, a large sample size does not guarantee adequate power, and high power alone does not ensure that results are informative. This section outlines ten complementary strategies for maximising experimental signal and reducing noise, and therefore increasing the information content of study data. Highlighted are strategies for reducing experimental variation before, rather than after, the experiment is conducted. Incorporating all ten strategies will also increase experimental efficiency – the ability of an experiment to achieve study objectives with minimal expenditure of time, money, and animals.

The ten strategies are as follows:

1. 'Well-built' research questions
2. Structured inputs (statistical study designs)
3. Reduce variation I: Process control
4. Reduce variation II: Research animals
5. Reduce variation III: Statistical control
6. Appropriate comparators and controls
7. Informative outcomes
8. Minimise bias
9. Think sequentially
10. Think 'right-sizing', not 'significance'

3.2 The 'Well-Built' Research Question

Once the investigator has identified an interesting clinical or biological research problem, the challenge is to turn it into an actionable, focused, and

A Guide to Sample Size for Animal-based Studies, First Edition. Penny S. Reynolds.
© 2024 John Wiley & Sons Ltd. Published 2024 by John Wiley & Sons Ltd.

testable question. A well-constructed research question consists of four concept areas: the study population or problem of interest, the test intervention, the comparators or controls, and the outcome. Format is modified according to study type (Box 3.1 and Figure 3.1).

Structuring the research question enables clear identification and discrimination of *causes* (factors that are manipulated or serve as comparators), *effects* (the outcomes that are measured to assess causality), and the *test platform* (the animals used to assess cause and effect). Breaking the research question into components allows the identification and correction of metrics that are otherwise poorly defined or unmeasurable.

A well-constructed research question is essential for effective literature searches. Comprehensive literature reviews provide current evidence-based assessments of the scientific context, the research gaps to be addressed, suitability of the proposed animal and disease model, and more realistic assessments of potential harms and benefits of the proposed research (Ritskes-Hoitinga and Wever 2018; Ormandy et al. 2019). Collaborative research groups such as CAMARADES (`https://www.ed.ac.uk/clinical-brain-sciences/research/camarades/about-camarades`) and SYRCLE (`https://www.syrcle.network/`) are excellent resources for certain specialities such as stroke, neuropathic pain, and toxicology, and provide a number of e-training resources and tools for assessing research quality. Construction of the research question in the PICOT framework was originally developed for evidence-based medicine. Information on constructing research questions and designing literature searches can be obtained from university library resources sections and the Oxford Centre for Evidence-based Medicine website.

The research question dictates formation of both the *research hypothesis* and related *statistical hypotheses*. These are often confused or conflated. The *research hypothesis* is a testable and quantifiable proposed explanation for an observed or predicted relationship between variables or patterns of events. It should be rooted in a plausible mechanism as to why the observation occurred. One or more testable predictions should follow logically from the central hypothesis ('If A happens, then B should occur, otherwise C'). A description of the scientific hypothesis provides justification for the experiments to be performed, why animals are needed, and rationale for the species, type or strain of animals, and justification of animal numbers.

The *statistical hypothesis* is a mathematically-based statement about a specific statistical population

---

BOX 3.1
The 'Well-Built' Research Question

Experimental/intervention studies: PICOT

   Population/Problem
   Intervention
   Comparators/Controls
   Outcome
   Time frame, follow up

Observational studies: PECOT

   Population/Problem
   Exposure
   Comparators
   Outcome
   Time frame, follow up

Diagnostic studies: PIRT

   Population/Problem
   Index test
   Reference/gold standard
   Target condition.

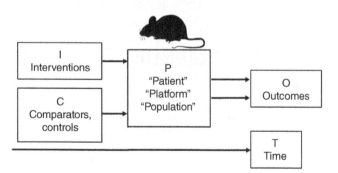

Figure 3.1: System diagram for the 'well-built' research question.

parameter. Hypothesis tests are the formal testing procedures based on the underlying probability distribution of the relevant sample test statistic. Choice of statistical test will depend upon the statistical hypothesis, the study design, the types of variables (continuous, categorical, ordinal time to event), designated as inputs or outcomes. Statistical hypotheses should be a logical extension of the research hypothesis (Bolles 1962). However, the research hypothesis may not immediately conform to any one statistical hypothesis, and multiple statistical hypotheses may be required to adequately test all predictions generated from the research hypothesis.

Example: Research Versus Statistical Hypotheses

Based on a comprehensive literature review, an investigator determined that ventricular dysrhythmia after myocardial infarction is associated with high risk of subsequent sudden cardiac arrest in humans (*clinical observation, clinical pattern*). The investigator wished to design a series of experiments using a mouse model of myocardial infarction to test the effects of several candidate drugs with the goal of reducing sudden cardiac death.

Scientific hypothesis. Pharmacological suppression of ventricular dysrhythmia should result in clinically important reductions in the incidence of sudden cardiac death.

Research question. In a mouse model of dysrhythmia following myocardial infarction (P), will drug X (I) when compared to a saline vehicle solution (C) result in fewer deaths (O) at four weeks post-administration (T)?

Quantified outcomes. Number of deaths (n) in each group and proportion of deaths (p) in each group.

Statistical hypothesis. The null hypothesis is that of no difference in the proportion of deaths for mice treated with drug X (p_X) versus the proportion of deaths for mice treated with control C (p_C) is H_0 $p_X = p_C$ or H_0: $p_X - p_C = 0$.

3.3 Structured Inputs (Experimental Design)

The design of an animal-based study will affect estimates of sample size. Good study designs are an essential part of the 3Rs (Russell and Burch 1959; Kilkenny et al. 2009; Karp and Fry 2021; Gaskill and Garner 2020; Eggel and Würbel 2021). Rigorous, statistically-based experimental designs consist of the formal arrangement and structuring of independent (or explanatory) variables hypothesised to affect the outcome. (Box 3.2). The optimum design will depend on the specific research problem addressed. However, to be fit for purpose, all designs must facilitate discrimination of signal from noise, by identifying and separating out contributions of explanatory variables from different sources of variation (Reynolds 2022). By increasing the power to detect real treatment differences, a properly designed experiment requires the use of far fewer time, money, and resources (including animals) for the amount of information obtained.

Well-designed studies start with a well-constructed research question and well-defined input and output variables. A good design also incorporates specific design features that ensure results are reliable and valid. These include correct specification of the unit of analysis (or experimental unit), relevant inclusion and exclusion criteria, bias minimisation methods (such as allocation concealment and randomisation; Section 3.10), and minimisation of variation (Addelman 1970). Useful designs for animal-based research include completely randomised design, randomised complete block designs, factorial designs, and split-plot designs (Festing and Altman 2002; Montgomery 2017; Festing 2020; Karp and Fry 2021).

BOX 3.2
Statistical Design of Experiments: Components

Study design. Formal structuring of input or explanatory variables according to statistically-based design principles

Study design features. Unit of analysis (experimental unit), inclusion/exclusion criteria, bias minimisation methods, sources of variation.

3.4 Reduce Variation I: Process Control

'Running a tight ship' to minimise process-related variation ensures high-quality research results (Box 3.3). Reliable and consistent results are achieved by a commitment to ongoing quality improvement (QI) and process control. The mechanism for process control is standardisation of procedures and human performance. Poor planning and lack of management can introduce additional uncontrolled and often unrecognised variation into experimental processes, and seriously jeopardise both research quality and reproducibility. When protocols have been previously approved by ethical oversight committees, deviations and non-compliances must be reported to institutional authorities and federal authorities.

Protocols should present a focused blueprint of work that could realistically be accomplished in the given time with resources, budget, and personnel actually available, and with best-practice care and welfare plans in place. It follows that investigators, technical staff, and other personnel involved with the study should be familiar with the protocol and standard operating procedures before the study begins, and routinely review procedures throughout.

Procedural variability is reduced by standardising operating procedures, using simple relevant performance benchmark metrics to evaluate procedural compliance, and incorporating quality control tools (such as checklists and visual aids) into the experimental workflow (Reynolds et al. 2019). Variation in operator performance, and hence measurement error, can be reduced by training all personnel up to predetermined best-practice standards, performing regular quality assurance (QA) checks for equipment calibration and drift, and frequent checks for data anomalies (Preece 1982, 1984).

Data management is an essential but often overlooked skill that contributes to the information power for a study. Datasets need to be properly structured and standardised to facilitate data collection, checking, cleaning, visualisation, and eventual analysis. Data entry should be formatted according to the rules of 'tidy data' (Wickham 2014; Box 3.3A). Incorrectly measured or recorded data and missing data reduce sample size and statistical power and invalidate results. Therefore sufficient time and resources must be allotted to QA methods for ensuring data are reliable, error-free, and complete. Plans must also be in place for appropriate data documentation, archiving, and security. Many journals now require all data necessary to replicate research findings are made available, either in appropriate public data repositories or as part of the published manuscript.

The data dictionary or variable key should be created on a separate spreadsheet. It should include names of all variables, a description, and definitions for each variable, together with measurement units and coding descriptions. All codes should be clearly

---

BOX 3.3
Process Control

Minimise process-related variation.

- Standardise protocol and operational procedures
- Standardise human performance; minimise human performance errors
- Regular quality checks
- Simple performance benchmarks
- Commit to upgrading and improving performance standards
- Good data management practices.

BOX 3.3A
Data Management Rules

1. One variable per column
2. One observation per row
3. One 'observational unit' per table
   - Each column header is a variable name.
   - Date and time are separate variables (therefore separate columns).
   - No special characters.
   - Do not code missing data as zero or leave cells empty.
   - Keep separate records of original raw data before manipulation.
   - Always create copies with multiple backups.
   - Always include a data dictionary.
   - Label auxiliary readouts with appropriate tracking information.

defined and interpretable. Regular quality checks should be performed to ensure that data required for the study are actually being collected, that data are not contaminated or missing due to technical artefacts, and recording and transcription errors are caught and corrected at an early stage. Ensure all samples and readouts can be matched to the correct subject and sample collection time (date, time, subject ID). Track and trace systems ensure data are reliable, enable troubleshooting, and may be required for regulatory compliance.

Formatting and organisation of data collection spreadsheets may evolve, so it is essential to test-drive first to find the most efficient methods for recording and quality review. Plans should also be in place for data archiving, security, and safeguarding of sensitive information. An excellent reference on electronic documentation of experimental procedures is Gerlach et al. (2019).

3.5 Reduce Variation II: Research Animals

Differences between animals are a major source of variation in experiments. Between-individual variability results from the characteristics of individual animals, the environment in which they are maintained, and conditions of handling and experimentation.

Choice of the animal model depends on the disease or injury condition to be simulated, or characteristics that best serve as a test platform for evaluating candidate therapeutics. Signalment, housing and husbandry, and care and welfare must be clearly described and scientifically justified (Lilley et al. 2020; Box 3.4).

Animal *signalment* is analogous to human demographic information. Signalment information includes species, strain, age, sex, weight, reproductive status, and source or vendor. These data provide information necessary for assessing model (construct) validity, generalisability (external validity), and similarity of group baseline characteristics for hypothesis-testing studies. Source and strain information is especially crucial for rodent models. Phenotype changes can occur as the result of random and quiet mutations, genetic drift, and even subtle differences in environmental factors (e.g. Mahajan

---

BOX 3.4
Research Animals

Minimise animal-related variation and unnecessary waste of animals.

Account for:

- Signalment
- Housing, husbandry, and routine handling

Have plans in place for:

- Care, housing, and husbandry
- Welfare and pain and stress management
- End-use disposition.

---

et al. 2016). Functional variants and sub-strain differences in phenotypic responses have been reported for the same strains even with the same vendor but from different barrier facilities (e.g. Simon et al. 2013). Misidentification, mislabelling, and/or cross-contamination of mouse strains and cell lines (Uchio-Yamada et al. 2017; Souren et al. 2022) are symptomatic of management failure. Investigators should be prepared to include genotyping and cell line authentication into protocol work descriptions and thoroughly document all strain- and cell-related authentication procedures (Souren et al. 2022).

It is a common misconception that between-animal variation can be reduced by 'standardising' housing and husbandry conditions by denying enrichment or opportunities for normal activity and behaviours. Such 'standardised' conditions have been shown to shorten rodent lifespan and adversely affect animal health and welfare. Animals kept in barren conditions cannot be considered representative of baseline or control states (Martin et al. 2010). Another misconception is that environmental enrichment compromises research data. Instead, there is abundant evidence to show refinements such as enrichment improve rather than compromise internal and external validity and reproducibility (Richter et al. 2009, 2010; André et al. 2018; Voelkl et al. 2018). Incorporating enrichment as 'planned heterogeneity' significantly reduces animal stress and minimises extraneous and uncontrolled variation, thus promoting more reliable results (Bailoo et al. 2018; Karp 2018; Eggel and

Würbel 2021). Habituation protocols (Reynolds et al. 2019; Ujita et al. 2021) and non-aversive handling (Gouveia and Hurst 2019) also reduce distress and anxiety in animals and make results less variable.

Unrelieved or inadequately treated pain, suffering, and stress modify animal behaviour and physiology and are, therefore, additional confounders of study outcomes. Protocols should include prospective and comprehensive welfare plans for recognition of pain and distress behaviours, best-practice palliation measures, choice of analgesic agents, and clearly defined humane endpoints, with criteria for deployment (Carbone 2011).

Careful end-use disposition planning is an important part of sample size reduction strategies (Smith et al. 2018; Reynolds 2021). Rodent breeding programmes in particular can involve considerable waste if animals are simply euthanised without being used; for example, if they age out of the study or are of an unwanted genotype. Strategies for minimising collateral losses and animal waste should incorporate preliminary estimates of the target number of animals required for experiments, the proportion of animals to be produced with both the desired and unwanted genotypes, and plans for reducing wasted animals (e.g. revised production schedule to minimise litter drop at any one time, transfer to other protocols, reduced study scope).

3.6 Reduce Variation III: Statistical Control

Extraneous nuisance or confounding factors increase variation or 'noise' in experiments and may obscure true effects. Statistical methods for reducing variation include controlling for known sources of variability and grouping similar experimental units together, and ensuring balanced sample sizes for groups (Box 3.5).

Blocking, stratification, and clustering used in statistics to reduce variability in experimental design and improve the precision of results. Known sources of variation are categorised as *nuisance variables*. Nuisance variables are not of direct interest for testing the scientific hypotheses but may be expected to potentially influence the outcome and increase variation.

BOX 3.5
Statistical Control

Minimise extraneous variation.

1. Controlling for known sources of variation

   Blocking
   Stratification
   Clustering

2. Sample size balance.

Blocks consist of similar experimental units or subjects, grouped into homogeneous subgroups or blocks, categorised by a nuisance variable. Randomisation is conducted independently for each block, so each block contains all treatments. Blocking variables may be physical (e.g. a cage of mice, a litter of piglets, a pen of calves, a tank of fish) or temporal (e.g. by week). The simplest blocking design is the randomised complete block design (RCBD). Permuted block randomisation is especially useful in preventing sample size imbalance and controlling for time dependencies when participants enter the study sequentially (Montgomery 2017).

Stratification involves dividing the experimental units into homogeneous subgroups based on characteristics of the subjects, such as age, sex, or baseline characteristics. Randomisation is conducted independently for each stratum (Kernan et al. 1999). Examples of stratification variables include age class (e.g. neonate, juvenile, adult), sex, body condition (e.g. below normal, normal, overweight, obese), clinical severity (e.g. IRIS stages of feline chronic kidney disease), tumour staging (Dugnani et al. 2018), and time. Stratification is common in survey research and may be useful for relatively small studies when responsiveness to a drug intervention or clinical prognosis will be strongly affected by a given confounding variable. However, if sample size is small, there is the possibility that some strata will have too few or zero experimental units, unless the number of strata is reduced, in which case baseline imbalances would not be correctable.

Clustering involves grouping experimental units into homogeneous subgroups based on a hierarchical relationship to a natural grouping variable, such that participants are contained or nested with clusters. Examples include litters clustered within breeding pairs, nestlings within nests, or participants clustered within clinics. The simplest cluster design has two levels, with individual subjects nested within clusters. Depending on the research question, randomisation can occur at either the cluster level (clusters are the experimental unit and clusters are randomised) or the individual level (subjects within clusters are the experimental unit and subjects within each cluster are randomised). Multilevel hierarchical designs can be considered (Aarts et al. 2015; Cook et al. 2016).

Sample size balance. Balanced allocation designs with equal sample sizes *n* per group have the highest precision and power for a given total sample size. Equal sample sizes may reduce the effect of chance imbalances in measured covariates and therefore minimising the potential for bias in the results. Unbalanced sample size allocation increases the variability of the sample, reducing precision and power and making it more difficult to detect true differences between groups.

3.7 Appropriate Comparators and Controls

Comparators and controls are used to establish cause-and-effect relationships between interventions and responses (Box 3.6). By providing a standard of comparison, controls act as a reference for evaluating if results are due to the effect of the independent or explanatory variables or are instead an artefact of some unknown confounder or time dependency in the data. Controls, therefore, provide

a method to distinguish experimental 'signal' from 'noise'. For valid comparisons, the experimental units or animals used as controls should always be as similar to those in the test groups as possible. To minimise bias, experiments with controls should also include formal randomisation and blinding (Moser 2019).

3.7.1 Types of Controls

Seven types of controls typically used to establish causality in animal research are positive, negative, standard, sham, vehicle (Johnson and Besselsen 2002), matching, and strain (Moser 2019). Baseline or 'own' controls are discussed separately.

Positive controls assess the validity of the test; that is, whether or not the response can actually be measured or distinguished from background noise. *Negative controls* assess the magnitude of response that would be expected if the test intervention had no effect. *Standard controls* are procedures or agents for which the effect is already known and verified (standard of care), to be compared to the new intervention. *Sham controls* are common in experiments where surgical procedures are required for instrumentation or to induce iatrogenic disease or injury. Shams account for inflammatory, morphological, and behavioural effects associated with surgical procedures, anaesthesia, analgesia, and post-surgical pain. They are used to discriminate systemic effects resulting from the procedures from those actually induced by the test intervention. However, sham procedures may be inappropriate controls for some models and can profoundly confound results (Cole et al. 2011). They may be unnecessary for certain models (see below).

Vehicle controls assess potential physiological effect of the active agent by accounting for the potential effects of the excipient, carrier, or solubilising solution in which the agent is suspended. Vehicles may not be physiologically inert and can radically affect experimental outcomes, so they must be chosen carefully (Moser 2019).

Matching is appropriate for both experimental and observational studies, especially when sample sizes are small and experimental units

---

BOX 3.6

Appropriate Comparators and Controls

Consider:

1. *Type*: positive, negative, standard, sham, vehicle, strain, and self.
2. *Allocation*: Concurrent versus historical.
3. *Necessity*: Is a specific type of control required?
4. *Non-redundancy*: Is a specific type of control informative?

differ considerably in baseline characteristics. Matching removes some of this bias by pairing subjects on specific baseline covariates, such as age, sex, sibling relationship, disease stage, and sample collection times. Regression-based models are often more powerful than simple matched-pairs comparisons (Greenland and Morgenstern 1990).

Strain controls. Several different control strains or sub-strains should be included in the experimental design (another example of planned heterogeneity; Åhlgren and Voikar 2019). Alternatively, wild-type littermates could be used as controls, for example, homozygous knockouts (Holmdahl and Malissen 2012) in blocked designs.

Selection of appropriate strain or species controls can be difficult. Availability and cost will affect sample size determinations. Comparisons of test strains with a single control or founder strain are highly problematic and not recommended (Garland and Adolph 1994). Observed differences between strains could be simply the result of random genetic drift (rather than the experimental intervention) and are also statistically confounded with strain membership. Different strains and sub-strains of rodents diverge both genetically and phenotypically, depending on source. For example, the C57BL/6 mouse strain is one of the most common and widely used inbred strains and is widely regarded as a 'gold standard' comparator. However, substantial differences in phenotype occurring between sub-strains and vendors contribute to contradictory results reported for common behavioural and physiological tests (Åhlgren and Voikar 2019) or disease conditions (Löscher et al. 2017).

Own control or *within-subject* designs have a baseline or covariate measurement on an animal that adjusts for subsequent measurements taken on the same animal. Designs with repeated measures are preferred to simple parallel-arm ('between-subjects') designs that do not incorporate covariates because they minimise the effects of between-subject variance (Karp and Fry 2021), and therefore have more power and require fewer animals. Methods of accounting for baseline differences are especially important when the experiment is small.

Examples of within-subject designs include paired, before-after, crossover, classic repeated measures, and longitudinal designs (Bate and Clark 2014; Hooijmans et al. 2014; Lazic 2016).

Concurrent controls are the most relevant comparators for determining treatment-related effects and most appropriate for ensuring internal validity. *Historical controls* are comparators for new experiments that consist of data from experiments completed in the past. Their use is occasionally suggested as a method to reduce animal numbers. However, historical controls are strongly discouraged for naïve hypothesis tests against new test data. First, randomisation is essential for valid statistical inference, but historical data cannot be integrated into randomisation protocols. Therefore, comparing new experimental data to historical data introduces serious biases that will compromise results. Second, historical controls will be valid only if stringent 'Pocock criteria' can be met (Hatswell et al. 2020). The Pocock criteria specify that all experimental conditions (animals, animal housing and husbandry, experimental procedures, study design, endpoints, and methods of measurement) must be identical to the historical condition, except for the treatment to be tested. However, these restrictions are unlikely to be met in practice (Kramer and Font 2017). In reality, historical controls will be subject to sources of variation not present in the new experiment, including changes in animal husbandry, laboratory techniques, personnel and personnel skill levels, strain, vendor or source, animal eligibility criteria, routes of agent administration, chemical reagents, and suppliers.

Historical control data can be extremely useful for planning new studies and informing study design, for example, for estimating variance and baseline means, or establishing plausible range of effect sizes (Kramer and Font 2017). They are also useful in oncology studies for interpreting rare tumour incidence, severity, and/or proliferation patterns or identifying evolving trends in tumour biology and behaviour (Keenan et al. 2009). Animal numbers can be reduced considerably if sample size estimates are based on Bayesian priors calculated from data obtained from previous experiments.

However, appropriate and informative estimates of priors require historical studies to be high-quality and well-powered to begin with. If prior data are biased and inadequate, priors should not be used, and sample sizes for the new experiment obtained from appropriate power calculations in the usual way (Bonapersona et al. 2021).

3.7.2 When Are Controls Unnecessary?

Inclusion (or omission) of specific types of control depend on study objectives and specific aims. Failing to include appropriate controls may result in an inconclusive experiment, resulting in the waste of all the animals used in the experiment. Conversely, use of large numbers of animals in unnecessary or redundant control groups is equally wasteful and should be discouraged. *Sham* controls may be unnecessary when the sham condition consists of animals left untreated after acute, well-established procedures with highly predictable experimental endpoints, and spontaneous recovery is not possible (Kramer and Font 2017). Examples include untreated spinal cord crush or transection (Kramer and Font 2017), severe haemorrhagic shock (Tremoleda et al. 2017), or sepsis (Zingarelli et al. 2019). Sham surgical controls may not be required if the only goal is to determine specific effects of the surgical procedure. Reducing variation in personnel technical skills, training up to a high and consistent performance level, and randomisation can eliminate the need for shams. *Redundant* controls occur if studies are designed as numerous one-factor-at-a-time (OFAT) two-group trials, each with its own 'control'. This approach is inefficient, has low precision, and cannot determine any potential interactions, such as synergisms or inhibitory effects which are usually of most interest. The result is the waste of large numbers of animals for limited information. More appropriate designs are available that enable screening of multiple predictors with small sample sizes and do not require multiple redundant control groups (Bate and Clark 2014; Lazic 2016).

Operational pilots used to determine, standardise, and finalise details of logistics, procedures, methods, resources, and processes will usually not need a comparator group (and may not need any animals at all). However, pilots designed to generate preliminary data for subsequent studies should include controls (and if possible be randomised and blinded) to strengthen the evidence for or against proceeding with the investigation. Study designs where animals act as their own control may not require a separate control group (see Bate and Clark 2014; Lazic 2016; Karp and Fry 2021).

3.8 Informative Outcome Variables

Experimental endpoints or outcome variables are the dependent or response variables used to assess if the test intervention 'works' (Box 3.7). Choice of variable used as the experimental outcome (continuous, binary, time to event) will affect estimates of effect size and sample. Specific outcomes will also determine the methods and procedures to be used in the experiments to obtain the data. Limitations in methods or technology may constrain the ability to measure the outcome. Humane endpoints and welfare intervention points must be accounted for when determining the choice of outcome variable.

Outcome variables must be clearly described, specific, and measurable ('What is being measured?'), and include frequency of measurements obtained per experimental unit ('How often will it be measured?'). Outcomes should be clinically or biologically relevances to the condition being studied. Outcomes measured in preclinical studies are often surrogates for clinical conditions, but if target biomarkers or molecular pathways are not clinically relevant, they will be impossible to translate. To maximise power and minimise sample size, continuous outcome variables are preferred to binary outcomes.

Many investigators collect data for large numbers of outcomes, presumably to maximise the amount

---

BOX 3.7
Informative Outcome Variables

- Can it be measured? Is it straightforward to measure?
- Is it biologically or clinically relevant?
- Consider choice of variable: Continuous rather than binary.
- Humane endpoints and welfare intervention points accounted for.

of information obtained from each animal. Choice of outcomes should be prioritised based on relevance and applicability to the central hypotheses, study goals, and objectives. Without such discrimination, studies will be overly large and unfocused, and therefore difficult to interpret (Vetter and Mascha 2017). Measurement and analysis of large numbers of irrelevant variables will greatly increase the probability of type I error and the number of false positives. Such studies are also liable to *P*-hacking (Head et al. 2015) and data dredging (Smith and Ebrahim 2002).

Outcomes should be prioritised by relevance and importance to study goals on a 'need to know' versus 'nice to know' basis. 'Need to know' or primary outcome variables are mission-critical and are the most important for addressing study objectives. The primary outcome drives power calculations and sample size estimates and is central to interpretation of results. Secondary outcomes ('nice to know') provide supporting or corroborating evidence for conclusions based on the primary outcome. Different objectives to be addressed in the same study may require very different outcome variables for separate experiments.

Choice and measurement of outcome variables must also account for *humane endpoints*, the study-relevant welfare indicators determining the point at which the animal will be removed from the study, and pain and distress terminated (for example, by euthanasia). Validity, reliability, and translation potential of the data collected depend on the health and welfare of the animals used in the study. It is also an ethical mandate to minimise pain, suffering, and distress of animals used in experiments. Appropriate *welfare indicators* are one or more predetermined, objective, and easily recognisable behavioural and physical criteria used to assess individual animal welfare and severity of animal distress. Multiple welfare indicators and appropriate monitoring schedules provide a complete welfare picture for each animal and minimise interpretation errors (Hawkins et al. 2010). Published guidelines provide information on choice of appropriate humane endpoints for specific research models (e.g. oncology: Workman et al. 2010; ischaemia: Percie du Sert et al. 2017a; sepsis: Zingarelli et al. 2019).

3.9 Minimise Bias

Bias is the systematic deviation of estimates based on observations from the true value of the population parameter. Bias can occur at all stages of the research cycle. However, most types of bias can be accounted for, or minimised, only during planning and design phases and cannot be removed statistically after data are collected. The SYRCLE Risk of Bias tool (Hooijmans et al. 2014) was developed for assessing methodological quality of published research for inclusion in systematic reviews and is extremely useful for identifying the most common types of bias in animal studies (Box 3.8).

Large sample size does not reduce bias. Only bias minimisation methods reduce bias. Essential best-practice methods are randomisation and allocation concealment (blinding). These should be incorporated early in the planning process to introduce and maintain research rigour (Kilkenny et al. 2009; Macleod et al. 2015; Bespalov et al. 2019; Karp and Fry 2021). Strategies are described in Bespalov et al. (2019). Randomisation ensures that a specific intervention is assigned with a given probability to any experimental unit. Randomisation minimises the effect of systematic bias, and, more crucially, is the cornerstone of inferential statistical hypothesis testing. Allocation concealment involves the coded relabelling of group identifiers, so the association between any specific intervention and the group is concealed from relevant personnel until after data collection or analysis. Allocation concealment minimises biases in selection or allocation, detection, and outcome assessment.

Unfortunately, routine incorporation of randomisation and allocation concealment into animal experiments is not standard practice (Kilkenny et al. 2009; Muhlhausler et al. 2013; Macleod et al. 2015). There is also considerable evidence that

---

BOX 3.8
Minimise Bias

- Randomisation
- Allocation concealment (blinding)
- Consider other sources of bias during planning – selection, performance, detection, attrition, reporting, and publication.

investigators confuse the technical probability-based meaning of 'random' with the lay meaning of 'unplanned' or 'haphazard' (Reynolds and Garvan 2020). Non-randomisation seriously compromises study validity, both by inflation of the false positive rate and invalidation of statistical hypothesis tests.

3.10 Think Sequentially

The most efficient and cost-effective experiments are iterative (Box 3.9). That is, results of previous experiments should inform the approach for the next set of experiments, allowing rapid convergence to an optimal solution (Box et al. 2005). In practice however, preclinical research is typified by numerous two-group, trial-and-error, OFAT comparisons. It is sometimes argued that the OFAT approach 'promotes creativity' and discovery of serendipitous results. However, there are several major disadvantages of OFAT. These include the inability to detect interactions between multiple factors (which results in most 'discovery by chance'), prioritising statistical significance over biological or clinical relevance, and extremely inefficient use of animals compared to statistically-based designs (Montgomery 2017; Festing 2020; Gaskill and Garner 2020).

Adaptive designs are one type of iterative design common in human randomised clinical trials for assessing efficacy, futility, or safety. They are characterised by planned changes based on observed data that allow refinement of sample size, dropping of unpromising interventions or drug doses, and early termination for futility, thus increasing power and efficiency and reducing costs (Pallmann et al. 2018). However, most adaptive design methods are unsuited for preclinical studies, because of different study objectives, complex methodology, and sample sizes that would be prohibitively large for animal-based experiments.

> *Sequential assembly* is an alternative approach that will accomplish many of the same objectives as adaptive designs and is more suited to

---

BOX 3.9
Think Sequentially

'Several small rather than one large'

Each phase informs the conduct and structure of the next: pilot, screening, optimisation, confirmatory.

---

animal-based studies (Box et al. 2005; Montgomery 2017). Experiments are designed to run in series, with results from the preceding stage informing the design of the next stage. A *pilot* phase may be necessary to determine logistics and standardise procedures. The next phase is focused on *factor reduction*: 'screening out the trivial many from the significant few'. In this phase, experiments consist of two-level factorial designs where a relatively large number of candidate inputs or interventions thought to affect the response are reduced to a few of the more promising or important. Subsequent experiments can then be designed to *optimise* responses based on the specific factors identified in the preceding phases. In many cases, only two or three iterations are required to obtain the best or optimal response. Advantages of the sequential assembly approach include overall reduction of sample sizes (and therefore number of animals), estimation of treatment effects with greater precision for a given N, a better understanding of relationships between variables (such as dose-response relationships), and improved decision-making (such as earlier termination of less-promising directions of enquiry). A further advantage of this structured approach, especially for exploratory animal-based research, is the increased chance of 'discovering the unexpected' or serendipitous results not possible with OFAT or trial-and-error (Box et al. 2005).

3.11 Think 'Right-Sizing', Not 'Significance'

Right-sizing experiments, or sample size optimisation, refer to the process of determining the most efficient and optimal sample size for a study to achieve its objectives with the least amount of resources (Box 3.10). This involves finding the balance between adequate effect size and study feasibility. Evidence of right-sizing is provided by a clear plan for sample size justification and reporting the number of all animals used in the study. In contrast, 'chasing statistical significance' refers to the experimental goal of obtaining a statistically significant ($P < 0.05$) result (Marín-Franch 2018).

Chasing significance is a particularly insidious and harmful practice contributing to the waste of countless number of animals in so-called 'negative'

BOX 3.10
Think 'Right-Sizing', Not 'Significance'

A 'right-sized' experiment has the most efficient and optimal sample size to ethically achieve scientific objectives with the least amount of resources and without wasting animals.

'Chasing statistical significance' has the sole goal of obtaining a statistically significant result. It leads to questionable research practices, selective reporting, and waste of large numbers of animals in 'negative' experiments.

experiments. Strength of conclusions is derived from model relevance, study design, bias control, and appropriate data collection methods, not *P*-values (Greenland et al. 2016). *P*-values do not necessarily provide support for the research hypothesis, have no clinical or biological meaning in themselves, and do not mean that results are scientifically important or meaningful. This is because small *P*-values can be an artefact of poor study design, inappropriate sample sizes, sampling variation, methodological errors, incorrect analyses, and bias.

Animals are obvious collateral damage resulting from the three immediate consequences of chasing significance: selective reporting, questionable research practices, and distortion of the evidence base. Perceived bias against the publication of negative results means that investigators probably do not report all experiments with all animals but only those with statistically-significant findings (Sena et al. 2010; ter Riet et al. 2012; Conradi and Joffe 2017; Wieschowski et al. 2019). Nearly all recent publications report statistically significant results (Chavalarias et al. 2016), indicating a high preponderance of selective reporting and discarding of so-called 'failed' experiments with negative results. The perception of research animals, especially mice, as disposable 'furry test tubes' (Garner et al. 2017) further contributes to non-reporting of results. Chasing significance encourages questionable research practices such as P-hacking and N-hacking. P-hacking is the manipulation of data and analysis methods to produce statistically significant results (Head et al. 2015). N-hacking is the selective manipulation of sample size to achieve statistical significance by increasing sample size, cherry-picking observations, or excluding outliers without justification (Szucs 2016). Statements such as 'We continuously increased the number of animals until statistical significance was reached to support our conclusions' are particularly egregious forms of N-hacking. Both P-hacking and N-hacking increase the false positive rate (meaning that results are essentially wrong), make reported *P*-values 'essentially uninterpretable', and violate standards of ethical scientific and statistical practice (Altman 1980; Wasserstein and Lazar 2016). Finally, publication bias resulting from selective reporting distorts the evidence base. Once the study is completed, the investigator has the ethical obligation to report all methods and results transparently, completely, and honestly (MacCallum 2010; Percie du Sert et al. 2020). Complete reporting is necessary to weigh the reliability and validity of contributions to the evidence base, and provide the basis for further pursuit of particular lines of scientific enquiry (Landis et al. 2012). Missing study data result in falsely optimistic estimates of efficacy (Sena et al. 2010), and make it impossible to assess the impact of test interventions on research animal welfare or to assess translation potential, especially the safety and efficacy of treatments intended for use in humans and other animals.

3.A Resources for Animal-Based Study Planning

Oxford Centre for Evidence-Based Medicine (CEBM) https://www.cebm.ox.ac.uk/resources
https://www.cebm.ox.ac.uk/resources/ebm-tools/finding-the-evidence-tutorial; https://www.cebm.ox.ac.uk/resources/data-extraction-tips-meta-analysis/no-intervention
FRAME https://frame.org.uk/resources/experimental-design/
EQIPD https://quality-preclinical-data.eu
PREPARE https://norecopa.no/PREPARE
NC3Rs Resource Library https://www.nc3rs.org.uk/
Percie du Sert, N., Bamsey, I., et al. (2017). The experimental design assistant. *PLoS Biology* 15(9): e2003779. https://doi.org/10.1371/journal.pbio.2003779

The Experimental Design Assistant. https://eda.nc3rs.org.uk/

Jackson Laboratories: Choosing appropriate controls for transgenic animals https://www.jax.org/jax-mice-and-services/customer-support/technical-support/breeding-and-husbandry-support/considerations-for-choosing-controls

References

Aarts, E., Dolan, C.V., Verhage, M., and van der Sluis, S. (2015). Multilevel analysis quantifies variation in the experimental effect while optimizing power and preventing false positives. *BMC Neuroscience* 16: 93. https://doi.org/10.1186/s12868-015-0228-5.

Addelman, S. (1970). Variability of treatments and experimental units in the design and analysis of experiments. *Journal of the American Statistical Association* 65 (331): 1095–1109.

Åhlgren, J., and Voikar, V. (2019). Experiments done in Black-6 mice: what does it mean? *Lab Animal (NY)* 48 (6): 171–180. https://doi.org/10.1038/s41684-019-0288-8.

Altman, D.G. (1980). Statistics and ethics in medical research: misuse of statistics is unethical. *BMJ* 281: 1182–1184.

André, V., Gau, C., Scheideler, A. et al. (2018). Laboratory mouse housing conditions can be improved using common environmental enrichment without compromising data. *PLoS Biology* 16 (4): e2005019. https://doi.org/10.1371/journal.pbio.2005019.

Bailoo, J.D., Murphy, E., Boada-Saña, M. et al. (2018). Effects of cage enrichment on behavior, welfare and outcome variability in female mice. *Frontiers in Behavioral Neuroscience* 26: https://doi.org/10.3389/fnbeh.2018.00232.

Bate, S.T. and Clark, R. (2014). *The Design and Statistical Analysis of Animal Experiments*. Cambridge: Cambridge University Press.

Bespalov, A., Wicke, K., and Castagné, V. (2019). Blinding and randomization. In: *Good Research Practice in Non-Clinical Pharmacology and Biomedicine*, Handbook of Experimental Pharmacology 257 (ed. A. Bespalov, M. Michel, and T. Steckler). Cham: Springer https://doi.org/10.1007/164_2019_279.

Bolles, R.C. (1962). The difference between statistical hypotheses and scientific hypotheses. *Psychological Reports* 11 (3): 639–645. https://doi.org/10.2466/pr0.1962.11.3.639.

Bonapersona, V., Hoijtink, H., and RELACS Consortium (2021). Increasing the statistical power of animal experiments with historical control data. *Nature Neuroscience* 24: 470–477.

Box, G.E.P., Hunter, J.S., and Hunter, W.G. (2005). *Statistics for Experimenters*, 2e. New York: Wiley.

Carbone, L. (2011). Pain in laboratory animals: the ethical and regulatory imperatives. *PLoS ONE* 6 (9): e21578. https://doi.org/10.1371/journal.pone.0021578.

Chavalarias, D., Wallach, J.D., Li, A.H., and Ioannidis, J.P. (2016). Evolution of reporting *P* values in the biomedical literature, 1990-2015. *JAMA* 315 (11): 1141–1148. https://doi.org/10.1001/jama.2016.1952.

Cole, J.T., Yarnell, A., Kean, W.S. et al. (2011). Craniotomy: true sham for traumatic brain injury, or a sham of a sham? *Journal of Neurotrauma* 28: 359–369.

Conradi, U. and Joffe, A.R. (2017). Publication bias in animal research presented at the 2008 Society of Critical Care Medicine Conference. *BMC Research Notes* 10 (1): 262. https://doi.org/10.1186/s13104-017-2574-0.

Cook, A.J., Delong, E., Murray, D.M. et al. (2016). Statistical lessons learned for designing cluster randomized pragmatic clinical trials from the NIH Health Care Systems Collaboratory Biostatistics and Design Core. *Clinical Trials* 13: 504–512. https://doi.org/10.1177/1740774516646578.

Dugnani, E., Pasquale, V., Marra, P. et al. (2018). Four-class tumor staging for early diagnosis and monitoring of murine pancreatic cancer using magnetic resonance and ultrasound. *Carcinogenesis* 39 (9): 1197–1206. https://doi.org/10.1093/carcin/bgy093.

Eggel, M. and Würbel, H. (2021). Internal consistency and compatibility of the 3Rs and 3Vs principles for project evaluation of animal research. *Laboratory Animals* 55 (3): 233–243. https://doi.org/10.1177/0023677220968583.

Festing, M.F.W. (2020). The 'completely randomised' and the 'randomised block' are the only experimental designs suitable for widespread use in pre-clinical research. *Scientific Reports* 10: 17577. https://doi.org/10.1038/s41598-020-74538-3.

Festing, M.F.W. and Altman, D.G. (2002). Guidelines for the design and statistical analysis of experiments using laboratory animals. *ILAR Journal* 4394): 244–258.

Garland, T. Jr. and Adolph, S.C. (1994). Why not to do two-species comparative studies: limitations on inferring adaptation. *Physiological Zoology* 67 (4): 797–828.

Garner, J.P., Gaskill, B.N., Weber, E.M. et al. (2017). Introducing therioepistemology: the study of how knowledge is gained from animal research. *Lab Animal (NY)* 46 (4): 103–113. https://doi.org/10.1038/laban.1223.

Gaskill, B.N., and Garner, J.P. (2020). Power to the people: power, negative results and sample size. *Journal of the American Association for Laboratory Animal*

Science (JAALAS) 59 (1): 9–16. https://doi.org/10.30802/AALAS-JAALAS-19-000042.

Gerlach, B., Untucht, C., and Stefan, A. (2019). Electronic lab notebooks and experimental design assistants. In: *Good Research Practice in Non-Clinical Pharmacology and Biomedicine*, Handbook of Experimental Pharmacology 257 (ed. A. Bespalov, M. Michel, and T. Steckler). Springer https://doi.org/10.1007/164_2019_287.

Gouveia, K. and Hurst, J.L. (2019). Improving the practicality of using non-aversive handling methods to reduce background stress and anxiety in laboratory mice. *Scientific Reports* 9: 20305. https://doi.org/10.1038/s41598-019-56860-7.

Greenland, S. and Morgenstern, H. (1990). Matching and efficiency in cohort studies. *American Journal of Epidemiology* 131: 151–159.

Greenland, S., Senn, S.J., Rothman, K.J. et al. (2016). Statistical tests, *P* values, confidence intervals, and power: a guide to misinterpretations. *European Journal of Epidemiology* 31 (4): 337–350. https://doi.org/10.1007/s10654-016-0149-3.

Hatswell, A., Freemantle, N., Baio, G. et al. (2020). Summarising salient information on historical controls: a structured assessment of validity and comparability across studies. *Clinical Trials* 17 (6): 607–616. https://doi.org/10.1177/1740774520944855.

Hawkins, P., Morton, D.B., Burman, O., et al. (2010). *A guide to defining and implementing protocols for the welfare assessment of laboratory animals*. Eleventh report of the BVAAWF/FRAME/RSPCA/UFAW Joint Working Group on Refinement. www.rspca.org.uk/sciencegroup/researchanimals/implementing3rs/refinement (accessed 2021).

Head, M.L., Holman, L., Lanfear, R. et al. (2015). The extent and consequences of *P*-hacking in science. *PLoS Biology* 13 (3): e1002106.

Holmdahl, R. and Malissen, B. (2012). The need for littermate controls. *European Journal of Immunology* 42: 45–47. https://doi.org/10.1002/eji.201142048.

Hooijmans, C.R., Rovers, M.M., de Vries, R.B. et al. (2014). SYRCLE's risk of bias tool for animal studies. *BMC Medical Research Methodology* 14: 43. https://doi.org/10.1186/1471-2288-14-43.

Johnson, P.D. and Besselsen, D.G. (2002). Practical aspects of experimental design in animal research. *ILAR Journal* 43 (4): 202–206. https://doi.org/10.1093/ilar.43.3.202.

Karp, N.A. (2018). Reproducible preclinical research—is embracing variability the answer? *PLoS Biology* 16 (3): e2005413. https://doi.org/10.1371/journal.pbio.2005413.

Karp, N.A. and Fry, D. (2021). What is the optimum design for my animal experiment? *BMJ Open Science* 5: e100126.

Keenan, C., Elmore, S., Francke-Carroll, S. et al. (2009). Best practices for use of historical control data of proliferative rodent lesions. *Toxicologic Pathology* 37 (5): 679–693. https://doi.org/10.1177/0192623309336154.

Kernan, W.N., Viscoli, C.M., Makuch, R.W. et al. (1999). Stratified randomization for clinical trials. *Journal of Clinical Epidemiology* 52 (1): 19–26. https://doi.org/10.1016/s0895-4356(98)00138-3.

Kilkenny, C., Parsons, N., Kadyszewski, E. et al. (2009). Survey of the quality of experimental design, statistical analysis and reporting of research using animals. *PLoS ONE* 4: e7824. https://doi.org/10.1371/journal.pone.0007824.

Kramer, M. and Font, E. (2017). Reducing sample size in experiments with animals: historical controls and related strategies. *Biological Reviews of the Cambridge Philosophical Society* 92 (1): 431–445. https://doi.org/10.1111/brv.12237.

Landis, S., Amara, S., Asadullah, K. et al. (2012). A call for transparent reporting to optimize the predictive value of preclinical research. *Nature* 490: 187–191. https://doi.org/10.1038/nature11556.

Lazic, S. (2016). *Experimental Design for Laboratory Biologists: Maximising Information and Improving Reproducibility*. Cambridge: Cambridge University Press.

Lilley, E., Stanford, S.C., Kendall, D.E. et al. (2020). ARRIVE 2.0 and the British Journal of Pharmacology: updated guidance for 2020. *British Journal of Pharmacology* 177 (16): 3611–3616. https://doi.org/10.1111/bph.15178.

Löscher, W., Ferland, R.J., and Ferraro, T.N. (2017). The relevance of inter- and intra-strain differences in mice and rats and their implications for models of seizures and epilepsy. *Epilepsy and Behaviour* 73: 214–235. https://doi.org/10.1016/j.yebeh.2017.05.040.

MacCallum, C.J. (2010). Reporting animal studies: good science and a duty of care. *PLoS Biology* 8: e1000413.

Macleod, M.R., Lawson McLean, A., Kyriakopoulou, A. et al. (2015). Risk of bias in reports of *in vivo* research: a focus for improvement. *PLoS Biology* 13 (11): e1002301.

Mahajan, V.S., Demissie, E., Mattoo, H. et al. (2016). Striking immune phenotypes in gene-targeted mice are driven by a copy-number variant originating from a commercially available C57BL/6 strain. *Cell Reports* 15 (9): 1901–1909.

Marín-Franch, I. (2018). Publication bias and the chase for statistical significance. *Journal of Optometry* 11 (2): 67–68. https://doi.org/10.1016/j.optom.2018.03.001.

Martin, B., J, S., Maudsley, S., and Mattson, M.P. (2010). 'Control' laboratory rodents are metabolically morbid: why it matters. *Proceedings of the National Academy of Sciences USA* 107 (14): 6127–6133. https://doi.org/10.1073/pnas.0912955107.

Montgomery, D.C. (2017). *Design and Analysis of Experiments*, 8e. New York: Wiley.

Moser, P. (2019). Out of control? Managing baseline variability in experimental studies with control groups. In: *Good Research Practice in Non-Clinical Pharmacology and Biomedicine, Handbook of Experimental Pharmacology*, 257 (ed. A. Bespalov, M. Michel, and T. Steckler). Cham: Springer https://doi.org/10.1007/164_2019_280.

Muhlhausler, B.S., Bloomfield, F.H., and Gillman, M.W. (2013). Whole animal experiments should be more like human randomized controlled trials. *PLoS Biology* 11 (2): e1001481. https://doi.org/10.1371/journal.pbio.1001481.

Ormandy, E.H., Weary, D.M., Cvek, K. et al. (2019). Animal research, accountability, openness and public engagement: report from an international expert forum. *Animals* 9 (9): 622. https://doi.org/10.3390/ani9090622.

Pallmann, P., Bedding, A.W., Choodari-Oskooei, B. et al. (2018). Adaptive designs in clinical trials: why use them, and how to run and report them. *BMC Medicine* 16: 29. https://doi.org/10.1186/s12916-018-1017-7.

Percie du Sert, N., Alfieri, A., Allan, S.M. et al. (2017a). The IMPROVE guidelines (Ischaemia models: procedural refinements of in vivo experiments). *Journal of Cerebral Blood Flow and Metabolism* 37 (11): 3488–3517. https://doi.org/10.1177/0271678X17709185.

Percie du Sert, N., Bamsey, I., Bate, S.T. et al. (2017b). The experimental design assistant. *PLoS Biology* 15 (9): e2003779. https://doi.org/10.1371/journal.pbio.2003779 [Web-based The Experimental Design Assistant. https://eda.nc3rs.org.uk/].

Percie du Sert, N., Hurst, V., Ahluwalia, A. et al. (2020). The ARRIVE guidelines 2.0: updated guidelines for reporting animal research. *PLoS Biology* 18 (7): e3000410. https://doi.org/10.1371/journal.pbio.3000410.

Preece, D.A. (1982). The design and analysis of experiments: what has gone wrong? *Utilitas Mathematica* 21A: 201–244.

Preece, D.A. (1984). Biometry in the Third World: science not ritual. *Biometrics* 40: 519–523.

Reynolds, P.S. (2021). Statistics, statistical thinking, and the IACUC. *Lab Animal (NY)* 50 (10): 266–268. https://doi.org/10.1038/s41684-021-00832-w.

Reynolds, P.S. (2022). Between two stools: preclinical research, reproducibility, and statistical design of experiments. *BMC Research Notes* 15 (1): 73. https://doi.org/10.1186/s13104-022-05965-w.

Reynolds PS, Garvan CW (2020). Gap analysis of swine-based hemostasis research: houses of brick or mansions of straw? *Military Medicine*, 185 (S1): 88–95. https://doi.org/10.1093/milmed/usz249

Reynolds, P.S., McCarter, J., Sweeney, C. et al. (2019). Informing efficient pilot development of animal trauma models through quality improvement strategies. *Laboratory Animal* 53 (4): 394–403. https://doi.org/10.1177/0023677218802999.

Richter, S.H., Garner, J.P., Auer, C. et al. (2010). Systematic variation improves reproducibility of animal experiments. *Nature Methods* 7 (3): 167–168. https://doi.org/10.1038/nmeth0310-167.

Richter, S.H., Garner, J.P., and Würbel, H. (2009). Environmental standardization: cure or cause of poor reproducibility in animal experiments? *Nature Methods* 6 (4): 257–261.

Ritskes-Hoitinga, M. and Wever, K. (2018). Improving the conduct, reporting, and appraisal of animal research. *BMJ* 360: j4935. https://doi.org/10.1136/bmj.j4935.

Russell, W.M.S. and Burch, R.L. (1959). *The Principles of Humane Experimental Technique*. London: Methuen and Co. Limited.

Sena, E.S., van der Worp, H.B., Bath, P.M., Howells, D.W., Macleod, M.R. (2010). Publication bias in reports of animal stroke studies leads to major overstatement of efficacy. *PLoS Biology*, 8(3):e1000343. https://doi.org/10.1371/journal.pbio.1000343.

Simon, M.M., Greenaway, S., White, J.K. et al. (2013). A comparative phenotypic and genomic analysis of C57BL/6J and C57BL/6N mouse strains. *Genome Biology* 14 (7): R82. https://doi.org/10.1186/gb-2013-14-7-r82.

Smith, A.J., Clutton, R.E., Lilley, E. et al. (2018). PREPARE: guidelines for planning animal research and testing. *Laboratory Animals* 52: 135–141. https://doi.org/10.1177/0023677217724823.

Smith, G.D. and Ebrahim, S. (2002). Data dredging, bias, or confounding. *BMJ* 325: 1437–1438.

Souren, N.Y., Fusenig, N.E., Heck, S. et al. (2022). Cell line authentication: a necessity for reproducible biomedical research. *EMBO Journal* 27: e111307. https://doi.org/10.15252/embj.2022111307.

ter Riet, G., Korevaar, D.A., Leenaars, M. et al. (2012). Publication bias in laboratory animal research: a survey on magnitude, drivers, consequences and potential solutions. *PLoS ONE* 7 (9): e43403. https://doi.org/10.1371/journal.pone.0043403.

Szucs, D. (2016). A tutorial on hunting statistical significance by chasing N. *Frontiers in Psychology* 7: 1443. https://doi.org/10.3389/fpsyg.2016.01444.

Tremoleda, J.L., Watts, S.A., Reynolds, P.S. et al. (2017). Modeling acute traumatic hemorrhagic shock injury:

challenges and guidelines for preclinical studies. *Shock* 48 (6): 610–623. https://doi.org/10.1097/SHK.0000000000000901.

Uchio-Yamada, K., Kasai, F., Ozawa, M., and Kohara, A. (2017). Incorrect strain information for mouse cell lines: sequential influence of misidentification on sublines. *In Vitro Cellular and Developmental Biology* 53 (3): 225–230. https://doi.org/10.1007/s11626-016-0104-3.

Ujita, A., Seekford, Z., Kott, M. et al. (2021). Habituation protocols improve behavioral and physiological responses of beef cattle exposed to students in an animal handling class. *Animals* 11 (8): 2159. https://doi.org/10.3390/ani11082159.

Vetter, T.R. and Mascha, E.J. (2017). Defining the primary outcomes and justifying secondary outcomes of a study: usually, the fewer, the better. *Anesthesia and Analgesia* 125 (2): 678–681. https://doi.org/10.1213/ANE.0000000000002223.

Voelkl, B., Vogt, L., Sena, E.S., and Würbel, H. (2018). Reproducibility of pre-clinical animal research improves with heterogeneity of study samples. *PLoS Biology* 16 (2): e2003693. https://doi.org/10.1371/journal.pbio.2003992.

Wasserstein, R.L. and Lazar, N.A. (2016). The ASA's statement on *P*-values: context, process, and purpose. *The American Statistician* 70: 129–133. https://doi.org/10.1080/00031305.2016.1154108.

Wickham, H. (2014). Tidy data. *Journal of Statistical Software* 59 (10): 1–23. https://doi.org/10.18637/jss.v059.i109.

Wieschowski, S., Biernot, S., Deutsch, S. et al. (2019). Publication rates in animal research. Extent and characteristics of published and non-published animal studies followed up at two German university medical centres. *PLoS ONE* 14 (11): e0223758. https://doi.org/10.1371/journal.pone.0223758.

Workman, P., Aboagye, E.O., Balkwill, F. et al. (2010). Guidelines for the welfare and use of animals in cancer research. *British Journal of Cancer* 102 (11): 1555–1577. https://doi.org/10.1038/sj.bjc.6605642.

Zingarelli, B., Coopersmith, C.M., Drechsler, S. et al. (2019). Part I: minimum quality threshold in preclinical sepsis studies (MQTIPSS) for study design and humane modeling endpoints. *Shock* 51 (1): 10–22. https://doi.org/10.1097/SHK.0000000000001243.

II

Sample Size for Feasibility and Pilot Studies

4 Why Pilot Studies?

CHAPTER OUTLINE HEAD

4.1 Introduction

Pilot studies are small preparatory studies meant to inform the design and conduct of the later main experiment (Box 4.1). In 1866, Augustus De Morgan wrote 'The first experiment already illustrates a truth of the theory, well confirmed by practice, whatever can happen will happen if we make trials enough.' Now enshrined as Murphy's three laws, problems of implementation and timing are integral to experimentation and research. Pilot studies are a tool to anticipate and to some extent circumvent these problems (Box 4.2). The focus of all pilot studies is on feasibility, process, and description, rather than statistical tests of inference (Lancaster et al. 2004; Thabane et al. 2010; Moore et al. 2011; Shanyinde et al. 2011; Abbott 2014; Lee et al. 2014; Kistin and Silverstein 2015).

The main reason for a pilot study is for early identification and correction of potential problems

---

BOX 4.1
Pilot Trials

Pilot trials are SMALL.

What: Small, preparatory studies for informing design and conduct of subsequent 'definitive' experiments.

Why: Essential tools to ensure:

- Methodological rigour
- Feasibility and trial viability
- Internal and external validity
- Problem identification and troubleshooting
- To minimise effect of Murphy's law

Pilot trials *should not* be used to:

- Test hypotheses
- Assess efficacy
- Assess safety

A Guide to Sample Size for Animal-based Studies, First Edition. Penny S. Reynolds.
© 2024 John Wiley & Sons Ltd. Published 2024 by John Wiley & Sons Ltd.

BOX 4.2
The Role of Pilot Studies

One or more pilot studies should be included in the research plan.

Murphy's First Law: Whatever can go wrong, will go wrong.

Pilots identify potential problems and enable development of safeguards and workarounds before beginning the definitive study.

Murphy's Second Law: Nothing is as easy as it looks.

Pilots are opportunities to road-test, improve, and standardise methods and identify potential problems before they occur.

Murphy's Third Law: Everything takes longer than originally thought.

Pilots enable identification and realistic assessment of timelines, task scheduling, and milestones.

one of the biggest contributors to unreproducible results. Unwanted variation arises in part from unreliability of measurement processes (Eisenhart 1952; Hahn et al. 1999). Once uncontrolled variation is introduced early in the study cycle, the resulting problems cannot be fixed by later statistical analyses (Sackett 1979; Altman 1980; Gelman et al. 2020). Therefore, pilots enable identification and control of unwanted variation from multiple sources (Hahn et al. 1999; Reynolds et al. 2019).

Reducing variation at all stages of the research process can greatly increase statistical power without increasing sample size, an important consideration for animal-based studies (Lazic 2016, 2018). Ensuring quality and reliability of research data also increases the probability that research results can be successfully translated from basic research to veterinary and human clinical practice (Davies et al. 2017; Vollert et al. 2020).

before considerable time, resources, money, and animals have been invested in a full study (Clayton and Nordheim 1991). When properly designed and conducted, pilot studies can be cost-effective and powerful tools for reducing experiment failures and increasing data quality. Thoughtful and well-conducted pilot studies coupled with sufficient planning can avoid futility and waste of animals in non-informative experiments and increase the probability that a true effect of the test intervention can be detected if it exists. Sample sizes are small, and use of animals may be completely unnecessary.

Pilot studies are not intended to formally test hypotheses or provide definitive evidence of benefit or safety. Because the purpose is to develop and refine subsequent experimental protocols, pilot studies are methodologically unstable. A single noisy estimate of effect size from a pilot will result in seriously underpowered definitive studies (Leon et al. 2011; Albers and Lakens 2018).

The main purpose of pilot studies is to provide methods for experimental quality control, improvement, and assurance. From a statistical point of view, unrecognised and uncontrolled variation is

4.2 Pilot Study Applications

Definitions and terminology for pilot studies can be confusing and may vary considerably between sources (Arain et al. 2010; Whitehead et al. 2014; Eldridge et al. 2016). In part, this may be because pilots are remarkably versatile. They can be designed to perform a range of functions vital to successful performance of the main experiment (Box 4.3), including:

Assessment of overall feasibility ('Can it work?', 'Does it work?', 'Will it work?').

Assessment of specific logistic, procedural, and resource requirements.

BOX 4.3
Pilot Study Applications

Feasibility assessment
Performance gap identification
Equipment function testing
Locating procedural bottlenecks
Identifying animal welfare concerns
Identifying stakeholder issues
Obtaining data for preliminary proof of concept.

Identification of performance gaps in personnel skill and experience.

Assessment of equipment function, calibration, drift, reliability, and integration with computational and information technology.

Obtaining data for preliminary proof of concept.

Refinement, standardisation, adjustment, and correction of specific tasks, procedures, and protocols.

Identification of potential bottlenecks, procedural hurdles, and human error.

Identification of major sources of variation in experimental processes and how they might be controlled.

Determining whether or not the outcome variables can actually be measured.

Estimating variation in outcome variables.

Identifying potential 'cultural' challenges faced by different stakeholders from implementation of processes and procedures.

Identifying animal welfare concerns, such as severity of the proposed procedures or interventions; effectiveness of remediation and palliative care measures; and definition of early humane endpoints.

Regulatory and ethical oversight bodies and funding agencies encourage preliminary pilots for all types of research investigations when:

Experimental procedures are novel, complex, or difficult to implement.

The 'best' or optimum sample size is difficult to determine.

The best or optimum subset of agents or interventions must be selected from a larger candidate pool.

Large numbers of animals are requested for definitive studies without clear justification.

There is a conflict between animal welfare expectations and the perceived goals of the experiment (for example, requests to exceed established humane endpoints 'because the experiment outcomes could not be measured otherwise').

4.2.1 The Role of Pilot Studies in Laboratory Animal-Based Research

The majority of preclinical laboratory-based studies are essentially exploratory rather than confirmatory (Kimmelman et al. 2014; Mogil and Macleod 2017). They typically consist of multiple factors to be investigated, coupled with very small sample sizes. Structurally and heuristically, they are more closely aligned with agricultural and industrial-type study designs pioneered by Fisher and Box, rather than the two-group, large N designs more typical of clinical trials. As a result, sample size guidelines designed for human clinical pilot trials are, for the most part, inappropriate for laboratory animal-based studies (Reynolds 2022a). However, before animals are used in experiments, all experimental methods, processes, and procedures should be standardised and operational variation minimised. This includes correction of study-specific performance issues and assurance that all technical personnel are trained to the highest possible standard and are fully compliant with experiment protocols.

4.2.2 The Role of Pilot Studies in Veterinary Research

Veterinary clinical trials (like human clinical trials) are generally intended to be confirmatory tests of efficacy. They typically consist of only two or a few treatment arms (usually a test intervention and a comparator), with treatments randomly assigned to many subjects representative of the target population (Reynolds 2022a). However, in practice, many published veterinary clinical trials are small, underpowered, and poorly designed (Wareham et al. 2017; Tan et al. 2019). Trials conducted in the clinical setting may involve considerable management logistics, especially when research protocols involve client-owned animals and require integration with individual case management and clinical decision-making processes (Morgan and Stewart 2002). The research protocol may be further complicated by difficulties in patient recruitment, client consent, compliance, and retention. Administration of test interventions, measurement of experimental outcomes, data collection, and other procedures or standard-of-care interventions can be difficult to implement, especially if they deviate from normal practice.

Pilot trial designs and rules of thumb used for human clinical trials may be appropriate for 'regulated' veterinary clinical trials, when veterinary clinical data is to be submitted to regulatory authorities (such as the FDA) for approval (Davies et al. 2017). Guidelines are available for pilots intended to inform human clinical trials (Lancaster et al. 2004; Arnold et al. 2009; Thabane et al. 2010), and these provide invaluable advice for planning.

4.2.3 Pilot Study Results Should Be Reported

It is an ethical obligation to make the best use of research and research animals by reporting all aspects of the research process, including pilot studies (van Teijlingen et al. 2001; Thabane et al. 2010; Friedman 2013). Many preclinical studies are never written up or published if experiments are perceived to have 'gone nowhere' because results were 'negative', not 'statistically significant', or otherwise unfavourable (Kimmelman and Federico 2017). Carefully performed pilots provide useful information on identification and description of problems, potential solutions, and best practices and will not only benefit other researchers but contribute to minimising waste of animals. Venues include dedicated journals and protocol description sites (e.g. Lancaster 2015; Macleod and Howells 2016) and are increasingly encouraged by mainstream journals (Dolgin 2013).

4.3 Pilot Studies: What They Are Not

Six categories of studies are often labelled incorrectly – and misleadingly – as 'pilot studies' (Box 4.4):

Throw-away studies lacking methodological rigour. These include trial and error studies to see what 'might work', studies conducted with little or no resources and funding, or studies conducted by personnel (usually students) without the requisite skills for meaningful research (Moore et al. 2011).

Misaligned studies, where the objectives of the pilot study consist of vague or undefined

---

> **BOX 4.4**
> *What a Pilot Study Is Not*
>
> 1. A throw-away/unfunded study (or 'student project')
> 2. A study lacking methodological rigour
> 3. A study misaligned with stated aims and objectives
> 4. A study without purpose
> 5. An underpowered, too-small study
> 6. A completed study that cannot reject the null hypothesis.

statements of 'feasibility', but the study goes on to present hypothesis tests of efficacy on numerous outcomes. Misaligned studies usually confuse or conflate the purpose of pilots with that for exploratory studies (see next).

Unsubstantiated studies that make vague statements of intent that any resulting data will be to inform future (funded) studies, but give no plans for subsequent action or how data will be used.

Futility loop studies that consist of an endless cycle of small experiments conducted to 'find something', but lack specific decision criteria or action plans that would enable assessment of progress (Moore et al. 2011).

Sometimes completed studies are retrospectively relabelled as 'pilots' because they are either underpowered or null. *Underpowered* studies did not have a large enough sample size in the first place to test a meaningful hypothesis. Underpowered studies can occur because of insufficient time, resources, funding, or merely because of poor planning (Arain et al. 2010). In contrast, *null* studies may not produce statistically significant results regardless of sample size or power (Abbott 2014). A completed study that is too small or cannot reject the null hypothesis is not *ipso facto* a pilot study.

4.3.1 Pilot Studies Differ from Exploratory and Screening Studies

A *pilot* is a small study acting as a preliminary to the main definitive study, and used to demonstrate feasibility, not provide results of hypothesis testing.

Exploratory studies are usually stand-alone heuristic studies, which may include identification of large-scale mechanisms and patterns of response (for example, in gene arrays).

A *screening study* is a type of exploratory study where the objective is to whittle down a large number of predictor variables, test agents, or genes to a smaller, more manageable, subset of the most promising candidates thought to influence a response (Montgomery and Jennings 2006).

Any study based on trial and error is inefficient and wasteful, producing results that are hard to interpret. Therefore, formal statistically based designs are strongly recommended for all experimental studies to provide maximal effectiveness and information content (Box and Draper 1987; Box et al. 2005; Montgomery and Jennings 2006; Montgomery 2012). Like definitive confirmatory studies, exploratory studies should also prioritise internal validity (randomisation, blinding, appropriate controls) and study sensitivity (control of variation), both to increase the probability of detecting an effect if it exists and minimise risk of false negatives.

4.4 Pilot Study Planning

Sound planning requires both understanding the general principles of pilot study design and having a clear justification for the information to be obtained from a specific pilot study. The focus of all pilots is on 'feasibility, process, and description' (Kistin and Silverstein 2015), rather than hypothesis testing (Lee et al. 2014). The four principles of pilot study planning are the 3Rs, utility, similitude, and proportionate investment, and justification is supported by comprehensive literature review and stakeholder requirements (Box 4.5).

4.4.1 Principles

The 3Rs. Before conducting any pilot, determine if animals are required at all. In the early phases of experimental process development, animal use can be eliminated entirely (e.g. equipment

---

BOX 4.5
Pilot Study Planning

Planning priorities for all pilot types include:

Principles

The 3Rs
Utility
Similitude
Proportionate investment

Justification

Literature reviews
Stakeholder requirements.

---

testing, personnel training) or minimised (e.g. using carcasses or culls). Determine if similar information can be obtained from *in vitro* sources (e.g. cell cultures), systematic literature reviews, surveys of academic or industry research groups, or updated reviews of current best practices established for the speciality.

Utility. A pilot trial should result in useful information for planning future experiments. There must be a clear description of the information to be obtained from the pilot and how it will be used to inform the conduct of later studies. An inadequate sample size or budget or non-significant results are not justifications for declaring an experiment to be a pilot trial.

Similitude. Because the pilot is in effect a rehearsal of the main trial, the pilot model for animals, disease, and/or process must be as similar as possible to the target model. Results obtained from dissimilar pilots will be misleading and essentially useless as evidence for potential feasibility or decisions to proceed with further experiments.

Proportionate Investment. Prepare to invest 10–20% of resources (time, personnel, budget, equipment) in a pilot or a small series of pilots. Unanticipated problems will occur in any experiment. It is better to identify and correct problems before time, money, resources, and animals are invested in the full study.

4.4.2 Justification

The choice of pilot design depends on the information to be obtained from that pilot and how that information is to be used. The information to be acquired by the pilot is determined by the scientific or procedural information gaps to be addressed, the specific information items required to address those gaps, the availability of current information, and the quality of that information. Justification is based on literature reviews and stakeholder input.

4.4.2.1 Literature Reviews

Literature reviews are international consensus best practice for study preparation (PREPARE) and research reporting (ARRIVE 2.0; Percie du Sert et al. 2020). Systematic literature reviews of previous research are used to identify specific gaps that the proposed study will be designed to address. One or several pilots will be justifiable if the literature and prior data present numerous diverse methodological approaches (e.g. computer-based, *in vitro, in vivo*), different animal and disease models, or if there is little or no relevant information on either mechanism or effect of test agents (De Wever et al. 2012).

Literature reviews have two additional goals. First, it is necessary to affirm that the proposed research is not an unnecessary duplication of previous research (Morgan 2011; Johnson et al. 2013). Second, a systematic and comprehensive description of 'lessons learned' prior to beginning a project is advisable for avoiding obvious design flaws or problems, and determine if the proposed methods are best practice or could be improved (Glasziou and Chalmers 2018). Information obtained from the literature must be carefully evaluated because it is certain to vary in quality and relevance to the proposed study.

4.4.2.2 Stakeholder Requirements

Different stakeholders have different roles to play in the project and therefore will require different kinds of information. Stakeholders include the investigators, technical personnel, lab managers, animal care staff, veterinarians, facilities and infrastructure staff, fiscal administrators, and the ethical oversight committee. For example, veterinarians are primarily concerned with animal health and welfare,

fiscal administrators with how much it is going to cost (and who pays), and technical and animal care staff with the amount of extra work and time that will be involved (Reynolds 2022b). It is frequently helpful to identify and confer with major stakeholders beforehand to ensure compliance with existing procedures, to communicate how project-specific operating procedures might affect normal operations (Smith et al. 2018), and determine the specific information that should be generated from the pilot.

4.5 What Kind of Pilot Trial?

Pilot studies are operationalised by clearly defining the research needs, objectives, and specific aims. The specific rationale determines pilot focus, design, and choice of assessment metrics. The objectives and aims will influence the larger strategy for execution; for example, if the pilot is to be conducted in iterative stages, or if each objective requires its own separate pilot (Robinson 2000; Moore et al. 2011).

There are three main categories of pilot studies: operational, empirical, and translational (Box 4.6). These categories have been broadly adapted from those proposed for study feasibility assessment in the public health arena (Bowen et al. 2009).

> *'Can it work?' Operational pilots* are used to determine and finalise details of logistics, procedures, methods, resources, and processes. Operational pilots may not require any animals at all (Chapter 6).

BOX 4.6
What Kind of Pilot?

'Can it work?'	Operational
'Does it work?'	Empirical, intervention-related
'Will it work?'	Translational, generalisation

All types of pilots require a protocol:

- Procedural documentation
- Clearly defined 'success' metrics
- Evaluation component
- Follow-on action plan

'Does it work?' Empirical pilots are used to determine if outcome measurements are fit for purpose and capable of meeting study objectives (Chapter 7).

'Will it work?' Translational pilots are used to determine generalisability, that is, if results are robust, consistent, and broadly applicable over a wider range of models and conditions. Proof of concept and proof of principle studies fit into this category (Chapter 7*)*.

All types of pilots require a fully articulated protocol that should include:

Clear and specific *documentation* of methods and procedures (What will be done, and how will it be done?).

Clearly defined, specific, and measurable *'success' criteria* (How will you know it 'works'?).

An *evaluation* component (Did it 'work'?).

An *action plan* with *decision criteria* (If the pilot is successful, what are the next steps?).

The type of pilot dictates the specific metrics used to determine whether or not the pilot works and how the pilot will inform the next step of the experimentation process. Measurable *decision criteria* ('go-no go') are used to decide if the experiment can proceed without modification or with only a few minor changes ('green light'), if major changes to protocol need to be made and what those changes would entail ('yellow light'), or the experiment should be terminated altogether ('red light'). For veterinary clinical trials, recruitment issues will be especially important to identify at an early stage. Recruitment metrics to be assessed will include projected subject recruitment and accrual rates, client consent rates, treatment compliance, and enrolment strategies (Hampson et al. 2018).

4.6 How Large a Pilot?

There are no hard and fast sample size guidelines for animal-based pilot studies. However, the chief guiding principle for animal-based pilot studies is that

> **BOX 4.7**
> *Pilot Study Sample Size*
>
> **a.** How large a pilot?
>   Pilot studies should be SMALL: Maximise the most information from the fewest possible animals
>   Zero-animal
>   Pragmatic
>   Precision-based
> **b.** How many pilots?
>   Several small versus one large

they are small: the goal is to maximise the most information from the fewest possible number of animals. Pilot studies should also be designed to be nimble. They should have sufficient flexibility to change scope, procedures, and direction as necessary to establish normal best-practice standard operating procedures. Formal power calculations are not appropriate because by design and intent, pilots are small-scale assessments of feasibility and therefore too small to obtain reliable estimates of effect size or variance or find 'statistical significance' (Kraemer et al. 2006; Lee et al. 2014). Multiple small pilots may be preferable to one larger study especially if different types of information are required.

Animal numbers for a pilot can be based on one or more of the following strategies: zero-animal, pragmatic, and precision-directed (Box 4.7).

4.6.1 Zero-Animal Sample Size

The principles of the 3Rs must always be considered before planning any experiment. *Pilot studies may not require any animals at all.* For personnel training, skill acquisition, and development of consistent levels of technical standardisation, simulators, carcasses, and/or *ex vivo* models should always be considered before live animals obtained for purpose. For example, medical residents trained in the repair of penetrating cardiac injury on either an *ex vivo* model or animals obtained for purpose had similar skill and management competencies at assessment (Izawa et al. 2016). Equipment testing for reliability (calibration, drift, measurement errors) should always be conducted on non-animal standard and test materials and confirmed before using animals.

Even 'proof of concept' studies may not necessarily require data from live animals. Systematic literature reviews and meta-analyses (Rice et al. 2008; Hooijmans et al. 2018; Soliman et al. 2020; Vollert et al. 2020) promote animal-based study validity and improve translation potential by synthesising all available external data. Therefore, these provide the best possible estimates of the likely true effect size and variation. The Collaborative Approach to Meta-Analysis and Review of Animal Experimental Studies (CAMARADES) research group specialises in systematic reviews of preclinical neurological disease models such as stroke, neuropathic pain, and Alzheimer's disease, and provides guidance on study design and current best practice (https://www.ed.ac.uk/clinical-brain-sciences/research/camarades/about-camarades). Targeted surveys of academic and/or industry research organisations can be an alternative to systematic literature reviews. Survey results can be used to determine range of likely outcomes or responses anticipated for candidate animal models, identify current best practice, and explore the range of study designs and methods that best minimise animal use (e.g. Chapman et al. 2012).

4.6.2 Pragmatic Sample Size

Pragmatic sample size is determined by the availability of time, budget, resources, and technical personnel. Pragmatic sample size justification may also apply to retrospective chart reviews, where the total sample size may be restricted by the relative rarity of the condition of interest, or the number of records with complete or relevant information.

Pragmatic sample size approximation does not require power calculations. *Arithmetic approximations* use basic arithmetic 'back of the envelope' calculations to confirm and cross-validate numbers for operational feasibility (*Chapter 8 Arithmetic*). Calculations based on *sequential probabilities* are useful when identification of subjects requires screening of large number of candidates, or if animals enter the trial sequentially rather than being randomised simultaneously (*Chapter 9 Probability).*

The amount of useable data obtained from each subject is inversely related to the number of subjects (Morse 2000). When resources are limited, animal-based pilot studies can be designed to be as information-dense as possible by maximising the amount of useable data from the fewest number of subjects. Designing a pilot study based on information power (Malterud et al. 2016) requires descriptions of the intended scope of the pilot, data collection methods, and study design. For example, before-after and longitudinal studies produce more data than do cross-sectional studies with only a single observation per subject. High-dimensionality technologies such as mass cytometry, immunohistochemistry, single-cell RNA sequencing, microfluidics, and bioinformatics can provide massive datasets for relatively few animals.

4.6.3 Precision-Based Sample Size

The emphasis of pilot trials should be on descriptions based on point estimates and confidence intervals (or some other measure of precision). Because 'significance' as such is relatively meaningless, Lee et al. (2014) recommend that precision intervals should be constructed to provide an estimated range of possible treatment effects. Simulations and sensitivity analyses are extremely valuable for estimating uncertainties associated with various projected sample sizes and exploring best-and worst-case scenarios for the statistical models, covariates, and analysis methods proposed for the definitive study (Bell et al. 2018; Gelman et al. 2020). Chapter 6 presents methods for evaluating pilot data based on preliminary estimates of precision and inspection of data plots.

Sample size calculations for external and internal pilot trials have been developed for application to human clinical trials (Julious 2005; Whitehead et al. 2014; Machin et al. 2018). External pilots are conducted prior to the definitive trial, and although data are used to inform subsequent trial design they are not included in later data analyses. Internal pilots are conducted to permit reassessment of sample size within the ongoing definitive trial. Sample sizes based on considerations of human clinical trials are usually prohibitively large for most lab-based animal studies and therefore not recommended. They may be appropriate for large-scale veterinary clinical trials.

References

Abbott, J.H. (2014). The distinction between randomized clinical trials (RCTs) and preliminary feasibility and pilot studies: what they are and are not. *Journal of Orthopaedic and Sports Physical Therapy* 44 (8): 555–558.

Albers, C. and Lakens, D. (2018). When power analyses based on pilot data are biased: inaccurate effect size estimators and follow-up bias. *Journal of Experimental Social Psychology* 74: 187–195. https://doi.org/10.1016/j.jesp.2017.09.004.

Altman, D.G. (1980). Statistics and ethics in medical research. Collecting and screening data. *British Medical Journal* 281 (6252): 1399–1401. https://doi.org/10.1136/bmj.281.6252.1399.

Arain, M., Campbell, M.J., Cooper, C.L., and Lancaster, G.A. (2010). What is a pilot or feasibility study? A review of current practice and editorial policy. *BMC Medical Research Methodology* 10: 67. https://doi.org/10.1186/1471-2288-10-67.

Arnold, D.M., Burns, K.E., Adhikari, N.K. et al. (2009). The design and interpretation of pilot trials in clinical research in critical care. *Critical Care Medicine* 37 (Suppl 1): 69–74. https://doi.org/10.1097/CCM.0b013e3181920e33.

Bell, M.L., Whitehead, A.L., and Julious, S.A. (2018). Guidance for using pilot studies to inform the design of intervention trials with continuous outcomes. *Clinical Epidemiology* 18 (10): 153–157. https://doi.org/10.2147/CLEP.S146397.

Bowen, D.J., Kreuter, M., Spring, B. et al. (2009). How we design feasibility studies. *American Journal of Preventive Medicine* 36 (5): 452–457. https://doi.org/10.1016/j.amepre.2009.02.002.

Box, G.E.P. and Draper, N.R. (1987). *Empirical Model-Building and Response Surfaces*. New York: Wiley.

Box, G.E.P., Hunter, W.G., and Hunter, J.S. (2005). *Statistics for Experimenters: An Introduction to Design, Data Analysis, and Model Building*, 2e. New York: Wiley.

Chapman, K.L., Andrews, L., Bajramovic, J.J. et al. (2012). The design of chronic toxicology studies of monoclonal antibodies: implications for the reduction in use of non-human primates. *Regulatory Toxicology and Pharmacology* 62 (2): 347–354. https://doi.org/10.1016/j.yrtph.2011.10.016.

Clayton, M. and Nordheim, E.V. (1991). [Avoiding statistical pitfalls]: comment. *Statistical Science* 6 (3): 255–257.

Davies, R., London, C., Lascelles, B., and Conzemius, M. (2017). Quality assurance and best research practices for non-regulated veterinary clinical studies. *BMC Veterinary Research* 13: 242. https://doi.org/10.1186/s12917-017-1153-x.

De Wever, B., Fuchs, H.W., Gaca, M. et al. (2012). Implementation challenges for designing integrated *in vitro* testing strategies (ITS) aiming at reducing and replacing animal experimentation. *Toxicology In Vitro* 26 (3): 526–534. https://doi.org/10.1016/j.tiv.2012.01.009.

Dolgin, E. (2013). Publication checklist proposed to boost rigor of pilot trials. *Nature Medicine* 19: 795–796. https://doi.org/10.1038/nm0713-79.

Eisenhart, C. (1952). The reliability of measured values: fundamental concepts. National Bureau of Standards Report, Publication Number 1600. Commerce Department, National Institute of Standards and Technology (NIST). https://www.govinfo.gov/content/pkg/GOVPUB-C13-78b559d7a3b0f60880eba04-f045e72f8/pdf/GOVPUB-C13-78b559d7a3b0-f60880eba04f045e72f8.pdf (accessed 2022).

Eldridge, S.M., Lancaster, G.A., Campbell, M.J. et al. (2016). Defining feasibility and pilot studies in preparation for randomised controlled trials: development of a conceptual framework. *PLoS ONE* 11 (3): e0150205. https://doi.org/10.1371/journal.pone.0150205.

Friedman, L. (2013). Commentary: why we should report results from clinical trial pilot studies. *Trials* 14: 14.

Gelman, A., Hill, J., and Vehtari, A. (2020). *Regression and Other Stories*. Cambridge: Cambridge University Press.

Glasziou, P. and Chalmers, I. (2018). Research waste is still a scandal—an essay by Paul Glasziou and Iain Chalmers. *BMJ* 363: https://doi.org/10.1136/bmj.k4645.

Hahn, G.J., Hill, W.J., Hoerl, R.W., and Zinkgraf, S.A. (1999). The impact of Six Sigma improvement—a glimpse into the future of statistics. *The American Statistician* 53 (3): 208–215.

Hampson, L.V., Williamson, P.R., Wilby, M.J., and Jaki, T. (2018). A framework for prospectively defining progression rules for internal pilot studies monitoring recruitment. *Statistical Methods in Medical Research* 27 (12): 3612–3627. https://doi.org/10.1177/0962280217708906.

Hooijmans, C.R., de Vries, R.B.M., Ritskes-Hoitinga, M. et al. (2018). Facilitating healthcare decisions by assessing the certainty in the evidence from preclinical animal studies. *PLoS ONE* 13: e0187271.

Izawa, Y., Hishikawa, S., Muronoi, T. et al. (2016). *Ex vivo* and live animal models are equally effective training for the management of a penetrating cardiac injury. *World Journal of Emergency Surgery* 11: 45. https://doi.org/10.1186/s13017-016-0104-3.

Johnson, J., Brown, G., and Hickman, D.L. (2013). Is 'duplicative' really duplication? *Lab Animal* 42 (3): 81–825.

Julious, S.A. (2005). Sample size of 12 per group rule of thumb for a pilot study. *Pharmaceutical Statistics* 4: 287–291.

Kimmelman, J. and Federico, C. (2017). Consider drug efficacy before first-in-human trials. *Nature* 542: 25–27. https://doi.org/10.1038/542025a.

Kimmelman, J., Mogil, J.S., and Dirnagl, U. (2014). Distinguishing between exploratory and confirmatory preclinical research will improve translation. *PLoS Biology* 12 (5): e1001863. https://doi.org/10.1371/journal.pbio.1001863.

Kistin, C. and Silverstein, M. (2015). Pilot studies: a critical but potentially misused component of interventional research. *JAMA* 314 (15): 1561–1562.

Kraemer, H.C., Mintz, J., Noda, A. et al. (2006). Caution regarding the use of pilot studies to guide power calculations for study proposals. *Archives of General Psychiatry* 63 (5): 484–489.

Lancaster, G.A. (2015). Pilot and feasibility studies come of age! *Pilot and Feasibility Studies* 1: 1. https://doi.org/10.1186/2055-5784-1-1.

Lancaster, G.A., Dodd, S., and Williamson, P.R. (2004). Design and analysis of pilot studies: recommendations for good practice. *Journal of Evaluation in Clinical Practice* 10 (2): 307–312. https://doi.org/10.1111/j.2002.384.doc.x.

Lazic, S.E. (2016). *Experimental Design for Laboratory Biologists: Maximising Information, Improving Reproducibility*. Cambridge: Cambridge University Press.

Lazic, S.E. (2018). Four simple ways to increase power without increasing the sample size. *Laboratory Animals* 52 (6): 621–629.

Lee, E.C., Whitehead, A.L., Jacques, R.M., and Julious, S.A. (2014). The statistical interpretation of pilot trials: should significance thresholds be reconsidered? *BMC Medical Research Methodology* 14: 41.

Leon, A.C., Davis, L.L., and Kraemer, H.C. (2011). The role and interpretation of pilot studies in clinical research. *Journal of Psychiatric Research* 45 (5): 626–629. https://doi.org/10.1016/j.jpsychires.2010.10.008.

Machin, D., Campbell, M.J., Tan, S.B., and Tan, S.H. (2018). *Sample Sizes for Clinical, Laboratory and Epidemiology Studies*, 4e. Wiley-Blackwell.

Macleod, M. and Howells, D. (2016). Protocols for laboratory research. *Evidence-Based Preclinical Medicine*. https://doi.org/10.1002/ebm2.21.

Malterud, K., Siersma, V.D., and Guassora, A.D. (2016). Sample size in qualitative interview studies: guided by information power. *Qualitative Health Research* 26 (13): 1753–1760. https://doi.org/10.1177/1049732315617444.

Mogil, J. and Macleod, M. (2017). No publication without confirmation. *Nature* 542: 409–411. https://doi.org/10.1038/542409a.

Montgomery, D.C. (2012). *Design and Analysis of Experiments*, 8e. New York: Wiley 752 pp.

Montgomery, D.C. and Jennings, C.L. (2006). An overview of industrial screening experiments, Chapter 1. In: *Screening: Methods for Experimentation in Industry, Drug Discovery, and Genetics* (ed. A. Dean and S. Lewis), 1–20. New York: Springer 332 pp.

Moore, C.G., Carter, R.E., Nietert, P.J., and Stewart, P.W. (2011). Recommendations for planning pilot studies in clinical and translational research. *Clinical and Translational Science* 4 (5): 332–337. https://doi.org/10.1111/j.1752-8062.2011.00347.x.

Morgan D (2011). Avoiding duplication of research involving animals. Occasional Paper No. 7, New Zealand National Animal Ethics Advisory Committee, 2011.

Morgan, D.G. and Stewart, N.J. (2002). Theory building through mixed-method evaluation of a dementia special care unit. *Research in Nursing and Health* 25 (6): 479–488. https://doi.org/10.1002/nur.10059.

Morse, J.M. (2000). Determining sample size. *Qualitative Health Research* 10 (1): 3–5.

Percie du Sert, N., Hurst, V., Ahluwalia, A. et al. (2020). The ARRIVE guidelines 2.0: updated guidelines for reporting animal research. *PLoS Biology* 18 (7): e3000410. https://doi.org/10.1371/journal.pbio.3000410.

Reynolds, P.S. (2022a). Between two stools: preclinical research, reproducibility, and statistical design of experiments. *BMC Research Notes* 15: 73. https://doi.org/10.1186/s13104-022-05965-w.

Reynolds, P.S. (2022b). Introducing non-aversive mouse handling with 'squnnels' in a mouse breeding facility. *Animal Technology and Welfare Journal* 21 (1): 42–45.

Reynolds, P.S., McCarter, J., Sweeney, C. et al. (2019). Informing efficient pilot development of animal trauma models through quality improvement strategies. *Lab Animal* 53 (4): 394–404. https://doi.org/10.1177/0023677218802999.

Rice, A.S.C., Cimino-Brown, D., Eisenach, J.C. et al. (2008). Animal models and the prediction of efficacy in clinical trials of analgesic drugs: a critical appraisal and call for uniform reporting standards. *Pain* 139: 243–247. https://doi.org/10.1016/j.pain.2008.08.017.

Robinson, G.K. (2000). *Practical Strategies for Experimentation*. Chichester: Wiley.

Sackett, D.L. (1979). Bias in analytic research. *Journal of Chronic Diseases* 32: 51–63. https://doi.org/10.1016/0021-9681(79)90012-2.

Shanyinde, M., Pickering, R.M., and Weatherall, M. (2011). Questions asked and answered in pilot and feasibility randomized controlled trials. *BMC Medical Research Methodology* 11: 117–117.

Smith, A.J., Clutton, R.E., Lilley, E. et al. (2018). PREPARE: guidelines for planning animal research and testing. *Laboratory Animals* 52 (2): 135–141. https://doi.org/10.1177/0023677217724823.

Soliman, N., Rice, A., and Vollert, J. (2020). A practical guide to preclinical systematic review and meta-analysis. *Pain* 161 (9): 1949–1954. https://doi.org/10.1097/j.pain.0000000000001974.

Tan, Y.J., Crowley, R.J., and Ioannidis, J.P.A. (2019). An empirical assessment of research practices across 163 clinical trials of tumor-bearing companion dogs. *Scientific Reports* 9: 11877. https://doi.org/10.1038/s41598-019-48425-5.

Thabane, L., Ma, J., Chu, R. et al. (2010). A tutorial on pilot studies: the what, why and how. *BMC Medical Research Methodology* 10: 1. http://www.biomedcentral.com/1471-2288/10/1.

Van Teijlingen, E.R., Rennie, A.M., Hundley, V., and Graham, W. (2001). The importance of conducting and reporting pilot studies: the example of the Scottish Births Survey. *Journal of Advanced Nursing* 34: 298–295. https://doi.org/10.1046/j.1365-2648.2001.01757.x.

Vollert, J., Schenker, E., Macleod, M. et al. (2020). Systematic review of guidelines for internal validity in the design, conduct and analysis of preclinical biomedical experiments involving laboratory animals. *BMJ Open Science* 4: e100046.

Wareham, K.J., Hyde, R.M., Grindlay, D. et al. (2017). Sponsorship bias and quality of randomised controlled trials in veterinary medicine. *BMC Veterinary Research* 13 (1): 234. https://doi.org/10.1186/s12917-017-1146-9.

Whitehead, A.L., Sully, B.G.O., and Campbell, M.J. (2014). Pilot and feasibility studies: is there a difference from each other and from a randomised controlled trial? *Contemporary Clinical Trials* 38: 130–133.

5

Operational Pilot Studies: 'Can It Work?'

5.1 Introduction

Operational pilots are a crucial preliminary to determine logistic feasibility of planned experiments ('Can it work?'). Quality can only be built into the experimental process, not added on after the experiment has been performed. Useful high-quality data can be obtained only if quality is built into the experimental process from the beginning, before animals are used. This requires understanding of the entire experimental process, knowledge of variation, organisation, and frequent evaluation and appraisal (Box 5.1).

The goal of operational pilot studies is to standardise all the logistics, procedures, tasks, and task sequences necessary to conduct the experiment. High-quality and reliable data can be obtained only if all operational and logistic aspects are identified, standardised, and mistake-proofed before the definitive experiments are performed. Operational issues are the largest contributor to uncontrolled variation in experiments Therefore, the pre-experiment stage is indispensable for ensuring the research is both rigorous and reproducible. Operational pilots will not require statistical significance testing (Deming 1987), and may not even need animals. However, researchers should be prepared to invest up to 20% of resources (time, personnel, budget, and equipment) in operational pilots.

There are four components to an operational pilot. The first step is *task identification* and the order in which procedures must be performed. *Benchmark, or 'success', metrics* are used to evaluate whether or not the operational pilot 'works' (Chambers et al. 2014). The success of operational pilots is determined by process and procedure stabilisation without large fluctuations in day-to-day performance and achievement of pre-defined target performance standards. *Deliverables* for an operational pilot include development of standardised operating procedures (SOPs), attainment of uniformly high and consistent skill levels by technical staff, and total compliance with all experimental protocols. SOPs should be in document form for reference purposes. They enable

A Guide to Sample Size for Animal-based Studies, First Edition. Penny S. Reynolds.
© 2024 John Wiley & Sons Ltd. Published 2024 by John Wiley & Sons Ltd.

BOX 5.1
'Can It Work?' Operational Pilots

If you can't describe what you are doing as a process, you don't know what you're doing.

– W E Deming

What are the operational goals and objectives?
What is the work process?
What will be measured?
Where do bottlenecks occur? How can they be fixed?
Where can the process be improved?

What: Workflow maps, checklists, run charts.

How will you know it 'works'? 'Success' metrics:

Outcome-neutral performance measures.

What next? Protocol decision criteria

Proceed as is? Amend? Terminate?

Is the process as good as it can be?
What are the next steps? Action plan

What do we achieve?

Deliverables:

- Standardised operating procedures
- Consistent high-quality technical skill levels and competencies
- 100% compliance with all protocol specifications

consistent performance over the course of complicated sets of experiments, minimise human performance error, and reduce miscommunication and non-compliance.

The operational pilot stage is an invaluable opportunity for research personnel to acquire best possible research practices and skills before animals are used. Technical implementation of experiments can be challenging. Experimental protocols are often complex, with many moving parts, and with high potential for deviation and error. Operator training, technical skill sets, and competencies may vary considerably among staff performing the same task, and on-the-job learning will result in large changes in technical ability and performance over time. Choice between different procedures and non-test interventions (e.g. analgesia, anaesthesia) can profoundly influence both the animal's baseline status and experimental endpoints. Systemic problems with study design and conduct may not become apparent until after experiments

are underway (Reynolds et al. 2019). Trial-and-error 'tweaking' or 'adjustment' of experiments already in progress will seriously compromise data integrity.

All animal-based research would benefit by incorporating best-practice quality improvement and, quality control methodologies into experimental protocols. Food and Drug Administration (FDA)-regulated veterinary trials are expected to be compliant with Good Clinical Practice (GCP) guidelines (VICH GL9. No 85) and other regulatory and QA standards to ensure data quality and patient safety. Davies et al. (2017) suggest several strategies (with checklists) for integrating similar best-practice quality management practices into basic research. There are no mandated best-practice quality requirements for non-regulated academic research. However, numerous quality methodologies, resources, and tools are available through the American Society for Quality (ASQ) and World Health Organisation (UNDP/World Bank/WHO 2006) that could be easily adapted to research protocols and experimental processes. All methods share the common goals of improving outcomes and achieving predictable results through standardisation of processes, reduction of variation, and use of iterative, data-driven strategies to drive improvement (Tague 2005, Reynolds et al. 2019, Backhouse and Ogunlayi 2020).

5.2 Operational Tools

There are three major tools used to identify and control variability in procedures: the *workflow (or process) map, checklists,* and *run charts*. These tools can also be used for day-to-day management of the main or definitive studies. They serve as memoranda or memory-joggers, ensuring that all personnel are compliant with the established protocol, and know what to do, and where, when, and how to do it (Tague 2005; Gygi et al. 2012). These tools are simple, user-friendly, and not highly technical. Above all, these tools can enable a complete understanding of the experiment as a dynamic system without formal data analyses.

5.2.1 Process, or Workflow, Maps

Improving data quality occurs by considering each experiment as a process system. The workflow map is a graphic showing the separate, in-sequence steps for all tasks in the study system (Box 5.2; Fig 5.1). Understanding, identifying, and mapping

BOX 5.2
Constructing Workflow Maps

Title: Define the process.

Landmarks: Where and when does the process start? End?

Brainstorm with personnel who actually do the work to identify all necessary activities, tasks, and decision points.

Write tasks on separate cards or sticky notes.

Arrange all tasks in proper sequence.

Review with all relevant personnel involved – is the process accurately represented?

Revise as necessary.

See `https://asq.org/quality-resources/flowchart;` Tague (2005).

The study workflow map identifies and describes all *tasks and task ordering* necessary to accomplish study objectives; specific *actions, resources, personnel, time, and decision*s to be made at each step; areas of unnecessary *complexity, bottlenecks,* and targets for process improvement; and provides *documentation* of study methods and processes.

Reporting guidelines, such as ARRIVE 2.0, CONSORT, STROBE, and STROBE-VET (`www.equator-network.org`), and many journals strongly suggest incorporation of workflow diagrams in submitted manuscripts. Workflow diagrams require far less explanation in the text, and enable complicated methodology and study design to be explained and evaluated more easily.

experiment components and dependencies beforehand enables a more complete understanding of variation and its causes, and how to control the process to maximise data quality and utility (Walton 1986, NHS Institute for Innovation and Improvement 2008, Trebble et al. 2010). Informative 'downstream' experimental results occur as the consequence of minimising 'upstream' errors (Schulman 2001).

Example: Surgical Instrumentation for Physiological Monitoring in a Swine Model

(Adapted from Reynolds et al. 2019). A swine model was developed to assess efficacy of certain fluid resuscitation interventions following acute shock trauma. The task map is shown in Fig 5.1. Anesthetised animals were surgically instrumented to enable measurement

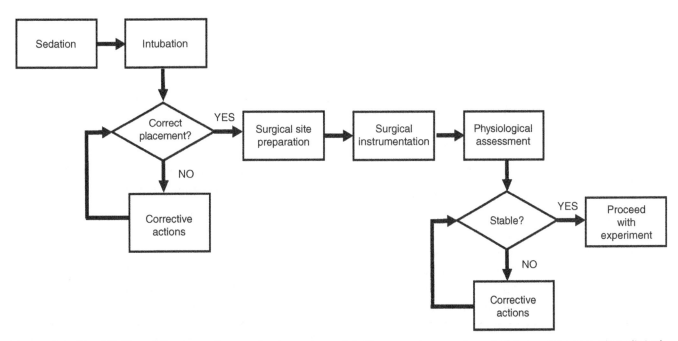

Figure 5.1: Simplified workflow map for a swine trauma model. There are two process decision points requiring clinical assessment of surgical quality and where corrective actions are performed if necessary.

Source: Adapted from Reynolds et al. (2019).

of multiple physiological variables. The experiment proper began with administration of experimental or control intervention. However, pre-intervention inconsistencies in procedures and uncorrected physiological disruptions (such as acidemia, hypothermia, and unstable hemodynamics) will result in unreliable and highly variable outcomes. Therefore, a priority of the pilot study was to establish procedures resulting in consistent and stable physiological baseline measurements. Two process decision points were identified where potential for physiological disruption was great, and where corrective actions could be easily performed if necessary. The first was confirmation of correct endotracheal tube placement with objective assessment of adequate oxygenation and ventilation. The second was after a 30-minute post-instrumentation rest period with blood gas measurements used to assess and correct pre-intervention physiological status before application of the experimental intervention.

5.2.2 Checklists

Checklists consist of simple, specific, precise, and convenient step-by-step list of actions or tasks for standard procedures or task groups. Checklists are not intended as a detailed list of procedures, but instead, act as an *aide memoire*. Checklists reduce errors of omission and memory lapse, ensure staff compliance with procedures and protocol, and enable personnel to maintain a consistent and high standard of performance and safety. Checklists also simplify management and oversight by increasing staff investment in the experimental process, so there is less reliance on the laboratory chief or principal investigator for routine protocol implementation and monitoring (Gawande 2009, Reynolds et al. 2019). Checklists can also be designed for other purposes, including troubleshooting (how to identify what went wrong and how to fix it), to-do lists (what needs to be done today), and consultation (what outside entities must be called in and when). Checklists are used routinely for mistake-proofing complex operational procedures in such diverse fields as industry, manufacturing, surgery, and aircraft operations (World Health Organization 2009, Gawande 2009).

Example: Animal Surgery Procedural Checklist

(Adapted from Reynolds et al. 2019). Table 5.1 shows a sample checklist of essential compliance items during surgical instrumentation and monitoring of anaesthetised swine. Technical staff were already familiar with the approved protocols and detailed standard operating procedures. Checklists were posted throughout the lab to remind personnel of tasks to be performed in complex procedures.

5.2.3 Run Charts, Process Behaviour Charts

Run charts are dot and line graphs of performance measurements plotted in time order. Run charts are useful for monitoring the improvement or deterioration of personnel skills over time, flagging any major changes or unusual events, pinpointing areas requiring remediation, or indicating if any protocol changes improved quality (Chakraborty and Tan 2012, Gygi et al. 2012).

Construction of a control chart requires a clearly defined set of measurable performance metrics and the time period for data collection. Formal control charts include a horizontal line indicating the average measure and upper and lower lines to indicate the control limits (usually derived from historical data). These may not be necessary if the goal is to achieve some pre-specified target value (e.g. zero errors by the end of pre-specified number of sessions)

Example: Process Chart for Technical Performance Improvement

(Adapted from Reynolds et al. 2019). Figure 5.2 shows a process chart of critical errors made over twenty sequential experiments. Fourteen specific critical errors were identified and defined *a priori* and in consultation with all study personnel. Critical procedural errors were associated with airway management, surgical procedure, and animal physiological monitoring. Errors per experiment were counted and plotted in time order.

Table 5.1: Sample checklist for surgical procedures and pre-intervention animal physiological status. All checks were to be documented in the surgical record for each animal.

Experimental phase	Task
Pre-surgical	Surgical skin site preparation × 3
	Sedation: correct dose, route, time to effect adequate?
Airway	Intubation attempt <60 s?
	Pulse oximetry readings >90%
	Bilateral breath sounds heard?
Surgery	Isoflurane, oxygen levels adequate?
	Pre-emptive analgesia given?
	Anaesthetic depth checks every 15 min
Baseline	Mean arterial pressure >70 mmHg?
	Arterial pCO_2 35–55 mmHg; pO_2 95–110 mmHg? Arterial lactate $\leq$2 mmol/L?
Monitoring: Animal	Pulse oximetry readings >90%
	Core temperature 36-40 °C
	Heart rate <120 bpm
	Surgical plane adequate? Check every 15 min
Monitoring: Procedural	Blood sampled at designated times?
	Catheters patent?
	Cardiac output signal present?
	Mechanical sigh breaths every 15 min

Source: Adapted from Reynolds et al. (2019).

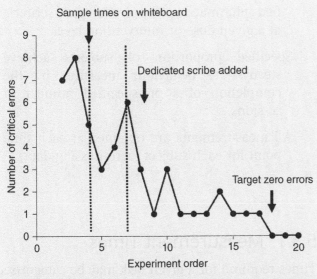

Figure 5.2: Process chart of critical errors made during twenty sequential experiments. Errors decreased after a dedicated data scribe was added to the personnel roster (*Source:* Adapted from Reynolds et al. 2019).

The target was 0/14 errors per experiment. The graph shows that the addition of a dedicated data scribe after the first six experiments was associated with increased protocol compliance and reduced protocol errors.

Run charts are also useful for assessing variation in animal behaviour. Animal behaviour studies often require performance stability after a period of habituation time. Habituation time is adequate if task performance times or learning trajectories stabilise by the end of a certain number of training sessions. Run charts are also useful for identifying floor-ceiling artefacts in behavioural tasks. Floor-ceiling artefacts occur if the task is either too difficult, such that most or all animals fail, or too easy, and all animals succeed.

Example: Process Chart for Animal Behaviour

(Data from Higaki et al. 2018). Morris water maze testing (MWM) is used to evaluate spatial learning, navigation, and reference memory in laboratory rodents. In this study, mice were tested over five consecutive days. Each session was terminated at 120 seconds of search time. Longitudinal data plots for four mice (Fig. 5.3) show increasing between-mouse variation in search times over five days and subject-specific patterns in learning trajectories. One mouse showed no improvement, and the remainder showed modest to substantial reductions.

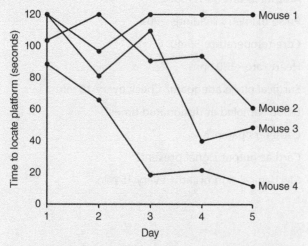

Figure 5.3: Sequential plot of platform location times for four mice tested over five days in a Morris Water Maze. Search times were truncated at 120 seconds. Between-mouse variation in search times increased with day. Mouse 1 show no improvement, whereas the remaining mice showed modest to substantial reductions

(*Source:* Data from Higaki et al. 2018).

5.3 Performance Metrics

Performance metrics are used to assess if the operational pilot has achieved pre-specified objectives (the operational pilot 'worked'). Performance metrics are outcome-neutral, that is, they are not response variables and do not test the research hypothesis. However, they must be relevant to major operational objectives. Metrics must be simple, quantifiable, and transparent to everyone involved with the project

(Gygi et al. 2012). Metrics that are not clearly defined will be inconsistently measured and collected and will not provide reliable information.

Choice of which specific metrics to use will require a preliminary process or workflow map to understand which personnel are involved with what parts of the project, what tasks must be completed, the amount of time taken to perform each task, and the resources needed for each task.

Examples: Training Metrics

All personnel to complete 100% of training requirements by a pre-specified time.

All personnel to complete all defined critical tasks with zero performance errors.

All subjects receive the study drug within a pre-specified time from study start.

All subjects are sampled within ± 2 minutes of pre-specified sampling times.

All necessary dose adjustments are made in response to pre-defined laboratory criteria.

Complete (100%) follow-up of 100% of enrolled subjects.

Specified proportion of subjects achieve specified pharmacological/physiological criteria at a given dose or intervention level.

Specified proportion of subjects achieve steady-state behaviour measures by the completion of a pre-specified number of sessions.

All measurements are obtained at each time point for each subject (zero missing data).

5.3.1 Measurement Times

Times required for a given task may be categorised as 'set' times, 'target' times, or 'untargeted' times.

Set times are protocol-driven, used for standardising operational procedures.

Examples: Set-Time Metrics for an Operational Pilot

Fixed sampling time points are specified for obtaining blood and tissue.

Pre-specified duration is defined for physiological stabilisation of an animal following surgical instrumentation and before experimental intervention.

Target times are mandated by best-practice requirements.

Example: Best-Practice Target Times

The definitive airway (intubation and confirmation of correct placement) for large-animal surgery models must be secured in <60 seconds to prevent injury and physiological compromise.

Untargeted times do not have rigid performance limits.

Example: No Pre-specified Performance Time

The primary objective of a successful surgical instrumentation of an experimental animal is completion of all procedures without physiological compromise. Total preparation time cannot be too rigidly fixed because different animals will respond differently to anaesthesia and surgery, and require different corrective measures and time to achieve stability. Personnel should not be compelled to sacrifice the quality of the preparation by rushing or taking shortcuts to meet an arbitrary time limit. However, if unduly prolonged preparation times result in physiological compromise, this is grounds for exclusion of that subject and its data (Reynolds et al. 2019).

5.3.2 Subjective Measurements

Some outcome variables are subjective. Assessors differ in skill levels, experience, and even interest. To ensure that all variables are collected and rated

BOX 5.3
Subjective Metric Applications

- Devising codes for variables and endpoints
- Classification of subjects into disease categories
- Behavioural research and behaviour outcomes
- Assessment of humane endpoints
- Subjective evaluations of human or animal performance
- Clinical decisions based on signs and symptoms
- Decisions based on score thresholds
- Retrospective chart reviews
- Questionnaire and survey instrument validation
- Qualitative research
- Systematic reviews and meta-analyses

consistently with minimal bias, an operational pilot will determine if measurements are *valid* (data and data assessments capture what they were intended to measure), *reliable* (data are consistent), and assessors show *agreement* (the frequency with which different assessors assign the same score to the same items). For high-priority items and welfare assessments, at least two data evaluators should be used and trained to a uniform standard (Box 5.3).

5.4 Pilots for Retrospective Chart Review Studies

In retrospective chart review studies, subject data have already been collected and are contained in patient medical records or registry databases. Medical record data are commonly used in veterinary medicine to assess clinical outcomes, treatment patterns, clinical care resources, and costs. Retrospective chart reviews may also be conducted for quality assurance and improvement assessments (Jones et al. 2019).

Chart review studies are highly susceptible to bias. To prevent wasting time and effort in a futile study, piloting a chart review study begins by assessing a very small number of records (say 5–10 cases) (Box 5.4). Preliminary scans of these records should show if there are enough case samples with

---

BOX 5.4
Piloting Retrospective Chart Reviews: Checklist

Research question and objectives defined
Identify case inclusion and exclusion criteria
Identify all databases and registries to be reviewed. Define search terms
Define and operationalise all variables (signalment, primary outcomes, secondary outcomes, predictor variables or covariates)
Determine methods of data collection (manual, automated)
Standardise data collection/entry templates
Assessors:
- Number of data extractors, assessors.
- Number of replicates used to establish agreement.
- Are ratings conducted independently?

Assessor training:
- Sufficient? (define performance metrics).
- Adequate monitoring and oversight?
- SOPS or procedure handbook?

Assessor performance:
- Reliable?
- Consistent?
- High agreement?

Identify methods for data management, screening, archiving and security.

---

characteristics that are representative of the population of interest, that the variables needed to address the research question are actually available in the records, and if there are enough complete records for later statistical analyses. Routinely-collected medical data may not include the variables necessary to answer the research study question.

Once it is established that a larger study is realistic, the second stage of a pilot is to develop a standardised and user-friendly data collection instrument and a standardised set of operational procedures. Pilot case records are used to define and operationalise variables for data collection, define inclusion and exclusion criteria, and develop consistent data extraction procedures. Variables that are not clearly defined, ambiguous, or inconsistently recorded in the medical records will be inconsistently measured and collected and cannot

provide reliable information (Gilbert et al. 1996, Vassar and Holzmann 2013, Kaji et al. 2014).

5.4.1 Standardising Performance

Reliable and unbiased data are obtained by developing clear rules for systematic data extraction and coding (*data dictionary*, *standardised operating procedures*), then rehearsing data extraction and coding procedures before the large study. To minimise misclassification bias, at least two people should perform data extraction and collection. Agreement between data collectors is essential for ensuring consistent and reliable data extraction and coding (Kaji et al. 2014). Agreement measures are usually required by journals as part of best-practice reporting for chart review studies, for example, RECORD (Benchimol et al. 2015) and STROBE (von Elm et al. 2007).

Data dictionary. This is a list of all variables to be measured and their corresponding definitions and measurement units. The data dictionary can also include brief descriptions of how variables will be measured or assessed, other accepted meanings, and guidance on interpretation. Data that are ambiguous or have conflicting identifiers (e.g. drug names) will require clear handling instructions.

Standardised operating procedures describe how variables will be interpreted, extracted from the database, rated, counted, or categorised. Inclusion and exclusion criteria must be clearly defined. Standardised and user-friendly data collection templates are essential to guide data collection and minimise abstraction errors.

After data collection methods have been agreed upon, test-drive procedures with a few representative records. Have all personnel extract and code data independently, then review data records for discrepancies, inconsistencies, misunderstandings, ambiguities, major or consistent errors, inaccuracies, and/or omissions. Discrepancies between extractors and assessors can be flagged electronically and examined for patterns (Appendix 5.A). Formal statistical tests of inter-rater agreement are unnecessary at this stage and will be unreliable. However, inter-rater agreement measures are strongly suggested for larger definitive studies (Altman 1991).

Revise and repeat. There should be frequent discussions with all personnel (including the study lead) to resolve ambiguities, discrepancies, questions, and problems. Areas of disagreement are resolved by consensus.

Finalise. Once standardised, the definitive data collection spreadsheets and SOPs can be constructed, and all data extractors and assessors can be instructed in use. Planning of the definitive study will require identification of the study design (matched case, case–control, cohort), matching criteria, grouping criteria, and formal sample size justification. Formal procedures must be put in place for data screening, cleaning, storage, and security (Benchimol et al. 2015).

5.5 Sample Size Considerations

Operational pilots do not require statistical significance testing. The size of operational pilot studies is determined by pragmatic considerations and logistic constraints. There may be no need for animals at all. For planning highly complex large experiments, researchers should be prepared to invest up to 20% of all resources (time, personnel, budget, and equipment). If animals must be used, bounds on animal numbers are computed relative to available resources, using simple arithmetic (*Chapter 7*) or probability-based (*Chapter 8*) approximations. To assess stability and consistency of processes and procedures, charts of counts, averages and ranges are constructed from at least two sequential groups of 2–10 measurements each (Gygi et al. 2012).

5.A Conducting Paired Data Assessments in Excel

1. Two raters X and Y score items independently and enter data in separate Excel spreadsheets. The sheet format must be the same for each assessor (variable names, order, number of rows and columns). The first row of each sheet is reserved for variable header names.

2. Copy separate spreadsheets into the same workbook. Each sheet is given an assessor ID (X, Y).

3. Cell-by-cell comparison: Create a third (empty) sheet with the same variable header row. Copy the Excel formula: $=IF(X!A2<>Y!A2,"X:"\&X!A2\&CHAR(10)\&" Y:"\&Y!A2,"")$ into the first cell A2. Drag and drop so all row-column dimensions are covered in the third sheet. Any discrepancies between the two assessors will be flagged with the assessor ID and entry values.

4. Examine original entries indicated by flags for the cause of the difference. Differences may be the result of transcription and entry errors (e.g. typos, stray characters, missed information) or because of assessor differences in interpretation.

5. Correct obvious coding errors. Discrepancies occurring because of ambiguities in data recording, interpretation or coding are resolved by discussion and consensus. Retraining of data collection personnel, and revision of the data dictionary and SOPs may be required.

References

Altman, D.G. (1991). *Practical Statistics for Medical Research*. London: Chapman and Hall.

American Society for Quality (ASQ) (2023). https://asq.org/quality-resources

Backhouse, A., Ogunlayi, F. (2020). Quality improvement into practice. *BMJ*, 368:m865. doi: https://doi.org/10.1136/bmj.m865.

Benchimol, E.I., Smeeth, L., Guttmann, A. et al. (2015). The reporting of studies conducted using observational routinely-collected health data (RECORD) statement. *PLoS Medicine* 12 (10): e1001885.

Chakraborty, A., Tan, K.C. (2012) Qualitative and quantitative analysis of Six Sigma in service organizations. In *Total Quality Management and Six Sigma* (Ed. Tauseef Aized). doi: https://doi.org/10.5772/46104; https://www.intechopen.com/chapters/38085

Chambers, C.D., Feredoes, E., Muthukumaraswamy, S.D., and Etchells, P.J. (2014). Instead of 'playing the game' it is time to change the rules: registered reports at AIMS Neuroscience and beyond. *AIMS Neuroscience* https://doi.org/10.3934/Neuroscience.2014.1.4.

Davies, R., London, C., Lascelles, B., Conzemius, M. (2017) Quality assurance and best research practices for non-regulated veterinary clinical studies. *BMC Veterinary Research*, 13:242. doi https://doi.org/10.1186/s12917-017-1153-x

Deming, W.E. (1987). On the statistician's contribution to quality. Invited paper, 46th session of the International Statistical Institute (ISI), Tokyo. https://deming.org/wp-content/uploads/2020/06/On-The-Statisticians-Contribution-to-Quality-1987.pdf

von Elm, E., Altman, D.G., Egger, M., Pocock, S.J., Gøtzsche, P.C., Vandenbroucke, J.P., STROBE Initiative (2007). The strengthening the reporting of observational studies in epidemiology (STROBE) statement: guidelines for reporting observational studies. *Epidemiology*;18:800–4. doi:https://doi.org/10.1097/EDE.0b013e3181577654

Gawande, A. (2009). *The Checklist Manifesto: How to Get Things Right*. New York: Picador.

Gilbert, E.H., Lowenstein, S.R., Koziol-McLain, J., Barta, D.C., Steiner, J. (1996). Chart reviews in emergency medicine research: where are the methods? *Annals of Emergency Medicine*, 27(3):305–8. doi: https://doi.org/10.1016/s0196-0644(96)70264-0

Gygi, C., William, Z.B., and DeCarlo, N. (2012). *Six Sigma for Dummies*, 4th ed. NJ: John Wiley & Sons.

Higaki, A., Mogi, M., Iwanami, J. et al. (2018). Recognition of early stage thigmotaxis in Morris water maze test with convolutional neural network. *PLoS ONE* 13 (5): e0197003. https://doi.org/10.1371/journal.pone.0197003.

Jones, B., Vaux, E., and Olsson-Brown, A. (2019). How to get started in quality improvement. *BMJ* 364: k5408. https://doi.org/10.1136/bmj.k543730655245.

Kaji, A.H., Schriger, D., and Green, S. (2014). Looking through the retrospectoscope: reducing bias in emergency medicine chart review studies. *Annals of Emergency Medicine* 64 (3): 292–298. https://doi.org/10.1016/j.annemergmed.2014.03.025.

NHS Institute for Innovation and Improvement (2008). *Improvement Leaders Guide: Process Mapping, Analysis and Redesign*. Warwick: NHS.

Reynolds, P.S., McCarter, J., Sweeney, C., Mohammed, B.M., Brophy, D.F., Fisher, B., Martin, E.J., Natarajan, R. (2019) Informing efficient pilot development of animal trauma models through quality improvement strategies. *Lab Animal*, 53(4):394–404. doi: https://doi.org/10.1177/0023677218802999.

Schulman, J. (2001) 'Thinking Upstream' to evaluate and to improve the daily work of the Newborn Intensive Care Unit. *Journal of Perinatology*, 21: 307–311. doi: https://doi.org/10.1038/sj.jp.7200528

Tague, N.R. (2005). *The Quality Toolbox*, Seconde, 584. ASQ Quality Press American Society for Quality.

UNDP/World Bank/WHO Special Programme for Research and Training in Tropical Diseases & Scientific Working Group on Quality Practices in Basic Biomedical Research (2006). *Handbook: Quality Practices in Basic Biomedical Research/Prepared for TDR by the Scientific Working Group on Quality Practices in Basic Biomedical Research*. World Health Organization. https://apps.who.int/iris/handle/10665/43512.

Trebble, T.M., Hansi, N., Hydes, T., Smith, M.A., Baker, M. (2010) Process mapping the patient journey: an introduction. *BMJ*, 341: c4078–4078. doi: https://doi.org/10.1136/bmj.c4078.

Vassar, M. and Holzmann, M. (2013). The retrospective chart review: important methodological considerations. *Journal of Educational Evaluation for Health Professions* 10: 12. https://doi.org/10.3352/jeehp.2013.10.12.

Walton, M. (1986). *The Deming Management Method*. New York: Perigree.

World Health Organization (2009). *WHO Guidelines for Safe Surgery 2009: Safe Surgery Saves Lives*. Geneva, Switzerland: WHO Press, World Health Organization.

6 Empirical and Translational Pilots

CHAPTER OUTLINE HEAD

6.1 Introduction

Empirical and translational pilot studies provide preliminary evidence of efficacy and translation potential. They are used to determine if further experimentation will be worthwhile (Box 6.1; Table 6.1).

Empirical pilot studies provide evidence that the intervention 'might work' (Bowen et al. 2009). Efficacy potential is assessed by the quality of the outcome, or response, variables. The response should be large enough to be measured (magnitude), be stable and consistent over multiple determinations (accuracy and precision), and behave as predicted by the research hypothesis (relevance). The goal is to obtain the largest possible signal from the data. Therefore, empirical pilots should be conducted under controlled and relatively ideal conditions, rather than attempting a more realistic replication of the clinical or disease model. Generalisation is not a consideration at this phase.

> **BOX 6.1**
> *Empirical and Translational Pilot Studies*
>
> Sample size is optimised for information density and information power
>
> *Empirical pilots* provide evidence of *potential efficacy* under ideal conditions
>
> Goal: Signal maximisation
>
> Is the response large enough to be measured? Stable enough to be reliable? Expected from the research hypothesis? Realistic compared to previously published data?
>
> *Translational pilots* provide evidence of *generalisability* for a variety of models and conditions.
>
> Goal: Signal robustness
>
> When assessed over a range of animal/disease models and operating conditions, are results Consistent? Reliable? Concordant?

A Guide to Sample Size for Animal-based Studies, First Edition. Penny S. Reynolds.
© 2024 John Wiley & Sons Ltd. Published 2024 by John Wiley & Sons Ltd.

Table 6.1: Comparison of empirical and translational pilot studies.

'Might it work?' Empirical pilots	'Will it work?' Translational pilots
Is there preliminary evidence of *intervention and outcome feasibility* (the intervention 'might work')? Are outcomes and endpoints ▪ large enough to be measured? Can they be measured easily? Are laboratory readouts within limits of detection? ▪ stable and consistent? (minimal trend, minimal variation) ▪ clinically or biologically meaningful (concept validity)? ▪ Does the intervention do what it was supposed to do? ▪ Is the intervention potentially cost-and time-effective to implement?	Is there preliminary evidence of *efficacy* and *concordance*? *Efficacy* ▪ Are models and endpoints clinically relevant? ▪ Are results clinically or biologically meaningful? *Concordance* ▪ Are results robust? ▪ Are results consistent and reproducible over multiple independent models/locations/ conditions?
Success metrics ▪ Outcome or outcome surrogates compared to *a priori* biologically relevant standard ('minimum difference to be detected')	*Success metrics* ▪ Outcome or outcome surrogates compared to *a priori* biologically relevant standard ('minimum difference to be detected')
Validity ▪ Internal validity ▪ Construct validity	*Validity* ▪ Internal validity ▪ External validity: Construct validity, face validity, predictive validity
Tools ▪ Small, nimble, designed, randomised controlled experiments.	*Tools* ▪ Appropriately designed experiments; adequate and appropriate sample sizes, randomisation, blinding, and controls ▪ Multi-batch experiments ▪ Systematic heterogeneity
Deliverables Progression decision metrics Effect size and confidence intervals	*Deliverables* Progression decision metrics Effect size and confidence intervals Proof of principle, proof of concept Statistical significance tests may be appropriate, depending on study objectives

Translational pilot studies ('Will it work?') provide preliminary evidence of *generalizability* (Bowen et al. 2009). Generalisable results are broadly applicable over a range of disease models, animal models, and operating conditions. Preclinical *proof of principle* (pPoP) experiments demonstrate effect of the intervention on the target disease process or pathophysiology. Preclinical *proof of concept* (pPoC) experiments demonstrate effect of the intervention in clinically relevant animal models, assessed by measured outcomes with a direct relationship to the clinical disease under study (Dolgos et al. 2016; Table 6.2). The goal of translational research is the evidence-based transition from 'bench to bedside' – the progression from basic science and/or mechanistic studies to clinically relevant animal models, clinical proof of concept studies, and eventually to randomised clinical trials (Rossello and Yellon 2016; Heusch 2017).

Animal-based laboratory studies are usually exploratory, designed to investigate multiple factors with small sample sizes. Therefore, sample sizes for preclinical pilot studies will be determined by on *information power* rather than statistical power.

Table 6.2: Proof of principle and proof of concept: definitions.

Preclinical proof of principle (pPoP)
- Demonstrates effect on the target disease process or pathophysiology
- Outcome variables are *surrogates* for the disease process.

Example: Evidence of potential efficacy of a new therapeutic agent for cancer is assessed by biomarkers for cell proliferation and apoptosis as surrogate measures of 'benefit' (Dolgos et al. 2016).

Preclinical proof of concept (pPoC)
- Demonstrates effect in disease-relevant preclinical animal models,
- Outcome variables have a *direct relationship* to the clinical disease

Example. Evidence of potential efficacy of a new therapeutic agent for cancer is assessed by tumour remission in patient-derived xenografts mouse models (Dolgos et al. 2016).

Clinical proof of concept (cPoC)
Early phase randomised clinical trials used to
- demonstrate product or device viability
- decide if confirmatory phase III trials can proceed ('go/no-go') and inform their planning.

Information power is the maximum amount of useful data that can be collected from each subject (*information density*) for the fewest total number of animals. In veterinary clinical research, pilot studies are usually a preliminary to large randomised controlled trials. Pilot studies are used to assess feasibility and pilot data to estimate effect size components. Methods for estimating sample sizes for each type of research are discussed separately.

Evidentiary strength of pilot data determined by study *validity*, with the emphasis on *internal validity* and reduction of systematic biases (Berkman et al. 2013). It is measured by effect size and precision, and assessed by data visualisation and data plots. *P-values alone are not evidence for whether or not the intervention 'works'.* *P*-values do not provide information about alternative hypotheses, an essential component of evidence assessment (Goodman and Royall 1988), and because they are based on small and usually methodologically unstable samples, *P*-values will be unreliable. Cautious use of *P*-values may be justifiable for larger replication studies and clinically–oriented pilot studies. Translation potential is assessed by generalisability or *external validity*. Generalizability is established by *consistency* across multiple lines of evidence, *concordance* with clinically relevant conditions and *replication* of model results across diverse operating conditions (Pitkänen et al. 2013; Kimmelman et al. 2014; Kimmelman and Federico 2017; Karp et al. 2020).

6.2 Building in Evidentiary Strength

Study strength and data quality are determined by study *validity* (Box 6.2). Validity is the qualitative assessment of 'the degree to which a result from a study is likely to be true and free from bias' (Higgins et al. 2022). The objective of a good experiment is to establish cause-and-effect between the experimental intervention and the study results. However, biases in study design, data collection, outcome assessments, and analysis methods will lead to misleading conclusions. Without validity, results will merely reflect the 'comparison of uncomparable groups', rather than a meaningful effect of the experimental intervention. For a study

---

BOX 6.2
Validity

Validity = 'the degree to which a result from a study is likely to be true and free from bias'

Internal validity

Methods-based ('truth within the study').
Purpose: Minimises systematic error (bias)

External validity

Model-based ('truth beyond the study').
Purpose: Increases study generalisability.

to produce high-quality data, methods that ensure study validity are far more important than either large sample sizes or experiment replication (Muhlhausler et al. 2013).

Bias is unrelated to sample size. However, minimising bias is essential for reducing animal use and the total number of animals used. Biased studies result in misleading claims of efficacy and translation potential. A major consequence of biased studies is the unintended waste of all animals used in those studies.

There are two main categories of validity: *internal validity* and *external validity*.

> *Internal validity* is methods-based. It refers to the extent of systematic error (bias) in the study and the methods used to minimise and control bias ('truth within the study').

> *External validity* is model-based. It refers to the extent to which the results obtained under the specific study conditions (the model) can be applied (generalised) to the larger population ('truth beyond the study')[1].

Internal validity is necessary for both empirical and translational pilots and an essential prerequisite for all hypothesis-testing studies. Internal validity alone does not confer external validity. However, without internal validity, studies cannot demonstrate external validity (Moher et al. 2010). Both types of validity must be built into the study design before experiments are conducted. They cannot be produced by statistical analyses after the fact (Bespalov et al. 2019).

6.2.1 Internal Validity

Internal validity refers to the risk of bias or systematic error inherent to a specific study (Box 6.3). All measurements incorporate two sources of error: random error and systematic error. Random error measures the precision of results and is determined by sample size. Precision is summarised by the standard deviation and standard error. In contrast,

---

> **BOX 6.3**
> *Internal Validity*
>
> *Internal validity* refers to the amount of bias incurred by flawed study design, conduct, and/or reporting
>
> Bias is minimised by
>
> > Appropriate comparators
> > Randomisation
> > Allocation concealment (blinding)
> > Predefined inclusion and exclusion criteria
> > Clearly defined outcome measures
>
> Bias is not lack of precision.
>
> An unbiased study does not necessarily have good external validity.

systematic error (bias) is unrelated to either sample size or precision. Bias is the consistent directional mismatch of observed results with the true effect, resulting from deficiencies in study design, conduct, and/or reporting (Eisenhart 1968; Higgins et al. 2011). Therefore, internal validity is a measure of confidence in the cause-effect relationship demonstrated by the study. Internal validity provides grounds for conclusions that treatment effects or observed differences between groups are real, and not an artefact of other potential explanations for the observed effects (Henderson et al. 2013; Pound and Ritskes-Hoitinga 2018).

Bias adversely affects both internal and external validity; however, lack of bias does not confer external validity. For example, a study may be unbiased because it was conducted with high internal validity safeguards, but it cannot demonstrate external validity if the study sample is not representative of the larger target population.

Bias occurs during subject selection, allocation of interventions to subjects, subject processing order, measurement and assessment of outcomes, and as a result of subject drop-out and missing data. The major bias domains are selection bias, performance bias, detection bias, and attrition bias (Huang et al. 2020). Bias is minimised by appropriate *controls* or *comparators*, *randomisation*, *allocation concealment (blinding)*, clearly defined *inclusion and exclusion criteria,* and clearly defined *outcome* measures (Higgins et al. 2011; Huang et al. 2020). These are described more fully in Chapter 3.

---

[1] https://courses.internal.vetmed.wsu.edu/jmgay/
clinical-epidemiology-evidence-based-medicine-
glossary/clinical-study-design-and-methods-
terminology

6.2.2 External Validity

External validity refers to the extent to which results generated by the study (the model system) are *representative* and *robust*, and therefore more widely applicable (*generalisable*) to the larger population (Box 6.4). Representativeness refers to how well the animal model captures essential features of the condition being modelled (model fidelity). Robustness refers to consistency of results obtained for diverse models, and how well models and constructs perform over a variety of conditions. Study replication contributes to assessment of robustness, and external validity to translation potential. If results cannot be replicated under different conditions ('generalising across') then translational success ('generalising to') is unlikely.

Unfortunately, successful translation of preclinical results to clinical applications have been rare (Henderson et al. 2013; Freedman et al. 2015; Pound and Ritskes-Hoitinga 2018; Errington et al. 2021a, b). This is primarily because the internal validity of most preclinical studies is poor (Henderson et al. 2013; Macleod et al. 2015). External validity is not possible without internal validity and will not be improved if methods used to increase study realism compromise internal validity (Lynch 1982). However, even high internal validity will not produce externally valid results if the model system is not sufficiently representative or clinically relevant. A non-representative animal model compromises external validity (Pound and Ritskes-Hoitinga 2018).

6.2.2.1 Representativeness

The animal model will not be representative if the model does not match the target population in essentials. Model signalment may not match demographics of the target population, the disease model may not match the relevant pathophysiology of the target condition, and the overall model may lack clinical relevance (model fidelity). Representativeness is assessed by *face validity*, *construct validity,* and *predictive validity*. These criteria were first described in 1969 as principles for the development of animal models of psychiatric affective disorders (McKinney and Bunney 1969). They were subsequently elaborated and refined by Willner (1984) and have since been extended to other basic science animal models (e.g. Markou et al. 2009).

Face validity is assessed by similarity of the animal model to the clinical or disease phenotype, and that model results are specific to the clinical condition being modelled ('Does it look right?')

Construct validity is assessed by the similarity of the animal model and its performance to the target clinical condition (the model is homologous), and that results are related theoretically, empirically, and unambiguously to the target condition ('Does it act right?').

Predictive validity is assessed by the similarity of the animal model to the target population, and the specificity of response or therapeutic outcome to the target outcome ('Do results from the animal model correctly predict performance in the clinical setting?'). In drug discovery and pharmacological intervention studies, predictive validity refers to similarity of responses of humans to the animal model ('human-animal correlation of therapeutic outcomes'; Belzun and Lemoine 2011). It also refers to the clinical relevance of the disease model or therapeutic biomarkers. Predictive validity is generally considered the most important validity item for translation (Belzun and Lemoine 2011; McGonigle and Ruggeri 2014; Tadenev and Burgess 2019).

BOX 6.4
External Validity

Extent to which results are representative, robust, and generalisable.

External validity is not possible without internal validity.

Representative ('generalising to')

  Face validity
  Construct validity
  Predictive validity

Replicable ('generalising across')

  Direct replication
  Conceptual replication.

There are no hard and fast criteria as to how to maximise external validity, or decide how much face or construct validity is needed for predictive validity. External validity criteria tend to be subjective and are therefore subject to bias in both interpretation and practical application. Trade-offs between the various types of external validity will be necessary for models of complex multifactorial diseases, such as oncological and neurodegenerative diseases (Pitkänen et al. 2013; Tadenev and Burgess 2019).

For a model to be 'good enough' for translational application, it needs to be informative and reasonably predictive of potential efficacy. A model can be useful without strict model fidelity, which in any case will be rarely possible (Pitkänen et al. 2013; Galanopoulou et al. 2017), nor always necessary (Russell and Burch 1959). The best option is to understand the general principles of external validity, and learn how to apply them on a study-by-study basis. A practical guide for ensuring construct validity was developed by Henderson et al. (2013). This is a 25-item checklist in four domains: the animal model, the clinical condition or disease to be modelled, experimental outputs, and experimental operations (Table 6.3).

For preclinical drug development models, Ferreira et al. (2019) developed a rigorous scoring system and score calculator for comparing animal models and identifying the model that will best fit specific research objectives. The calculator is based on 22 validation criteria in eight domains: epidemiological, symptomology and disease natural history, genetic, biochemical, aetiological, pharmacological, and endpoints.

6.2.2.2 Sex as a Biological Variable

Sex is an essential variable for ensuring construct validity. Sex differences in basic biology, pathophysiology, pharmacokinetics, and response to interventions have been reported for numerous diverse animal models (Karp et al. 2017; Wilson et al. 2022). However, female animals are still markedly under-represented in much of biomedical research (Will et al. 2017; Karp and Reavey 2019). There is no evidence that females are inherently more variable than males, a common reason for justifying male-only studies (Beery and Zucker 2011; Beery 2018). As it is, many published animal-based studies are based only on a single sex or are too underpowered to detect potential interaction of sex with other factors.

Table 6.3: Construct validity checklist.

Animal model	Clinical/disease model	Outputs	Experimental/ operational
Do features of the animal model represent those of the clinical patient population?	Does the lab-based disease model represent the clinical disease syndrome?	Do response variables represent clinically relevant outcomes?	Are they appropriate and correctly performed?
Signalment Age (juvenile, adult, aged) Sex Body mass	Mechanistic pathways Signs & symptoms Pathophysiology Chronicity (Acute or chronic) Severity	Biomarkers Physiological response Allometric scaling	Appropriate controls Appropriate methodology Technical skill Experimental confounds
Baseline characteristics: Behaviour, physiology, other phenotype markers	Clinical treatment: How defined? Co-interventions		Location
Comorbidities	*Treatment delivery*: Timing Route/method Duration, exposure		
Inclusion and exclusion criteria	Clinical setting		

Source: Adapted from Henderson et al. (2013).

Because sex bias in preclinical research greatly limits translation, National Institute of Health best-practice standards strongly encourage inclusion of female animals in addition to males (Clayton and Collins 2014 and incorporation of sex as a biological variable in study design (Miller et al. 2017; Honarpisheh and McCullough 2019). Regulatory guidelines covering chemical and drug toxicity testing may mandate testing of both sexes at each dose level (e.g. OECD Test Guidelines `https://www.oecd.org/`).

Sample size considerations. Incorporating both sexes into the study design does not require doubling of sample size, nor does it need to cost more money (Clayton 2015). Including sex as a biological variable is handled by appropriate statistical study design. If sex is not of interest as a primary explanatory (independent) variable, variation due to sex differences can be controlled by statistically based methods, such as stratified randomisation on sex and sample size balance. If sex is included as an independent variable, statistical power for detecting sex effects is increased by factorial and repeated-measures designs, and further improved by incorporating sex-related covariates, such as weight.

6.2.2.3 Body Size and Allometric Scaling

Body size differences between model and target species are a major challenge to construct validity. Commonly, allometric relationships are determined for cross-sectional data over a range of body sizes from multiple species. Allometric models are used for scaling pharmacokinetics, risk, drug dosages (Caldwell et al. 2004; Huang and Riviere 2014), tumour growth (Pérez-García et al. 2020), and physiological responses. They can also be used to estimate the likely range of responses if specific data for a given species are unavailable (Lindstedt and Schaeffer 2002).

The relationship between body size for the reference and target species is

$$Y = \left(\frac{X_{target}}{X_{ref}}\right)^b$$

where X is a measure of body size, such as body mass, and b is the scaling coefficient. The scaling coefficient b is obtained by nonlinear regression $Y = aX^b$, or more commonly, by least-squares regression on log-transformed variables: $\log(Y) = \log(a) + b \cdot \log(X)$.

Ratios (for example, mg/kg body mass) should be used only with considerable caution. Ratios implicitly assume the response Y is directly proportional to body size ($b = 1$). If b is not equal to 1, then ratio data will result in considerable statistical bias, spurious correlation, and greatly reduced precision. Body size differences should be accounted for by regression-based methods, such as analysis of covariance or multiple regression (Tanner 1949; Kronmal 1993; Packard and Boardman 1988, 1999; Jasieński and Bazzaz 1999). Body mass is useful for laboratory animal-to-human drug dosage conversions if exponents are verified by current evidence or appropriate pharmacological models. However, body surface area (e.g. mg/cm³) has been long discredited as an imprecise, flawed, and obsolete metric without any biological basis and should not be used (Tang and Mayersohn 2011; Blanchard and Smoliga 2015).

Allometric equations may be extremely unreliable when used for prediction, especially for drug dosages and pharmacokinetics. General scaling relationship approximations can provide initial approximations for expected forms of the relationship between size and a given physiological response (Calder 1984; Schmidt-Nielsen 1984). However, extrapolation to other species based on inappropriate conversion assumptions and insufficient evidence of efficacy can result in catastrophic outcomes (Leist and Hartung 2013; Kimmelman and Federico 2017; Van Norman 2019). Validity of scaling exponents depends on the type of comparison (interspecific or intraspecific), the data used to construct the relationship, and validity of the allometric model itself (choice of the best-fit line, data distribution, influence of outliers, measurement error). The allometric models assume that the response is determined by size-related factors alone and do not account for other potential determinants, such as age, species or strain, body composition, and sex (Eleveld et al. 2022). Sex effects in mice may be trait-specific; case-by-case evaluation of drug dosage scaling for mice is recommended (Wilson et al. 2022).

Sample size considerations. For a simple allometric regression based on two variables, a rule of thumb sample size is $N \geq 25$ if variance is expected to be high (Jenkins and Quintana-Ascencio 2020). Sample size can be determined more formally by iteration for a pre-specified precision of the regression coefficients b. Precision is calculated from the two-sided $100(1 - \alpha)\%$ confidence interval, $b \pm t_{1-\alpha/2, n-2} \cdot SE(b)$. For multiple regression models, sample size can be determined from Cohen's effect size $f^2 = R^2/(1 - R^2)$ (Chapter 15). Disadvantages of this method are the requirement for preliminary estimates of R^2, and the assumption of a straight-line fit. More advanced methods for sample size determination based on the joint estimation of the intercept and slope and estimation of the non-centrality parameter have been described (Colosimo et al. 2007; Jan and Shieh 2019). Large formal analyses involving multiple strains or species should consider phylogenetic regression methods (Cooper et al. 2016).

6.3 Sample Size Determination

Sample size features: Statistical study design, sequential testing strategy, internal validity

Sample size for preclinical pilots are determined by *information density* and *information power* rather than statistical power (Box 6.5). The goal of pilot studies is to obtain the maximum amount of useable, scientifically productive data from the fewest number of animals. This contrasts with sample size for definitive or confirmatory experiments that must be based on predefined primary outcome, effect size, and power.

The concept is borrowed from qualitative research studies and is based on the concept that the more relevant information contained in the sample, the fewer number of subjects will be needed (Malterud et al. 2016).

6.3.1 Information Density

Information density is the maximum amount of useable, scientifically productive data that can be obtained from each animal. However, measuring as many outcome variables as possible is not the same as maximising the amount of useable information. There is a trade-off between technical efficiency and effectiveness (or thoroughness; Hollner 2009). *Efficiency* refers to procedural leanness: an efficient experiment minimises investment of animals, resources, time and workload to obtain information required to meet study objectives. *Effectiveness (or thoroughness)* refers to task quality: an effective study produces data that are correct, accurate, timely, and of uniformly high quality. An effective study will also include recognition and prompt correction of mistakes or errors in protocol implementation.

Design strategies for information density will involve trade-offs between animal welfare measures (*refinement*) versus overall reduction of animal numbers (Nunamaker et al. 2021). Emphasis on technical speed and efficiency comes at the expense of data quality. Measuring too many outcome variables can produce results that are too difficult to interpret (Vetter and Mascha 2017), but measuring too few may miss important relationships or not properly capture the test effect. More data-intensive procedures per subject will reduce the number of animals that can be processed in a given time. On the other hand, as complexity increases, procedural times also increase, together with the potential for increased suffering and risk of adverse events (such as the animal dying before completion of the procedure). After methodology and protocol have been standardised, experimental performance can shift to an emphasis on efficiency, although frequent monitoring of task quality throughout the research cycle is advised. Emphasis on task quality over task completion may necessitate a shift in lab culture

---

BOX 6.5
Information Density and Power

Information density
  Maximum amount of useable, scientifically productive data obtained from each animal

  Sample size features: Trade-off between *efficiency* and *effectiveness*.

Information power
  Maximum amount of useable, scientifically productive data from the fewest number of animals.

and will require adequate oversight, communication, and support from investigators (Reynolds et al. 2019).

Sample size for information density. Pilot studies should prioritise thoroughness over efficiency. Therefore, operational constraints (time, skilled technical personnel, budget, resources) will determine the number of endpoints measured on each animal and therefore the number of animals that can be processed in a given time (Chapter 5). The number of endpoints will be determined by procedural needs per task and variable priority ('need to know' mission-critical variables' versus less important 'nice to know' variables). When surgical instrumentation is required, the number of procedures that can be realistically completed per animal is assessed by placing strict limits on maximum permissible surgical duration and time under anaesthesia.

6.3.2 Information Power

Information power is the maximum amount of useable, scientifically productive data obtained from the fewest number of animals (*sensu* Malterud et al. 2016). Even for pilot studies, formal statistical experimental designs are strongly recommended. Compared to trial and error experimentation, experimental designs make the variance of the responses as small as possible, and avoid or minimise bias in estimates of effects. Bias is further reduced by randomisation and blinding. Designed pilot experiments greatly increase the chances of detecting a difference in responses between test interventions.

An experiment involves changing one or more explanatory variables (*factors*) and observing the effects on one or more response variables. With exploratory animal-based studies, there will be a large number of potential factors and very little prior knowledge of how they might be expected to affect response or each other. The conventional approach is experiments limited to two-group and one-factor-at-a-time (OFAT) comparisons, with sample size calculated as multiples of the number of experimental 'groups' (Box and Draper 1987; Czitrom 1999). This approach is slow, inefficient, misses potentially useful information, and wastes

large number of animals. In contrast, designed experiments, especially factorial-type designs, permit simultaneous assessment of multiple input variables and are flexible, efficient, and very economical. They can therefore contribute to large savings in animal numbers for much more information. In general, designed experiments involve selection of the most likely set of candidate factors, deciding on a small fixed number of levels for each factor (usually a high and low value), then conducting the experiment on all possible combinations of those factors and factor levels. Each run combination is randomly assigned to an experimental unit. Genuine replicates made at the same run combination and addition of centre points provide an estimate of the variance for the effect (usually a regression coefficient). Therefore, sample size is determined by the number of replicate runs, not by 'group size' (Box and Draper 1987; Box et al. 2005; Chapter 19).

The recommended experimental strategy is sequential testing in three phases (Trutna et al. 2012):

1. *Factor reduction.* Screening designs are used narrow down a large number of candidate factors to a smaller subset of the most important. Factor 'importance' is assessed by the size of main effects and two-way interactions. Classic screening designs include fractional factorials and Plackett-Burman designs (Montgomery and Jennings 2006). Modern, computer-generated definitive screening designs are more efficient and less biased than older designs, resulting in more precise and reliable results. Definitive screening designs should be used only in the early stages of experimentation, and if there are four or more factors to be assessed. Factors are usually continuous variables, but designs can accommodate a few two-level categorical factors. It may be necessary to conduct separate experiments on each level of a multi-level categorical factor to avoid problems with interpretation (Jones and Nachtsheim 2011).

2. *Factor importance.* After a smaller subset of the most important factors has been identified, experiments are conducted on these

few remaining factors to determine significance of interaction effects and identify regions of the optimal or target response. Typical designs include full factorial and response surface designs. Responses are quantified by linear or polynomial regression and ANOVA table statistics (Box et al. 2005; Montgomery 2017).

3. *Factor testing*. Optimising experimental conditions will maximise the experimental signal if it exists. Therefore definitive experiments can proceed with greater certainty of success.

Results from each experiment provide feedback on whether or not the experiment is converging to the target solution predicted by the research hypothesis. Sequential feedback means that the study factors can be changed strategically and sensibly as information from each stage becomes available, rather than having to resort to trial and error tweaking (Czitrom 1999; Box et al. 2005). Because the factorial approach allows a much greater amount of precision for a given number of experimental runs, it is much more likely that a true effect will be detectable and not swamped by experimental error and unknown variation.

Sample size for information power. Sample size recommendations for multi-batch studies are a minimum of three animals per 'group' (Karp et al. 2020). However, with screening designs, fewer animals can be used because sample size is determined, not by group size, but by the number of replicate runs. Statistical power increases both with the size of regression coefficients and coefficient hierarchy, that is, more power is associated with main effects and less power with interactions (Jones and Nachtsheim 2011). Dean and Lewis (2006) is an excellent introduction to screening designs. Customised screening designs can be generated in commercially available packages such as SAS, SAS JMP Pro, and R (e.g. Lawson 2020).

Example: Information Power: Conventional Versus Screening Designs

Investigators wished to evaluate efficacy potential of a trial vaccine in mice. They thought that measurements of efficacy would be most strongly affected by mouse strain, vaccine dose, challenge-killing interval, mouse age at vaccination, and mouse sex. The initial study plan was to evaluate six mouse strains, three vaccine doses, three challenge-killing intervals, and three ages on both male and female mice. The goals were to determine what combination of factors contributes to the 'best' response, and quantify any possible sex differences.

Conventional approach. The investigators proposed a series of experiments designed as a series of two-group comparisons or 'one-way ANOVAs' on combinations of the major factors. They intended to use five mice per group 'because that was necessary for statistical significance' and was 'what everyone else does'. They concluded that at least 1620 mice would be required for total of 324 experimental 'groups' with five animals per group: Strain $\times$ dose $\times$ interval $\times$ age $\times$ sex $= 6 \times (3 \times 3 \times 3 \times 2) = 324 \times 5 = 1620$.

This piecemeal approach has serious disadvantages. It is likely to miss the most important factors or optimal factor levels giving the 'best' response and cannot estimate interactions between factors. It is highly unlikely that an experiment involving over a thousand animals could be processed in a reasonable period of time. Animals will be wasted as they age out of the study before they can be used. Finally, a large un-designed experiment is difficult to control and risks potentially large, unmanageable, and undetectable sources of variation that will swamp true experimental signals (Reynolds 2021).

Definitive screening design. The same study could be more efficiently and economically designed as a screening study, using far fewer animals and increased probability of detecting a real effect. Strain and sex are categorical factors. Dose, interval, and age are continuous factors with three levels each.

A reasonable approach for a screening study is to run separate definitive screening experiments on

Table 6.4: Example of a definitive screening experiment for assessing vaccine efficacy in mice. The design was generated in SAS JMP Pro 16. Experiments are conducted on separate strains in two blocks on the factors of sex, dose, interval, and age. This type of design requires less than 10% of the animals originally requested.

Run	Block	Dose	Interval	Age	Sex
1	1	−1	−1	1	Female
2	1	−1	−1	1	Male
3	1	0	1	1	Female
4	1	0	−1	−1	Male
5	1	−1	1	0	Male
6	1	1	−1	0	Female
7	1	0	0	0	Male
8	1	1	1	−1	Male
9	1	0	0	0	Female
10	1	1	1	−1	Female
11	2	−1	0	−1	Male
12	2	1	1	1	Male
13	2	1	−1	1	Male
14	2	−1	−1	−1	Female
15	2	−1	1	1	Male
16	2	−1	1	−1	Female
17	2	1	0	1	Female
18	2	1	−1	−1	Female

6.3.3 Veterinary Clinical Trials

Data from empirical clinical pilots can be used to assess feasibility and also to obtain estimates of sample size parameters (such as outcome variance, event rates, and effect size) for the later clinical trial (Chapter 16.4). Suggested sample size for a clinical trial pilot is a minimum of 12 subjects per arm, for a total of 24 subjects (Julious 2005). Up to 70 subjects per arm may be necessary depending on the effect size to be estimated and the amount of precision required for the estimate of the variance (Teare et al. 2014).

6.3.4 A Note on Safety and Tolerability

Empirical and translation pilot studies are usually too small to reliably assess safety or tolerability of a clinical intervention, especially if the safety metric is occurrence of an adverse event (yes/no). Adverse events are usually rare, and zero events in a small study do not mean that the intervention is safe. The sample size N to detect at least one adverse event can be approximated if the probability of detection α and the expected prevalence p of adverse events are known (or can be guessed). Sample size required for a safety study can be approximated by probability-based feasibility calculations (Chapter 8).

6.4 Assessing Evidentiary Strength

Strength of evidence means having confidence in study results. There is no sense continuing a certain line of enquiry or method of experimentation if only poor or unreliable results are produced. Strong evidence is obtained by consistency of results across multiple lines of experimentation. Exploratory data analysis by data visualisation methods enables rapid assessment of size, direction, and precision of results. The most promising findings are verified by sequential testing and replication, first with different models of the same syndrome, followed by independent replication within and across laboratories or locations (Box 6.6).

each strain, with each strain experiment run in blocks so that experimental effort can be distributed over two sessions. Table 6.4 is an example of a definitive screening design to be run in two blocks. For each trial, 18 runs for each strain are performed in random order. Nine males and nine females are required for each experiment. Each factor level is replicated seven times on each low and high value, and four times on the intermediate levels. The centre points (0, 0, 0) provide the variance estimates for assessing significance of main effects. The total number of mice required is now (6 strains × 18 runs) = 108 mice. This design requires fewer than 10% of the number of animal originally proposed.

Definitive studies based on factorial designs will require formal simulation-based power analyses, especially if interaction effects are of primary interest (Chapter 19).

BOX 6.6
Assessing Strength of Evidence

A. Causal appraisal
   1. Consistency
   2. Exploratory data analysis (data visualisation, graphs, and plots)
B. Sequential testing
   1. Preliminary evidence of efficacy
      a. Input factor screening and reduction
      b. Tests of efficacy
   2. Replication:
      a. Different models of the same syndrome
      b. Independent replication across multiple laboratories.

BOX 6.7
Exploratory Data Analysis

Always plot your data

Routine assessments

Why: Assessing data patterns, anomalies, outliers, distributional assumptions
What: Basic graphics (dot plots, histograms, boxplots, scatterplots, etc.)

Evidentiary strength

Why: Assessment of effect size, direction, and precision
What: Coverage plots, profile plots, half-normal plots.

Simple preliminary appraisal checks are easily performed by plotting the data, examining patterns, then informally scoring observed patterns as consistent with the scientific hypothesis, inconsistent, or undetermined (Berkman et al. 2013; US EPA 2017). Data plots can be used to assess effect strength, direction, heterogeneity, and transitivity. Evidentiary strength guidelines were first proposed for observational epidemiological studies in the 1960s and have been updated (Howick et al. 2009). The Causal Analysis/Diagnosis Decision Information System (CADDIS; US EPA 2017) is a useful guide to performing causal assessments and can be readily adapted for laboratory studies.

6.4.1 Exploratory Data Analysis

Exploratory data analysis (EDA) with graphs and data plots is an indispensable first step for all studies, including pilots. Graphics provide easy and rapid visualisation of effect size, direction, and precision without formal statistical testing (Box 6.7).

Routine visual assessments of raw data with simple graphics (histograms, boxplots, scatterplots, cumulative distribution function graphs, etc.) are essential for identifying patterns, anomalies, and outliers, and checking distributional assumptions prior to analyses (Tukey 1977; Filliben and Heckert 2012).

Evidentiary strength is evaluated by specialised plots showing direction, magnitude, and precision of results. These include *coverage plots* (means and confidence intervals), *profile plots* (means and simultaneous confidence intervals) and *half-normal plots* (regression coefficients). Detailed descriptions methods for calculating confidence intervals and other measures of precision are described in Part III.

6.4.2 Coverage Plots

Coverage plots are graphs of means and corresponding confidence intervals relative to the null hypothesis and the target, or minimum biologically important, difference (MBID) to be detected (Clayton and Hills 1993). These plots allow simultaneous assessment of observed effect size, precision, confidence, and power, and allow immediate evaluation and interpretation of size, direction, and variation associated with the observed effect (Steidl et al. 1997; Cumming 2012; Rothman and Greenland 2018).

The target difference is a measure of effect size and drives subsequent power calculations. It is the measured difference in the primary outcome variable between interventions and/or controls. It must be determined *a priori* and clearly defined, and the biological importance of the difference must be scientifically justified (Cook et al. 2015, 2017). The target difference and rationale should be

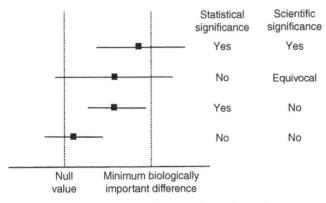

Figure 6.1: Coverage plot showing hypothetical mean effects and associated confidence intervals. Evidence for a biologically important effect is suggested if confidence intervals cross or exceed the predefined target difference to be detected ('yes'), 'equivocal' results if neither the target difference nor the expected null difference can be excluded, and unpromising if confidence intervals include the null but exclude the target difference ('no').

reported in the methods section of protocols and manuscripts. For the study to have sufficient power to detect the target difference, the lower limit of the confidence interval must be greater than the upper limit for null hypothesis value. However, intervals containing the null value may also contain values that are not statistically significant but may be of practical or biological importance (Figure 6.1). Therefore evaluation of pilot study results consists of a three-part decision tree:

1. Does the confidence interval for the observed difference cross or exceed the target difference and exclude the null value?

   If 'Yes', results are promising. Sample sizes for subsequent definitive studies can be obtained by working backwards from the confidence intervals to compute the sample size. Plotting confidence interval widths indicates how subsequent studies may be powered to increase precision of the mean difference, and confidence interval widths indicates how subsequent studies may be powered to increase precision.

2. Does the confidence interval cross both the target difference and the null value?

   If 'Yes', results are equivocal. Such large confidence intervals cannot preclude the possibility of either no true difference at all or a difference larger than target. Study design, study procedures, and experimental endpoints should be revisited and revised as needed. A new study may be necessary, or the current study redesigned.

3. Does the upper confidence limit *exclude the target difference*?

   If the confidence interval does not cover the *target difference* ('No'), a new study may be necessary. There is insufficient evidence of effect even if the confidence interval excludes the null value.

6.4.3 Sample Size From Confidence Intervals and Standard Deviation

Sample size greatly affects variability but does not affect estimates of the expected mean of the sample (Bishop et al. 2022). Lee et al. (2014) recommend that strength of preliminary evidence can be assessed by comparing the effect to null and reference values with confidence intervals of different widths (e.g. 95%, 90%, 80%, 75%). This information can then be used to calculate appropriately powered sample sizes by working backwards from a given confidence interval to compute the sample size. A visual approach simplifies evaluation of responses relative to some specific reference or target value. To obtain 80% power, the sample size must be large enough so that 80% of all possible estimates will be at least 1.96 standard errors from the reference point. If the observed effect does not attain statistical significance, the sample size in a new study must be increased by a factor calculated as the square of the ratio of the current value of the standard deviation to that value required to obtain a desired power with specified confidence (Clayton and Hills 1993).

Example: Calculating Sample Size

The target difference in a hypothetical experiment is 0.5. The standard deviation (SD) observed in a pilot study with sample size of 30 is 0.3. However, the standard deviation required to detect the target difference with confidence of 95% and power 80% is

$$SD = 0.5/(1.96 + 1.282) = 0.154$$

Therefore, to detect a difference of 0.5, sample size for the new study must be increased by a factor of $(0.3/0.154)^2 = 3.8$. The sample size in the new study would have to be

$$N = 3.8 \times 30 \cong 114 \text{ subjects.}$$

6.4.4 Profile Plots

Profile plots are useful for the simultaneous assessment of multiple variables in a study. If the variables have different units of measurement, they must be standardised t otherwise confidence intervals are meaningless and cannot be interpreted. Simultaneous confidence intervals SCI are constructed to assess multiple outcome variables to minimise false positive rates. The simultaneous confidence bands $100(1 - \alpha)\%$ for each standardised variable $\bar{r}_j$ are

$$\bar{r}_j \pm \frac{s_{zj}}{\sqrt{n}} \cdot \sqrt{\frac{p(n-1)}{n-p} \cdot F_{\alpha,p,n-p}}$$

(see Chapter 10 for details). Transformed means and corresponding confidence bands for all variables are plotted on the same graph. Confidence intervals that do not intersect the pre-specified null value suggest evidence for the existence of an effect. No evidence is provided against the null hypothesis if confidence intervals contain the null value. Appendix 10.A provides sample SAS code for calculating simultaneous confidence bands and generating profile plots.

Example: Nutrient Screening of Obese Dogs in a Weight Loss Study

---

(Data from German et al. 2015). Daily intake of 20 essential nutrients was measured for 27 obese dogs on an energy-restricted weight loss diet. The goal was to establish if average daily nutrient intakes complied with recommended dietary levels. Observations for each nutrient were first standardised to its respective specific minimum recommended daily allowance value (RDA). For example, crude protein intake is standardised by dividing each observation by 3.28, the NRC recommended level for crude protein. The reference ratio across all standardised variables equals 1. Computation details are given in Chapter 10.

Figure 6.2a shows standardised mean values and simultaneous confidence intervals for all 20 nutrients. Confidence intervals for fat-soluble vitamins A, D_3, and E, and folic acid and manganese are well above 1, with mean intakes between 10–35 times above recommended levels. When these extreme values are excluded from the plot (Figure 6.2b), it is seen more clearly that confidence intervals for eight nutrients are above recommended levels, those for magnesium and selenium below recommended levels, and those for the remaining six nutrients suggesting conformity with recommended intakes.

6.4.5 Half-Normal Plots

Complex experiments with multiple factors can be modelled by multiple or polynomial regression, and regression coefficients displayed in a half-normal plot. The regression coefficients are the effect sizes. The half-normal plot allows rapid visual assessment of the strength of all main effects and interactions, and discrimination of the most promising factors to include in subsequent experiments. The magnitude of each effect is assessed by the position of the coefficient relative to a reference line constructed from all the effects closest to zero. Effects furthest from zero on the x-axis have greater magnitude and are 'significant' (Daniel 1959). A factor is important if it has a large main effect or is involved in a large two-factor interaction (furthest from line of zero effect).

There are six steps in the construction of a half-normal plot:

1. Calculate all k regression coefficients for main effects and two-way interactions
2. Sort coefficients in ascending rank order from smallest to largest, $i = 1$ to N.

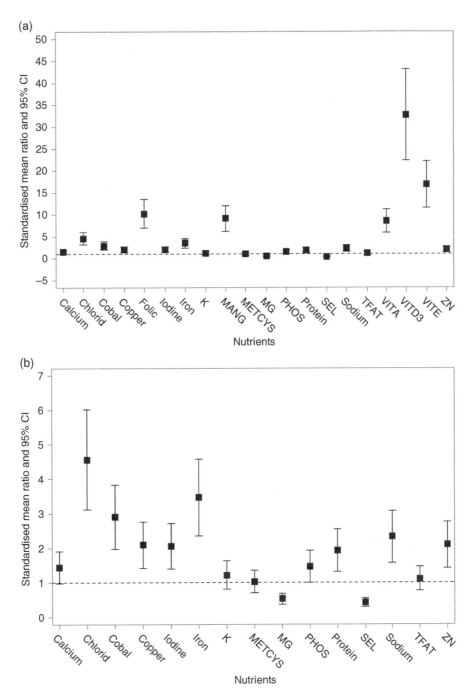

Figure 6.2: Profile plot. Mean daily intake for 27 obese dogs of essential nutrients and 95% simultaneous confidence intervals. Values are standardised to a common ratio based on recommended daily allowance levels.

Source: Data from German et al. (2015). (a) Standardised mean values and confidence intervals for 20 nutrients. Dotted line is the line of equality for the standardised recommended intake. (b) Mean observed values and confidence intervals for the subset of nutrients with standardised means less than 10.

3. Calculate median order statistic for each of m values: $i_m = (i - 0.5)/k$
4. Obtain the z value for each order statistic i_m from the standard normal distribution.
5. Plot the z-transformed order statistics against the effect sizes (regression coefficients).
6. Fit a reference line to the effects clustered most closely to zero.

Example: Effects of Environmental Toxins on Mouse Growth

(Data from Porter et al. 1984). An experiment was designed to simultaneously assess interactions of food and water restriction (factors 1 and 2), an immunosuppressant chemical (factor 3), an infectious agent (factor 4), and a diet supplement (factor 5) on growth rates of Swiss-Webster white mice. The study design was a five-factor half-fraction factorial study design at two levels. This design used $2^{5-1} = 16$ runs replicated twice, totalling 32 female mice. Growth rates were measured on pups from a single litter from each mouse.

There were $k = 15$ regression coefficients obtained from the 5 main effects and 10 interactions (Table 6.5). The half-normal plot (Figure 6.3) shows the most important factors were food restriction (factor 1), water restriction (factor 2), and the infectious agent (factor 4). Diet supplement (factor 5) and interaction effects were not important. Factor 5 was dropped from subsequent experiments and a second environmental contaminant chemical was substituted.

Table 6.5: Environmental toxins and mouse growth: screening for effect size. Regression coefficients obtained from polynomial regression on five factors and all two-way interactions.

Factor ID	Regression coefficient	Median order statistic	z
1	0.100	0.967	0.8331
2	0.053	0.900	0.8159
3	−0.008	0.233	0.5922
4	−0.033	0.033	0.5133
5	0.008	0.633	0.7367
1 × 2	0.011	0.700	0.7580
1 × 3	0.015	0.833	0.7977

(*continued*)

Table 6.5: (continued)

Factor ID	Regression coefficient	Median order statistic	z
1 × 4	−0.017	0.100	0.5398
1 × 5	0.013	0.767	0.7784
2 × 3	0.001	0.500	0.6915
2 × 4	−0.005	0.367	0.6431
2 × 5	−0.010	0.167	0.5662
3 × 4	−0.007	0.300	0.6179
3 × 5	−0.001	0.433	0.6676
4 × 5	0.003	0.567	0.7145

Source: Data from Porter et al. (1984).

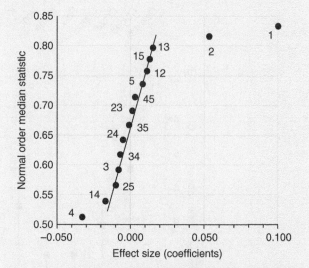

Figure 6.3: Half-normal plot. The experiment screened five environmental stressors thought to affect mouse pup growth. The plot displays regression coefficients (effect sizes) computed for the five factors and all two-way interactions.

Source: Data from Porter et al. (1984).

6.4.6 Interaction Plots

Interaction plots are a simple and rapid method for assessing generalizability. In general, the interaction plot shows the means of one independent categorical variable on the x-axis and displays the means of a second categorical variable as separate lines, showing how the means change with the levels of the first variable (Chapter 19). Interaction plots used to assess generalizability plot values of the treatment variables on the y-axis against categorical confounder variables on the x-axis.

Interactions are negligible if the line is horizontal parallel to the *x*-axis. Alternatively, if the interaction is significant, the line is not horizontal and the mean treatment response is not consistent across all the levels of the confounder. This suggests that treatment main effects are dominated by the confounder, may not be biologically important or are too variable (Lynch 1982).

Lynch (1982) suggests four potential sources of confounding interactions with treatment:

Subject group by treatment interactions, when the experimental units comprise distinct groups or cohorts. For example, young adult mice will not be representative of aged mice.

Block by treatment interaction. Lynch defines these as 'situational specifics' that include potential blocking variables such as location, lighting, noise, treatment administration, investigator, timing of measurement, etc.

Time by treatment interactions are suggested if cause-effect relationships differ between baseline and subsequent measurement periods.

Replicate × treatment interactions. Consistency between replicates suggests the replicate × treatment effect is negligible and provides evidence of the robustness of results.

Example: Interaction Plots Across Replicate Experiments

(From von Kortzfleisch et al. 2020). Investigators compared effectiveness of a conventional experimental design and a multi-batch design for evaluating behavioural and physiological differences between four mouse strains. Strain differences were evaluated across four replicate experiments. The overall effect size was the mean strain difference. Mean strain differences were estimated from the replicate experiments for each paired strain comparison. Evidence of reproducibility was suggested by consistency of the effect of mouse strain across replicate experiments; that is, the (strain x replicate) interaction is statistically negligible. They found that the multi-batch approach improved reproducibility, and increased discovery of treatment effects. Figure 6.4 shows an exemplary interaction plot presenting mean differences between strains and confidence intervals; the first plot shows poor repeatability across replicates, and the second plot good repeatability with negligible effect of replicate,

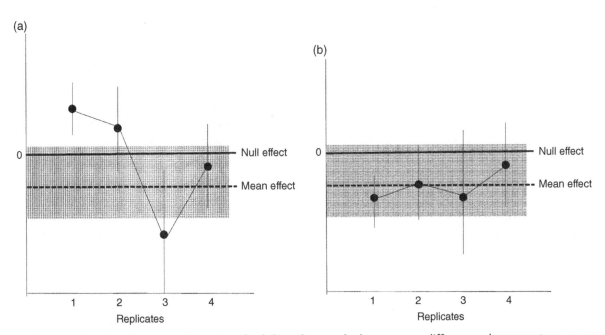

Figure 6.4: Interaction plots for assessing generalizability. The graph shows mean differences between two mouse stains and 95% confidence intervals across four experimental replicates.

Source: Adapted from von Kortzfleisch et al. (2020). (a) Poor repeatability across replicates; evidence for generalizability is poor. (b) Good repeatability across replicates; evidence for generalizability is good as the data suggest treatment effects are relatively consistent and independent of replicate effects

6.4.7 Replication

Replication studies consist of multiple experiments conducted to assess and validate the robustness of research results (Box 6.8). They are considered the ultimate test of external validity (Drude et al. 2021). However, there is considerable misunderstanding of how replication is defined, and how replication studies are to be designed and conducted. Frequently, investigators think a 'replication study' consists of two or more repetitions of an entire experiment over identical conditions (Fitts 2011; Frommlet and Heinze 2021). However, repeating an unrepresentative, biased, and poor-quality study does not provide evidence of external validity or translation potential, no matter how many times the study is repeated.

Pitkänen et al. (2013) recommend a three-phase experimental approach

1. Preliminary studies designed to show evidence of efficacy.
2. Independent replication studies in a different laboratory guided by data from the first study in a given model.
3. Confirmatory testing in another model of the same syndrome.

Replication studies should be conducted only after there is substantive preliminary evidence of efficacy. Animal use oversight committees expressly discourage unnecessary duplication of animal experiments. Justification for replication studies must be based on consistency of evidence across multiple sources (including a comprehensive literature review), evidentiary strength of prior data (especially strong internal validity), and clinical relevance of the models and outcome variables (Drude et al. 2021). Initial studies are followed by testing in another animal and or disease model of the same syndrome, and culminating in independent replication studies in different laboratories (Landis et al. 2012; Henderson et al. 2013; Pitkänen et al. 2013; Dirnagl et al. 2021; Drude et al. 2021).

Table 6.6 distinguishes the different types of replication experiments. The main components of a replication study are *location* (single laboratory, multiple laboratories), *study conditions* (animal environment, lab environment, equipment, reagents, personnel, etc.), *model system* (animal, animal model, disease model) and *variables measured* (explanatory variables, response variables).

Direct replications are multiple experiments that are the same across all study components, and performed by the same laboratory (Fraser et al. 2019). Simple direct repeats of experiments ('repeating an experiment in triplicate') by a single laboratory are not recommended unless internal validity is high and the experimental protocol is rigorously defined (Würbel, 2017; Drude et al. 2021; Frommlet and Heinze 2021). Simple repeats can only estimate differences due to sampling error between repeats, and thus estimate only the overall measurement error of the experiment, not the validity of experimental results. Moreover, if variation is non-systematic and uncontrolled, simple direct repeats are expensive and waste animals without increasing information. In contrast, if variation is systematically introduced with appropriate study designs (e.g. multi-batching), direct replication studies can provide a preliminary assessment of robustness across experiments, and inform decisions to proceed with replication between laboratories (Richter 2017; Drude et al. 2021).

Conceptual replications are multiple experiments that differ in one or more key inputs that influence the effect through different mechanisms

---

BOX 6.8
Replication

Multiple experiments conducted to assess robustness of research findings

Direct replications: Multiple experiments similar across all study components, and performed by the same laboratory.
Conceptual replications: Multiple experiments differing in one or more key inputs: Location, model, operating conditions, variables.

Repeating unrepresentative, biased, and poor-quality studies does not provide evidence of external validity or translation potential.

Table 6.6: Six types of replication studies defined by number of experimental locations, 'conditions', model system (animal or disease model) and variables measured (explanatory and/or response). Replication studies can be the same across all components (*direct replication*), or differ in one or more components (*conceptual replication*).

Location	Experimental conditions	Model system	Variables	What is evaluated?
Direct replication				
Same	Same	Same	Same	Single laboratory study with identical repeats (*direct internal replication*). Can assess sampling error, quality standards, mistakes, fraud, etc. *Measurement error, precision* *External validity* (limited) with multi-batch designs (planned heterogeneity)
Differ	Same	Same	Same	Multi-laboratory study (*independent replication*). Assesses effects of different lab procedures, personnel, equipment, environments, etc. in different labs *Robustness (between laboratories), external validity*
Conceptual replication				
Same	Differ	Same	Same	Systemic variation in experimental operating conditions (*planned heterogeneity*) between replications. Conditions that are varied include environment, batch, days, time of day, operators, suppliers, vendors, etc. *Robustness (procedural/process replication)*
Same	Same	Same	Differ	Systematic changes in explanatory factors, response variables, and/or methods for measuring or quantifying the response. Assesses operationalisation of the research question (e.g. 'benefit' = symptom relief *versus* improved function *versus* improved survival). *Robustness (outcome replication); external validity*
Same	Same	Differ	Same	Systematic changes in the animal model/species/strain, and/or disease model. Tests generalisation and robustness of results in a new model system related to measures of efficacy or effect tested *Robustness (animal and disease model); external validity*
Same or differ	Same or differ	Same or differ	Same or differ	Multiple simultaneous systematic controlled changes in any or all components of study design and operations *Robustness, external validity*

Source: Adapted from Fraser et al. (2019).

or pathways. These include test of treatment effect as a function of different animal models (species, strain), disease models, disease severity levels, age groups, treatment interventions, and/or experimental outcomes. Conceptual replications are an example of planned heterogeneity. Independent replication experiments performed across multiple laboratories are regarded as the gold standard for external validity and reproducibility (Karp 2018; Voelkl et al. 2018; Usui et al. 2021). However, single-laboratory replications consisting of a series of mini-experiments separated over time can be as effective as a multi-lab study, as well as more economical (von Kortzfleisch et al. 2020).

Efficiency-effectiveness trade-offs are inherent to replication studies. The relative efficiency of repeating experiments in different laboratories compared to a single laboratory can be judged by criteria such as greater statistical power, relative reduction in animal numbers to detect a given effect size, and improved coverage probability of the true effect size for a fixed number of animals (von Kortzfleisch et al. 2020). Replications across as few as two laboratories can produce substantial improvements in power and predictive validity, as long as protocols are standardised and internal validity is ensured (Karp 2018; Voelkl et al. 2018; Drude et al. 2021).

6.4.8 Design and Sample Size for Replication

Any statistically based experimental design appropriate for the study can be used for replication studies (factorial, split-plot, randomised complete block, etc.). Internal validity must be high, with clearly specified methods for randomisation and allocation concealment (blinding), and defined inclusion and exclusion criteria (Drude et al. 2021).

Sample size is greatly reduced by adoption of multi-batch designs rather than the conventional approach of replicating independently powered experiments. Multi-batch designs consist of several small independent experiments conducted at separate time points by the same laboratory or even in different laboratories. They are in effect large-scale randomised complete block experiments where each block is batch or replicate. Results are combined to assess intervention effects. The suggested minimum is three batches (Karp et al. 2020; von Kortzfleisch et al. 2020).

References

Beery, A.K. (2018). Inclusion of females does not increase variability in rodent research studies. *Current Opinion in Behavioral Sciences* 23: 143–149. https://doi.org/10.1016/j.cobeha.2018.07.016.

Beery, A.K. and Zucker, I. (2011). Sex bias in neuroscience and biomedical research. *Neuroscience and Biobehavioral Reviews* 35: 565–572. https://doi.org/10.1016/j.neubiorev.2010.07.002.

Belzun, C. and Lemoine, M. (2011). Criteria of validity for animal models of psychiatric disorders: focus on anxiety disorders and depression. *Biology of Mood & Anxiety Disorders* 1 (1): 9. https://doi.org/10.1186/2045-5380-1-9.

Berkman, N.D., Lohr, K.N., Ansari, M., et al. (2013). Grading the strength of a body of evidence when assessing health care interventions for the effective health care program of the agency for healthcare research and quality: an update. Methods guide for comparative effectiveness reviews (Prepared by the RTI-UNC Evidence-based Practice Center under Contract No. 290-2007-10056-I). AHRQ Publication No. 13(14)-EHC130-EF. Rockville, MD: Agency for Healthcare Research and Quality. www.effectivehealthcare.ahrq.gov/reports/final.cfm (accessed 2022).

Bespalov, A., Wicke, K., and Castasgné, V. (2019). Blinding and randomization. In: *Good Research Practice in Non-Clinical Pharmacology and Biomedicine*, Handbook of Experimental Pharmacology, 257 (ed. A. Bespalov, M. Michel, and T. Steckler). Springer Cham. https://doi.org/10.1007/164_2019_279.

Bishop, D.V.M., Thompson, J., and Parker, A.J. (2022). Can we shift belief in the 'Law of Small Numbers'? *Royal Society Open Science* 9: 211028. https://doi.org/10.1098/rsos.211028.

Blanchard, O.L. and Smoliga, J.M. (2015). Translating dosages from animal models to human clinical trials – revisiting body surface area scaling. *FASEB Journal* 29 (5): 1629–1634. https://doi.org/10.1096/fj.14-269043.

Bowen, D.J., Kreuter, M., Spring, B. et al. (2009). How we design feasibility studies. *American Journal of Preventive Medicine* 36 (5): 452–457. https://doi.org/10.1016/j.amepre.2009.02.002.

Box, G.E.P. and Draper, N.R. (1987). *Empirical Model-Building and Response Surfaces*. New York: Wiley.

Box, G.E.P., Hunter, W.G., and Hunter, J.S. (2005). *Statistics for Experimenters: An Introduction to Design, Data Analysis, and Model Building*, 2e. New York: Wiley.

Calder, W.A. III (1984). *Size, Function and Life History*. Cambridge, MA: Harvard University Press.

Caldwell, G.W., Masucci, J.A., Yan, Z., and Hageman, W. (2004). Allometric scaling of pharmacokinetic parameters in drug discovery: can human CL, V_{ss} and t1/2 be predicted from in-vivo rat data? *European Journal of Drug Metabolism and Pharmacokinetics* 29: 133–143.

Clayton, D. and Hills, M. (1993). *Statistical Models in Epidemiology*. Oxford: Oxford University Press.

Clayton, J.A. (2015). Studying both sexes: a guiding principle for biomedicine. *FASEB Journal* 30 (2): 519–524. https://doi.org/10.1096/fj.15-279554:10.1096/fj.15-279554.

Clayton, J.A. and Collins, F.S. (2014). Policy: NIH to balance sex in cell and animal studies. *Nature* 509 (7500): 282–283. https://doi.org/10.1038/509282a.

Colosimo, E.A., Cruz, F.R.B., Miranda, J.L.O., and Van Woensel, T. (2007). Sample size calculation for method validation using linear regression. *Journal of Statistical Computation and Simulation* 77 (6): 505–516. https://doi.org/10.1080/00949650601151729.

Cook, J.A., Hislop, J., Altman, D.G. et al. (2015). Specifying the target difference in the primary outcome for a randomised controlled trial: guidance for researchers. *Trials* 16: 12. https://doi.org/10.1186/s13063-014-0526-8.

Cook, J.A., Julious, S.A., Sones, W. et al. (2017). Choosing the target difference ('effect size') for a randomised controlled trial – DELTA2 guidance protocol. *Trials* 18 (1): 271. https://doi.org/10.1186/s13063-017-1969-5.

Cooper, N., Thomas, G.H., and FitzJohn, R.G. (2016). Shedding light on the 'dark side' of phylogenetic comparative methods. *Methods in Ecology and Evolution* 7 (6): 693–699. https://doi.org/10.1111/2041-210X.12533.

Cumming, G. (2012). *Understanding the New Statistics: Effect sizes, Confidence Intervals, and Meta-Analysis.* New York: Routledge.

Czitrom, V. (1999). One-factor-at-a-time versus designed experiments. *The American Statistician* 53: 126–131.

Daniel, C. (1959). Use of half-normal plots in interpreting factorial two-level experiments. *Technometrics* 1 (4): 311–341.

Dean, A. and Lewis, S. (2006). *Screening Methods for Experimentation in Industry, Drug Discovery, and Genetics.* Springer, 330 p. https://link.springer.com/content/pdf/10.1007%2F0-387-28014-6.

Dirnagl, U., Bannach-Brown, A., and McCann, S. (2021). External validity in translational biomedicine: understanding the conditions enabling the cause to have an effect. *EMBO Molecular Medicine* 14 (2): 1757–4676. https://doi.org/10.15252/emmm.202114334.

Dolgos, H., Trusheim, M., Gross, D. et al. (2016). Translational Medicine Guide transforms drug development processes: the recent Merck experience. *Drug Discovery Today* 21 (3): 517–526. https://doi.org/10.1016/j.drudis.2017.01.003.

Drude, N.I., Gamboa, L.M., Danziger, M. et al. (2021). Science Forum: improving preclinical studies through replications. *eLife* 10: e62101. https://doi.org/10.7554/eLife.62101.

Eisenhart, C. (1968). Expression of the uncertainties of final results. *Science* 160: 1201–1204. https://doi.org/10.1126/science.160.3833.1201.

Eleveld, D.J., Koomen, J.V., Absalom, A.R. et al. (2022). Allometric scaling in pharmacokinetic studies in anesthesiology. *Anesthesiology* 136 (4): 609–617. https://doi.org/10.1097/ALN.0000000000004115.

Errington, T.M., Denis, A., Perfito, N. et al. (2021a). Reproducibility in cancer biology: challenges for assessing replicability in preclinical cancer biology. *eLife* 10: e67995. https://doi.org/10.7554/eLife.67995.

Errington, T.M., Mathur, M., Soderberg, C.K. et al. (2021b). Investigating the replicability of preclinical cancer biology. *eLife* 10: e71601. https://doi.org/10.7554/eLife.71601.

Ferreira, G.S., Veening-Griffioen, D.H., Boon, W.P.C. et al. (2019). A standardised framework to identify optimal animal models for efficacy assessment in drug development. *PLoS ONE* 14 (6): e0218014. https://doi.org/10.1371/journal.pone.0218014.

Filliben, J.J. and Heckert, A. (2012). Exploratory data analysis. In: *NIST/SEMATECH e-Handbook of Statistical Methods* (ed. C. Croakin and P. Tobias). http://www.itl.nist.gov/div898/handbook/2012. https://doi.org/10.18434/M32189.

Fitts, D.A. (2011). Ethics and animal numbers: informal analyses, uncertain sample sizes, inefficient replications, and Type I errors. *Journal of the American Association for Laboratory Animal Science* 50: 445–453.

Fraser, H., Barnett, A., Parker, T.H., and Fidler, F. (2019). The role of replication studies in ecology. *Ecology and Evolution* 10: 5197–5207.

Freedman, L.P., Cockburn, I.M., and Simcoe, T.S. (2015). The economics of reproducibility in preclinical research. *PLoS Biology* 13: e1002165.

Frommlet, F. and Heinze, G. (2021). Experimental replications in animal trials. *Laboratory Animals* 55 (1): 65–75. https://doi.org/10.1177/0023677220907617.

Galanopoulou, A.S., Pitkänen, A., Buckmaster, P.S., and Moshé, S.L. (2017). What do models model? What needs to be modeled? In: *Models of Seizures and Epilepsy*, 2e (ed. A. Pitkänen, P.S. Buckmaster, A.S. Galanopoulou, and S.L. Moshé), 1107–1119. Elsevier.

German, A.J., Holden, S.L., Serisier, S. et al. (2015). Assessing the adequacy of essential nutrient intake in obese dogs undergoing energy restriction for weight loss: a cohort study. *BMC Veterinary Research* 11: 253. https://doi.org/10.1186/s12917-015-0570-y.

Goodman, S.N. and Royall, R. (1988). Evidence and scientific research. *American Journal of Public Health* 78 (12): 1568–1574. https://doi.org/10.2105/ajph.78.12.1568.

Henderson, V.C., Kimmelman, J., Fergusson, D. et al. (2013). Threats to validity in the design and conduct of preclinical efficacy studies: a systematic review of guidelines for *in vivo* animal experiments. *PLoS*

Medicine 10 (7): e1001489. https://doi.org/10.1371/journal.pmed.1001489.

Heusch, G. (2017). Critical issues for the translation of cardioprotection. *Circulation Research* 120 (9): 1477–1486. https://doi.org/10.1161/CIRCRESAHA.117.310820.

Higgins, J.P., Altman, D.G., Gøtzsche, P.C. et al. (2011). The Cochrane collaboration's tool for assessing risk of bias in randomised trials. *BMJ* 343: d5928. https://doi.org/10.1136/bmj.d5928.

Higgins, J.P.T., Savović, J., Page, M.J. et al. (2022). Chapter 8: Assessing risk of bias in a randomized trial. In: *Cochrane Handbook for Systematic Reviews of Interventions, version 6.3 (updated February 2022)* (ed. H. JPT, J. Thomas, J. Chandler, et al.). Cochrane www.training.cochrane.org/handbook.

Hollner, E. (2009). *The ETTO Principle: Efficiency-Thoroughness Trade-Off: Why Things That Go Right Sometimes Go Wrong*. Taylor & Francis Group, ProQuest Ebook Central https://ebookcentral.proquest.com/lib/ufl/detail.action?docID=438714.

Honarpisheh, P. and McCullough, L.D. (2019). Sex as a biological variable in the pathology and pharmacology of neurodegenerative and neurovascular diseases. *British Journal of Pharmacology* 176 (21): 4173–4192. https://doi.org/10.1111/bph.14675.

Howick, J., Glasziou, P., and Aronson, J.K. (2009). The evolution of evidence hierarchies: what can Bradford Hill's 'guidelines for causation' contribute? *Journal of the Royal Society of Medicine* 102 (5): 186–194. https://doi.org/10.1258/jrsm.2009.090020.

Huang, Q. and Riviere, J.E. (2014). The application of allometric scaling principles to predict pharmacokinetic parameters across species. *Expert Opinion on Drug Metabolism & Toxicology* 10: 1241–1253.

Huang, W., Percie du Sert, N., Vollert, J., and Rice, A.S.C. (2020). General principles of preclinical study design. In: *Handbook of Experimental Pharmacology*, vol. 257, 55–69. https://doi.org/10.1007/164_2019_277.

Jan, S.L. and Shieh, G. (2019). Sample size calculations for model validation in linear regression analysis. *BMC Medical Research Methodology* 19: 54. https://doi.org/10.1186/s12874-019-0697-9.

Jasieński, M. and Bazzaz, F.A. (1999). The fallacy of ratios and the testability of models in biology. *Oikos* 321–327.

Jenkins, D.G. and Quintana-Ascencio, P.F. (2020). A solution to minimum sample size for regressions. *PLoS ONE* 15 (2): e0229345. https://doi.org/10.1371/journal.pone.0229345.

Jones, B. and Nachtsheim, C.J. (2011). A class of three level designs for definitive screening in the presence of second-order effects. *Journal of Quality Technology* 43 (1): 1–15. https://doi.org/10.1080/00224065.2011.11917841.

Julious, S.A. (2005). Sample size of 12 per group rule of thumb for a pilot study. *Pharmaceutical Statistics* 4: 287–291.

Karp, N.A. (2018). Reproducible preclinical research—Is embracing variability the answer? *PLoS Biology* 16 (3): e2005413. https://doi.org/10.1371/journal.pbio.2005413.

Karp, N.A., Mason, J., Beaudet, A.L. et al. (2017). Prevalence of sexual dimorphism in mammalian phenotypic traits. *Nature Communications* 8: 15475. https://doi.org/10.1038/ncomms15475.

Karp, N.A. and Reavey, N. (2019). Sex bias in preclinical research and an exploration of how to change the status quo. *British Journal of Pharmacology* 176 (21): 4107–4118. https://doi.org/10.1111/bph.14539.

Karp, N.A., Wilson, Z., Stalker, E. et al. (2020). A multi-batch design to deliver robust estimates of efficacy and reduce animal use – a syngeneic tumour case study. *Scientific Reports* 10 (1): 6178. https://doi.org/10.1038/s41598-020-62509-7.

Kimmelman, J. and Federico, C. (2017). Consider drug efficacy before first-in-human trials. *Nature* 542: 25–27. https://doi.org/10.1038/542025a.

Kimmelman, J., Mogil, J.S., and Dirnagl, U. (2014). Distinguishing between exploratory and confirmatory preclinical research will improve translation. *PLoS Biology* 12 (5): e1001863. https://doi.org/10.1371/journal.pbio.1001863.

Kronmal, R.A. (1993). Spurious correlation and the fallacy of the ratio standard revisited. *Journal of the Royal Statistical Society Series A (Statistics in Society)* 156 (3): 379–392. https://doi.org/10.2307/2983064.

Landis, S.C., Amara, S.G., Asadullah, K. et al. (2012). A call for transparent reporting to optimize the predictive value of preclinical research. *Nature* 490: 187–191. https://doi.org/10.1038/nature11556.

Lawson J (2020). daewr: design and analysis of experiments with R. R package version 1.2-5. http://www.r-qualitytools.org (accessed 2022).

Lee, E.C., Whitehead, A.L., Jacques, R.M., and Julious, S.A. (2014). The statistical interpretation of pilot trials: should significance thresholds be reconsidered? *BMC Medical Research Methodology* 14: 41. https://doi.org/10.1186/1471-2288-14-41.

Leist, M. and Hartung, T. (2013). Inflammatory findings on species extrapolations: humans are definitely not

70-kg mice. *Archives of Toxicology* 87: 563–567. https://doi.org/10.1007/s00204-013-1038-0.

Lindstedt, S.L. and Schaeffer, P.J. (2002). Use of allometry in predicting anatomical and physiological parameters of mammals. *Laboratory Animals* 36 (1): 1–19. https://doi.org/10.1258/0023677021911731.

Lynch, J. (1982). On the external validity of experiments in consumer research. *Journal of Consumer Research* 9 (3): 225–239. https://doi.org/10.1086/208919.

Macleod, M.R., Lawson McLean, A., Kyriakopoulou, A. et al. (2015). Risk of bias in reports of *in vivo* research: a focus for improvement. *PLoS Biology* 13 (10): e1002273. https://doi.org/10.1371/journal.pbio.1002273.

Malterud, K., Siersma, V.D., and Guassora, A.D. (2016). Sample size in qualitative interview studies: guided by information power. *Qualitative Health Research* 26 (13): 1753–1760. https://doi.org/10.1177/1049732315617444.

Markou, A., Chiamulera, C., Geyer, M. et al. (2009). Removing obstacles in neuroscience drug discovery: the future path for animal models. *Neuropsychopharmacology* 34: 74–89. https://doi.org/10.1038/npp.2008.173.

McGonigle, P. and Ruggeri, B. (2014). Animal models of human disease: challenges in enabling translation. *Biochemical Pharmacology* 87 (1): 162–171. https://doi.org/10.1016/j.bcp.2013.08.007.

McKinney, W.T. and Bunney, W.E. (1969). Animal model of depression: I. Review of evidence: implications for research. *Archives of General Psychiatry* 21 (2): 240–248. https://doi.org/10.1001/archpsyc.1969.01740200112015.

Miller, L.R., Marks, C., Becker, J.B. et al. (2017). Considering sex as a biological variable in preclinical research. *FASEB Journal* 31 (1): 29–34. https://doi.org/10.1096/fj.201600781R.

Moher, D., Hopewell, S., Schulz, K.F. et al. (2010). CONSORT 2010 explanation and Elaboration: updated guidelines for reporting parallel group randomised trials. *BMJ* 340: c869. https://doi.org/10.1136/bmj.c869.

Montgomery, D.C. (2017). *Design and Analysis of Experiments*, 8th ed. New York: Wiley 752 pp.

Montgomery, D.C. and Jennings, C.L. (2006). Chapter 1: An overview of industrial screening experiments. In: *Screening: Methods for Experimentation in Industry, Drug Discovery, and Genetics* (ed. A. Dean and S. Lewis), 1–20. New York: Springer 332 pp.

Muhlhausler, B.S., Bloomfield, F.H., and Gillman, M.W. (2013). Whole animal experiments should be more like human randomized controlled trials. *PLoS Biology* 11: e1001481.

Nunamaker, E.A., Davis, S., O'Malley, C.I., and Turner, P.V. (2021). Developing recommendations for cumulative endpoints and lifetime use for research animals. *Animals (Basel)* 11 (7): 2031. https://doi.org/10.3390/ani11072031.

Packard, G. and Boardman, T. (1988). The misuse of ratios, indices, and percentages in ecophysiological research. *Physiological Zoology* 61: 1–9. https://doi.org/10.1086/physzool.61.1.30163730.

Packard, G. and Boardman, T. (1999). The use of percentages and size-specific indices to normalize physiological data for variation in body size: wasted time, wasted effort? *Comparative Biochemistry and Physiology, Part A* 122 (1): 37–44.

Pérez-García, V.M., Calvo, G.F., Bosque, J.J. et al. (2020). Universal scaling laws rule explosive growth in human cancers. *Nature Physics* 16: 1232–1237.

Pitkänen, A., Nehlig, A., Brooks-Kayal, A.R. et al. (2013). Issues related to development of antiepileptogenic therapies. *Epilepsia* 54 (Suppl 4): 35–43. https://doi.org/10.1111/epi.12297.

Porter, W.P., Hinsdill, R., Fairbrother, A. et al. (1984). Toxicant-disease-environment interactions associated with suppression of immune system, growth, and reproduction. *Science* 224 (4652): 1014–1017.

Pound, P. and Ritskes-Hoitinga, M. (2018). Is it possible to overcome issues of external validity in preclinical animal research? Why most animal models are bound to fail. *Journal of Translational Medicine* 16 (1): 304. https://doi.org/10.1186/s12967-018-1678-1.

Reynolds, P. (2021). Statistics, statistical thinking, and the IACUC. *Lab Animal* 50: 266–268. https://doi.org/10.1038/s41684-021-00832-w.

Reynolds, P.S., McCarter, J., Sweeney, C. et al. (2019). Informing efficient pilot development of animal trauma models through quality improvement strategies. *Laboratory Animals* 53 (4): 394–404. https://doi.org/10.1177/0023677218802999.

Richter, H. (2017). Systematic heterogenization for better reproducibility in animal experimentation. *Lab Animal (NY)* 46: 343–349. https://doi.org/10.1038/laban.1330.

Rossello, X. and Yellon, D.M. (2016). Cardioprotection: the disconnect between bench and bedside. *Circulation* 134: 574–575. https://doi.org/10.1161/circulationaha.116.022829.

Rothman, K.J. and Greenland, S. (2018). Planning study size based on precision rather than power. *Epidemiology* 29: 599–603.

Russell, W.M.S. and Burch, R.L. (1959). *The Principles of Humane Experimental Technique.* London: Methuen & Co.

Schmidt-Nielsen, K. (1984). *Scaling: Why Is Animal Size So Important?* Cambridge: Cambridge University Press.

Steidl, R.J., Hayes, J.P., and Schauber, E. (1997). Statistical power analysis in wildlife research. *Journal of Wildlife Management* 61 (2): 270–279.

Tadenev, A.L.D. and Burgess, R.W. (2019). Model validity for preclinical studies in precision medicine: precisely how precise do we need to be? *Mammalian Genome* 30 (5-6): 111–122. https://doi.org/10.1007/s00335-019-09798-0.

Tang, H. and Mayersohn, M. (2011). Controversies in allometric scaling for predicting human drug clearance: an historical problem and reflections on what works and what does not. *Current Topics in Medicinal Chemistry* 11 (4): 340–350. https://doi.org/10.2174/156802611794480945.

Tanner, J.M. (1949). Fallacy of per-weight and per-surface area standards, and their relation to spurious correlation. *Journal of Applied Physiology* 2 (1): 1–15. https://doi.org/10.1152/jappl.1949.2.1.1.

Teare, M.D., Dimairo, M., Shephard, N. et al. (2014). Sample size requirements to estimate key design parameters from external pilot randomised controlled trials: a simulation study. *Trials* 15: 264. https://doi.org/10.1186/1745-6215-15-264.

Trutna, L., Sapgon, P., Del Castillo, E. et al. (2012). Process improvement. In: *NIST/SEMATECH e-Handbook of Statistical Methods.* https://doi.org/10.18434/M32189.

Tukey, J. (1977). *Exploratory Data Analysis.* Reading: Addison-Wesley.

US EPA (Environmental Protection Agency) (2017). *Causal Analysis/Diagnosis Decision Information System (CADDIS).* Washington: Office of Research and Development https://www.epa.gov/caddis-vol1/consistency-evidence.

Usui, T., Macleod, M.R., McCann, S.K. et al. (2021). Meta-analysis of variation suggests that embracing variability improves both replicability and generalizability in preclinical research. *PLoS Biology* 19 (5): e3001009. https://doi.org/10.1371/journal.pbio.3001009.

Van Norman, G.A. (2019). Limitations of animal studies for predicting toxicity in clinical trials: is it time to rethink our current approach? *JACC: Basic to Translational Science* 4 (7): 845–854. https://doi.org/10.1016/j.jacbts.2019.10.008.

Vetter, T.R. and Mascha, E.J. (2017). Defining the primary outcomes and justifying secondary outcomes of a study: usually, the fewer, the better. *Anesthesia & Analgesia* 125 (2): 678–681. https://doi.org/10.1213/ANE.0000000000002224.

Voelkl, B., Vogt, L., Sena, E.S., and Würbel, H. (2018). Reproducibility of preclinical animal research improves with heterogeneity of study samples. *PLoS Biology* 16 (2): e2003693. https://doi.org/10.1371/journal.pbio.2003693.

von Kortzfleisch, V.T., Karp, N.A., Palme, R. et al. (2020). Improving reproducibility in animal research by splitting the study population into several 'mini-experiments'. *Scientific Reports* 10 (1): 16579. https://doi.org/10.1038/s41598-020-73503-4.

Will, T.R., Proaño, S.B., Thomas, A.M. et al. (2017). Problems and progress regarding sex bias and omission in neuroscience research. *eNeuro* 4 (6): eneuro.0278-17.2017.

Willner, P. (1984). The validity of animal models of depression. *Psychopharmacology* 83: 1–17. https://doi.org/10.1007/BF00427414.

Wilson, L.A.B., Zajitschek, S.R.K., Lagisz, M. et al. (2022). Sex differences in allometry for phenotypic traits in mice indicate that females are not scaled males. *Nature Communications* 13: 7502. https://doi.org/10.1038/s41467-022-35266-6.

Würbel, H. (2017). More than 3Rs: the importance of scientific validity for harm-benefit analysis of animal research. *Lab Animal* 46: 164–167. https://doi.org/10.1038/laban.1220.

7 Feasibility Calculations: Arithmetic

CHAPTER OUTLINE HEAD

7.1 Introduction

Operational feasibility is determined by sufficient money, space, equipment, time, and properly trained and competent technical personnel to meet study objectives (Box 7.1). Typically, formal assessments of operational feasibility are required by various stakeholders, such as funding agencies, product investment stakeholders, and ethical oversight committees. Funders and investors may require some projection of costs and resources to determine if the research is worth their investment. Animal care and use oversight committees need to be assured that animals will not be wasted in unfeasible and impractical studies that have little chance of completion.

Power calculations and hypothesis tests are not appropriate for operational feasibility. Simple 'back of the envelope' calculations using basic arithmetic may be all that are required to confirm that the projected number of subjects makes sense or that the study is even feasible. Even for studies where power calculations are necessary, basic arithmetic should be used to cross-validate estimates.

BOX 7.1
Operational Feasibility

Operational: Are current work practices, procedures, and trained personnel sufficient to support the project?

Budgetary: Are financial resources sufficient to support the project? How many samples can be processed for a fixed amount of money?

Time or scheduling: How long will it take to accomplish each task/all necessary tasks? Can the project be completed in the allotted time?

Subjects: How many subjects can be processed, given operational constraints?

7.2 The Process

There are numerous examples of simple arithmetic approximation for solving problems in physics, engineering, economics, claims validation, ecology, and tests of critical thinking (Weinstein and Adams 2008). The process is sometimes known as Fermi estimation, named after the physicist Enrico Fermi,

A Guide to Sample Size for Animal-based Studies, First Edition. Penny S. Reynolds.
© 2024 John Wiley & Sons Ltd. Published 2024 by John Wiley & Sons Ltd.

BOX 7.2
Approximation Process

1. *Problem structuring*: What information is needed?
2. *Calculations:* What are the initial sample size approximations?
3. *'Reality checks':* Do the numbers make sense? Do initial approximations align with available resources?
4. *Revision:* What protocol changes are required so numbers align with available resources?

who had an extraordinary aptitude for finding quick and accurate answers to practical problems when data were sparse or absent, and without using sophisticated mathematics (Weinstein and Adams 2008; Reynolds 2019).

Arithmetic approximation is a four-step cyclical process (Box 7.2) consisting of:

1. *Formulation* of the estimation problem through the logical structuring of the research question and identification of the quantitative elements.
2. *Calculation* of numbers for each element with simple arithmetic.
3. *'Reality checks'* to determine if the estimates obtained by arithmetic approximation are both sensible and feasible.
4. *Revision* of the estimates if necessary (Reynolds 2019).

7.2.1 Problem Structuring

The first step is to specify the approximation problem, the sub-problems or problem sub-components, and the items in each sub-problem that need quantification. If information is not readily available, then a few common-sense assumptions can be made to complete the calculations. A brief justification of the problem should be included.

7.2.2 Calculations

Simple arithmetic is used to 'guesstimate' numbers for each item of information. Any formula used should include the variable names for any quantity to be approximated and estimated. Calculations involve both approximations and any relevant

information are substituted into the equations. The total is obtained by adding all component parts together. Because totals are approximate, it is recommended that a reasonable range of estimates are calculated by bounding the initial total by minimum and maximum numbers along the lines of best and worst-case scenarios (Weinstein and Adams 2008).

To facilitate calculations and troubleshooting, measurement units for each variable should be retained, and all calculations should be checked for unit balance and conversion errors. Variables without formal specified units should be assigned some sort of unit based on context (Chizeck et al. 2009).

7.2.3 Reality Checks

Feasibility or 'reality checks' confirm that estimates align with available resources, that the proposed work practices, procedures, and personnel are sufficient to support the project, and that the project can be completed in the allotted time and within the budget. Estimates should also make sense and should be within a reasonable range of possible starting values. Tracking units helps to find simple maths errors or incorrect unit conversions.

7.2.4 Refinement

If the proposed number of animals is too large for study resources, then sample sizes will need to be reformulated or refined to meet more realistic operational capabilities.

7.3 Determining Operational Feasibility

Resources should be sufficient to complete the study. These include the number of trained technical personnel, duration of each task, procedures required to collect data, costs per procedure, cost for processing each sample, total study duration, and study budget (Box 7.3).

7.3.1 Basic Science/Laboratory Studies

The total animal sample size will be constrained by the availability of resources, number of trained

BOX 7.3
Operations and Logistics

Number of trained personnel
Duration of each task, procedure
Cost per procedure
Cost per subject
Total study duration
Total study budget.

personnel, estimated task and procedure duration, costs per procedure, total study duration, and study budget. Number of animals should be adjusted to numbers that can actually be processed in a reasonable amount of time.

Example: Laboratory Processing Capacity

An investigator requested 500,000 mice for a three-year project. Two technicians were listed on the protocol. The experimental surgeries to be performed on each animal took 30–60 minutes per mouse. Is the number of mice realistic?

Calculations. Even if it is assumed that, *at most,* each technician could work 350 days per year for 8 hours per day, the total procedure time is:

2 persons × 350 days/year × 3 year = 2100 person − days
2100 person − days × 8 hours/day = 16,800 person − hours

The number of mice that can be processed is approximately 500,000 animals/2100 person-days $\cong$ 238 animals per person per day.

Assuming an 8-hour work day, 500,000 animals/16800 person-hours $\cong$ 30 animals per person per hour, or one mouse approximately every 2 minutes.

Alternatively, the total procedural time required for 250,000 mice per technician is

1 mouse/30 min × 60 min/hr × 8 hr/day

= 16 mice per day

16 mice/day × 350 days/year

= 5600 mice/year

Or 45 years for 250,000 mice per technician

Reality check. Even if the realities of staff welfare and animal housing are ignored, procedures take 30–60 minutes per mouse. The projected numbers are unrealistic.

Example: Time Available for Large Animal Experiments

An investigator requested 120 swine for a three-year project. A single experiment on one animal required a full working day for five people. An additional 8–16 hours per week was allocated for equipment setup and takedown, supply inventory, data management, and related tasks. To prevent staff fatigue and burnout, the investigator considered two experiments per week was a reasonable performance goal. Before the study could be initiated, it was anticipated that at least four to six months would be required for protocol oversight paperwork to be filed and approved and to obtain necessary supplies and equipment. Can this number of animals realistically be processed in three years?

Calculations. Total time = 3 yr × 12 months/yr = 36 months

Available time = 36 months – 6 months

= 30 months

Time required

= 120 animals/(2 animals/week × 4 weeks/months)

= 15 months

Reality check. The investigator has allowed adequate time to complete the project, without compromising the well-being of staff or study quality, and with an adequate time cushion in case of unforeseen problems or breaks in the workflow.

7.3.2 Veterinary Clinical Trials

Veterinary clinical trials must consider both availability of resources and availability of eligible subjects, especially if these are client-owned animals

---

BOX 7.4
Veterinary Clinical Trials

Availability of resources
Clinic intake rate per week, month
Number and proportion of eligible subjects
Client consent rate
Cost per procedure, per subject
Total study duration
Total study budget.

---

entering the study sequentially and at variable intervals. For the trial to have a chance of success, anticipated enrolment rates must agree with numbers obtained from power-based sample size calculations (Box 7.4).

Example: Recruitment of Client-Owned Companion Animals

Investigators determined from power calculations that at least 100 subjects would be required to detect a meaningful difference between two intervention groups, with 50 subjects per comparison arm (healthy versus diseased). Clinic records indicated that a total of 90 subjects meeting study eligibility criteria had visited the clinic in the previous year and included 20 subjects with the disease of interest. However, only 50% of owners consented to participate in previous trials at the institution. How long will the trial have to run to enrol sufficient subjects for this new trial?

Information required. Size of the patient recruitment pool, anticipated client consent rate, and expected number of patients with the disease.

The expected number that can be recruited from the total pool of eligible subjects (N_T) is approximated as

$$N = N_T \cdot p_E$$

where p_E is the proportion of subjects enrolled. The number of subjects that can both be enrolled and have the condition of interest (N_D) is then

$$N_D = N \cdot p_D$$

Calculations. The expected number recruited is $N = 90(0.50) = 45$ per year. The number expected to have the disease is $N_D = 45(20/90) \cong 10$ subjects. To obtain the minimum number of subjects with the condition of interest, the study would have to run for

$$\frac{50 \text{ subjects required}}{\sim 10 \text{ subjects/year}} \cong 5 \text{ years}$$

Example: Trial Size for a Grant Proposal

A researcher wished to apply for a grant proposal to study a novel disease biomarker in client-owned companion animals. The plan was to develop a clinical prediction model featuring an additional 11 predictor variables already confirmed in the primary literature to be associated with severity of that disease. The study was intended to be a single-centre clinical prospective cohort study. How many animals will need to be recruited for a two-year study?

To develop an adequately robust clinical prediction model using 12 predictors (including the biomarker of interest), formal sample size calculations (Riley et al. 2020) suggested at least 260–300 subjects are required. Is this number of subjects a feasible recruitment goal?

Information required. Number of eligible subjects visiting the clinic per week or per month, anticipated client consent rate, total operating budget, processing costs per patient, and total study duration.

Clinic volume. Based on prior clinical record data, approximately 2–3 eligible animals visited the clinic per week. Fully subsidised clinical and laboratory costs encouraged high client consent rates, which were expected to average about 80%.

Anticipated intake $= 2 - 3$ animals/week $\times 50$ weeks
$= 100 - 150$ animals per year

Anticipated enrolment $= 100 (0.8) - 150 (0.8)$
$= 80 - 120$ animals per year

Budget. Laboratory and clinical work-ups were approximately $1500 per animal. The grant allowed a total operating budget of $120,000.

Maximum number processed

= $120,000/($1500/cat) = 80 animals in two years.

Reality check. Sample size based on power calculations does not align with either anticipated enrolment or budget constraints.

Refinement. Reducing the number of laboratory tests performed cut costs from $1500 to $1000. The revised sample size is now $120,000/($1000/cat) = 120 animals. However, this still does not align with the power-based sample size estimates in the original proposal. The proposal was revised in the direction of more modest research goals that could be accommodated by the budget. The investigator prioritised the marker of interest plus the top four candidate markers, with priority based on clinical relevance and importance. The reduced subset of five candidate predictors would allow reasonably precise estimates to be obtained with an anticipated enrolment of 100–120 animals and remain within budget.

7.3.3 High Dimensionality Studies

These studies typically involve the harvest and processing of cells or tissues from multiple animals. Typically, large volumes of output information are produced per subject. Examples include DNA/RNA microarrays, biochemistry assays, biomarker studies, gene expression, proteomic, metabolomic, and inflammasome profiles.

For identification of differentially expressed genes, sample size refers to either the number of arrays or the number of biological replicates, depending on the research question. For array-based studies, sample sizes are determined by the desired fold change to be detected, the number of replicate arrays, investigator-specified sensitivity (proportion of detected differentially expressed genes), number of expected false positives, the correlation between expression levels of different genes, and the number of sampling time points (Jung and Young 2012; Lin et al. 2010).

The number of animals required is less easy to estimate. In general, the number of animals will be determined by the total amount of tissue or cells required to perform the assays (M) and the amount of viable tissue or cells per animal (m/animal)

$$N = M/(m/\text{animal})$$

Example: Number of Mice Required for a Microarray Study

Suppose a cell culture requires $M = 10^6$ cells per plate to obtain a sufficient amount of RNA for analysis. However, only $m = 2.5 \times 10^5$ relevant cell types can be isolated per mouse.

The number of mice required

= 10^6 cells$/(2.5 \times 10^5$ cells per mouse$)$

= 4 mice/plate.

The evolution of microarray and RNA-seq technology means that very small amounts (for example, <~1 pg) of mRNA can be extracted from tissue and even single cells for gene expression profiling (Amit et al. 2009; Shalik et al. 2013; Ye et al. 2018). Therefore very few animals, or only one, maybe all that are necessary. Variability in array preparation and background determination will determine the number of technical replicates required. However, technical replicates only affect measurement precision; they do not contribute to reducing variance of the overall effect.

If the research question involves mapping heterogeneity at the subject level, the study will have to be designed to accommodate the true experimental unit, which is the whole animal. If the number of biological replicates is too small, statistical power for detecting differentially expressed genes will be too low, and false positive rates will be high. For detecting differential gene expression between two groups (e.g. knockout versus wild-type mice; tumour versus non-tumour tissue), sample size is determined by power calculations. Sample size will depend on the fold change to be detected, power to detect that change (the true positive rate, or power $1 - \beta$), confidence α (Type I error probability), and the variance in the sample. The variance can be estimated from the 75th percentile of the standard deviation of log ratio of expression levels (the variance for the 75% least variable genes in the array). Sample size can then be estimated as usual for a two-sample t-test, or iteratively using the formula for power and the non-centrality parameter (Wei et al. 2004).

7.3.4 Training, Teaching, Skill Acquisition

Animal numbers for training and skill acquisition are determined primarily by the number of repetitions of the task required to meet predetermined competency standards. For teaching purposes, the trainer:trainee ratio, number of assessors, and availability of teaching-related resources must be factored into estimates (Box 7.5).

Competency in essential clinical skills and surgical techniques is of major importance for preclinical laboratory experiments and veterinary and human medicine. On-the-job training is usually not sufficient to acquire competency (Bergmeister et al. 2020). High-quality and consistent skill sets are developed and maintained only with structured training and sufficient practice. Poor technique and inconsistent and unstandardised procedures will contribute to considerable variability in experiments, potentially hiding any effect of the experimental intervention.

Use of live animals for surgical skills training is still part of many medical and veterinary training programmes (DeMasi et al. 2016).

Use of live animals obtained for purpose should be considered a last resort, and only after critical skillsets have been developed on non-animal and/or *ex vivo* models. Computer simulations, virtual models, and simulators are rapidly replacing animals. Carcasses and culls can be used for skills acquisition (Baillie et al. 2016). *Ex vivo* models can be as effective as animals obtained for purpose, as well as considerably cheaper (e.g. Izawa et al. 2016). Additional refinement measures include supplementary educational materials (lectures, text, webinars, etc.) for orientation to equipment, techniques, and procedures. Detailed standard operating procedures and training plans should be devised. Instructional plans should describe how specific critical skills and skill sets will be taught before live animals are used. Specific and measurable proficiency metrics and competency assessments must be identified. Training must be species- and procedure-specific. Strategies for development of animal-based training programs are described by Conarello and Shepherd (2007).

Example: Surgical Training Laboratory

A teaching lab requested swine to be used for surgical training of residents in multiple invasive procedures. Students were to have received several weeks of intensive training on basic skill acquisition on simulators and excised organs prior to this lab. There are 16 trainees. Eight trainees can be accommodated per training session. One animal can be allocated to every two trainees. Four instructors are available. How many animals are required?

Calculations. Number of sessions = 16 trainees/ (8 trainees/session) = 2 sessions.

Number of animals/session = (8 trainees/session) × (1 animal/2 trainees) = 4 animals/session.

Total number of animals = (4 animals/session) × 2 sessions = 8 animals.

7.3.5 Rodent Breeding Production

Breeding protocols are required for in-house production of research animals that are too expensive

---

BOX 7.5
Training and Teaching

- Number of sessions required for competency
- Resource availability
- Number of trainees
- Number of trainees per session
- Number of instructors, assessors
- Time available per session
- Assessor availability.

to purchase in sufficient quantity or cannot be obtained commercially. Examples include the creation of new transgenic, knockout, or other genetically modified animals; back-crossing of genetically modified lines; or production of prenatal or early neonate subjects.

The projected total pup production and the anticipated number of pups for a specific genotype depend on the number of breeding adults, litter size, number of litters over the productivity lifespan, and weaning success (Box 7.6). Reasonable predictions of pup production can be based on breeding colony records, current breeding stock numbers, facility and personnel capability, and projections from past demand. For genetic analyses involving mice, estimating the number of individuals and breeding pairs per line can be approximated by simple rules of thumb (Table 7.1). Number of animals subjected to embryonic or foetal manipulations must be included in the total number of animals requested for a given study.

Scientific justification must be provided for breeding protocols and plans for disposition of unused animals. When target genotypes are of research interest, potentially large number of animals will be euthanised because they are the unwanted genotype, or if they are produced in such quantities that they age out or otherwise cannot be used in the experimental protocol. Justifications for

Table 7.1: Rules of Thumb for Estimating Mouse Numbers Required for Genetic Analyses.

Purpose	Number of breeding pairs per line	Number required
Maintenance and characterisation of transgenic or knockout line	up to 5	80–100
Strain construction with congenic genotyping	10–12	750–1200
Quantitative trait loci analysis	4–6 pairs inbred parental strains	500–1000 F2
	2–4 reciprocal F1 hybrid pairs	
Gene mapping	10–12	1200

Source: Adapted from Pitts (2002).

breeding protocols must include end-use disposition plans to minimise unacceptably large collateral losses (Reynolds 2021).

Example: Estimating Number of Mice With a Desired Genotype

From preliminary power calculations, an investigator determined that 50 homozygous knockout (KO $-/-$) and 50 homozygous wild-type (WT; $+/+$) mouse pups of a certain strain are required to study a specific disease. Only KO and WT mice were to be used for experiments.

Ten pairs of heterozygous breeder mice (HET $+/-$) were available, 10 males and 10 females. From past breeding records and vendor specifications, the expected average litter size for this strain was approximately 5 pups, and each pair was expected to produce 6 litters over 6 months of the productivity lifespan. Genotype distribution was thought to be approximately Mendelian; that is, approximately 50% of pups will be HET, 25% of pups KO, and 25% WT. However, perinatal losses of KO were 15–20%. How many litters will

---

BOX 7.6
Rodent Breeding Production

How many pups can be produced (N)?
How many breeding adults are required to produce N pups?
How many pups are born?
How many pups are successfully weaned?

Information required

Initial number of breeders
Sex ratio (number of females per breeding male)
Litter size = (number of pups/litter/female)
Estimated total number of pups produced
Proportion of pups lost (attrition)
Expected genotype distribution
Proportion of desired genotype required
Time frame (pups produced per unit time).

need to be produced, and how long will it take to get the requisite number of pups?

Calculations: Total number of pups: $N = 10$ females $\times$ (5 pups/litter/female) $\times$ (6 litters/female/6 months) = 300 pups in 6 months. Assuming a Mendelian 1.2.1 distribution of genotypes, then the proportions of each genotype are $p_{WT} = p_{KO} = 0.25$, and $p_{HET} = 0.50$.

The anticipated perinatal loss of KO pups is $p = 0.15–0.20$. Then the number of surviving KO pups is $p_{KO} \cdot N \cdot (1 - p) = 0.25 (300) \cdot (1 - 0.20) = 60$ pups to $0.25(300) \cdot (1 - 0.15) = 63$ pups. The number of WT pups = $(0.25)(300) = 75$, and the number of HET pups = $(0.50)(300) = 150$.

Therefore, for 10 adult pairs, we expect the production of 60–63 viable knockout pups and 75 wild-type pups every 6 months. These are sufficient to meet the objectives of the study. The total number of mice required is 20 adults + 300 pups = 320 mice, with 'waste' of approximately 150 heterozygotes. Instead of being euthanised, heterozygotes could be transferred to another protocol or used to replace the breeding stock.

References

Amit, I., Garber, M., Chevrier, N. et al. (2009). Unbiased reconstruction of a mammalian transcriptional network mediating pathogen responses. *Science* 326 (5950): 257–263. https://doi.org/10.1126/science.1179050.

Baillie, S., Booth, N., Catterall, A., et al. (2016). *A guide to veterinary clinical skills laboratories.* https://www.researchgate.net/publication/296964846 (accessed 2022).

Bergmeister, K.D., Aman, M., Kramer, A. et al. (2020). Simulating surgical skills in animals: systematic review, costs & acceptance analyses. *Frontiers in Veterinary Science* 7: https://doi.org/10.3389/fvets.2020.570852.

Chizeck, H.J., Butterworth, E., and Bassingthwaighte, J. B. (2009). Error detection and unit conversion: automated unit balancing in modeling interface systems. *IEEE Engineering in Medicine & Biology* 28 (3): 50–57.

Conarello, S.L. and Shepherd, M.J. (2007). Training strategies for research investigators and technicians. *ILAR Journal* 48 (2): 120–130.

DeMasi, S.C., Katsuta, E., and Takabe, K. (2016). Live animals for preclinical medical student surgical training. *Edorium Journal of Surgery* 3 (2): 24–31.

Izawa, Y., Hishikawa, S., Muronoi, T. et al. (2016). Ex-vivo and live animal models are equally effective training for the management of a penetrating cardiac injury. *World Journal of Emergency Surgery* 11: 45. https://doi.org/10.1186/s13017-016-0104-3.

Jung, S.-H. and Young, S.S. (2012). Power and sample size calculation for microarray studies. *Journal of Biopharmaceutical Statistics* 22 (1): 30–42.

Lin, W.-J., Hsueh, H.-M., and Chen, J.J. (2010). Power and sample size estimation in microarray studies. *BMC Bioinformatics* 11: 47.

Pitts, M. (2002). *Institutional Animal Care and Use Committee Guidebook*, 2e. Bethesda: Office of Laboratory Animal Welfare, National Institutes of Health.

Reynolds, P.S. (2019). When power calculations won't do: fermi approximations of animal numbers. *Lab Animal* 48: 249–253.

Reynolds, P.S. (2021). Statistics, statistical thinking, and the IACUC. *Lab Animal (NY)* 50 (10): 266–267. https://doi.org/10.1038/s41684-021-00832-w.

Riley, R.D., Ensor, J., Snell, K.I.E. et al. (2020). Calculating the sample size required for developing a clinical prediction model. *BMJ* 368: m441. https://doi.org/10.1136/bmj.m441.

Shalik, A., Satija, R., Adiconis, X. et al. (2013). Single-cell transcriptomics reveals bimodality in expression and splicing in immune cells. *Nature* 498: 236–240. https://doi.org/10.1038/nature12172.

Wei, C., Li, J., and Bumgarner, R.E. (2004). Sample size for detecting differentially expressed genes in microarray experiments. *BMC Genomics* 5: 87. https://doi.org/10.1186/1471-2164-5-87.

Weinstein, L. and Adams, J.A. (2008). *Guesstimation: Solving the World's Problems on the Back of a Cocktail Napkin*. Princeton, NJ: Princeton University Press.

Ye, Y., Song, H., and Shi, S. (2018). Understanding the biology and pathogenesis of the kidney by single-cell transcriptomic analysis. *Kidney Diseases* 4: 214–225. https://doi.org/10.1159/000492470.

8

Feasibility: Counting Subjects

CHAPTER OUTLINE HEAD

8.1 Introduction

Many studies require preliminary screening of large numbers of potential subjects in order to capture a much smaller target number of subjects with a specific trait or characteristic (Box 8.1). This type of sample size problem is essentially like a series of

---

BOX 8.1

Examples of Probability-Based Sampling

Determining the total number of subjects to be screened to ensure the inclusion of a pre-specified number with a given trait or condition.

Determining the number of subjects required to observe at least one specified event.

Estimating risk of adverse events when none were observed during the preceding sequence of procedures.

Determining the number of batches and batch size to estimate disease prevalence.

'experiments' analogous to flipping a coin many times. Each experiment has one of two possible outcomes: 'success' representing the selection of a subject either with the condition or 'failure' if the subject is without the condition. The outcome is therefore a binomial variable with only two possible outcomes, success s or failure f. The probability of having the condition is the proportion p of successes s in the total number of selected subjects N, and $(1 - p)$ is the expected proportion without the condition ('failures').

The conventional approach to determining sample size is to use standard sample size formula based on the normal distribution. As sample size increases towards infinity, distributions of count data will converge to approximately normal distribution (large-scale approximations). However, sample size estimates tend to be inaccurate and biased when applied to discrete data based on small samples. Better estimates can be obtained with *exact methods*. These are based on specific discrete probability distributions enabling an explicit model of the

A Guide to Sample Size for Animal-based Studies, First Edition. Penny S. Reynolds.
© 2024 John Wiley & Sons Ltd. Published 2024 by John Wiley & Sons Ltd.

BOX 8.2
Determining Feasibility of Counts by Exact Methods

Sample size formula based on the normal distribution are inaccurate and biased when applied to count data.

1. Choose the appropriate discrete probability distribution.
2. Choose the appropriate *cumulative distribution function* (is the probability of possible outcomes equal to, less than, or greater than the specified number).

probability of obtaining a pre-specified number of successes. They are valid regardless of sample size. Although computationally intensive (Newman 2001), this is trivial consideration with the availability of computer programmes. Worked examples are provided with exemplary SAS code in each section and in Appendices 8.A and 8.B.

The families of discrete probability distributions considered here are the binomial, geometric, negative binomial, and hypergeometric distributions (Box 8.2). Choice of sampling distribution is determined by feasibility and sampling objectives. The binomial distribution is used to determine the total number of successes s or prevalence p in the population for a fixed number of selections N. The geometric distribution is used to determine the number of selections N required to obtain the first success ($s = 1$) after which sampling stops. The negative binomial distribution is used to determine the number of trials N needed to obtain a specific number of successes $s = x$. The hypergeometric sampling is used when the objective is to determine the number of samples to be screened to obtain a pre-specified number of successes in a small finite population.

After identifying the distribution family most appropriate for addressing feasibility objectives, the form of the *cumulative distribution function*

must be specified (Table 8.1). The cumulative distribution function (cdf) of a random variable X, evaluated at x, is the probability function that X will take a value equal to, less than, or greater than x. That is, to determine the target number of subjects s ('successes'), we need to first decide on whether the problem requires a given probability of obtaining exactly s, or s that is more or less than some predetermined limit X. In practice, the probability of choosing exactly s subjects is too restrictive. It is usually more feasible to determine the probability of getting at least as many or more than s subjects.

Finally, sample size is calculated using the appropriate exact method as follows:

1. Specify a range of candidate sample sizes n
2. Specify the form of the cumulative probability distributions. For example, if the target value for the expected number of success was to be at least 10 in a sample of N subjects, then $\Pr(X \geq 10)$
3. Compute the probability y for each n using the cumulative distribution function and/or the probability density function of the relevant distribution family. Computation methods and SAS codes are given in Appendix 8.A.
4. Select the sample size that equals or exceeds the target confidence $1 - \alpha$.

8.2 Normal Distribution

Conventional sample size formulas are based on the normal distribution. The large-sample approximation is

$$N = \left(z_{1-\alpha/2}\right)^2 \cdot p(1-p)/d^2$$

where N is the total sample size, $p =$ the expected proportion of the condition of interest or prevalence, d is the target precision, and z is the standard

Table 8.1: Cumulative Probabilities, Interpretation, Representation, and Computation.

Interpretation	Representation	Computation
Probability of getting *exactly x*	$\Pr(X = x)$	
Probability of getting *fewer than x*	$\Pr(X < x)$	$\Pr(X = 0) + \Pr(X = 1) + \dots + \Pr(X = x - 1)$
Probability of getting *at most x*	$\Pr(X \leq x)$	$\Pr(X = 0) + \Pr(X = 1) + \dots + \Pr(X = x)$
Probability of getting *more than x*	$\Pr(X > x)$	$\Pr(X = x + 1) + \dots + \Pr(X = n)$
Probability of getting *at least x*	$\Pr(X \geq x)$	$\Pr(X = x) + \dots + \Pr(X = n)$

score based on the standard normal distribution and pre-specified confidence level. For example, for a two-tailed test with 95% confidence, $\alpha = 0.05$ and $z = 1.96$. The expected number of subjects with the condition (n) is obtained by multiplying total sample size N by prevalence p:

$$n = N \cdot p$$

Although simple to perform, this method will give only poor approximations for most practical purposes (Newcombe 1998). Bias correction adjustment methods (such as the Agresti-Coull method) are described in Chapter 10.

8.3 Binomial (Exact) Distribution

The binomial distribution is the probability distribution of a binomial random variable. It is used to determine the number of successes s in N selections. It is applicable when the population is very large relative to the target sample, and the N selections are independent; that is, the outcome on one selection does not affect the outcome on other selections, and the probability of success p is the same on every trial (that is, $p = 1/2$ or 0.5). The binomial probability is the probability that a binomial 'experiment' results in exactly s successes. In practice, the cumulative binomial probability will be applicable to most studies. This is the probability that a binomial experiment results in s successes that occur within a specified range (either greater than or equal to a given lower limit or less than or equal to a given upper limit).

The binomial distribution is given by

$$\Pr(X = s) = \binom{N}{s} p^s (1-p)^{N-s}$$

where N is the total sample size (number of 'trials' or selections), s is the number of subjects with the condition ('successes'), $p =$ the expected proportion of successes, and $(1 - p)$ is the expected proportion without the condition or event ('failures'). The probability of success p is the same on every trial.

Because the binomial variable is the sum of N independent binomial variables, the mean $\mu = N \cdot p$, and the variance $\sigma^2 = (N \cdot p)(1 - p) = \mu(1 - p)$

Sample size is calculated using the binomial exact method:

1. Specify the range of candidate sample sizes n.
2. Specify the form of the cumulative binomial probability. For example, if the target value for the expected number of success was to be at least 10 in a sample of N subjects, then $\Pr(X \geq 10)$.
3. Compute the probability y for each n using the cumulative distribution function and/or the probability density function of the binomial distribution (Appendix 8.A).
4. Select the sample size that equals or exceeds the target confidence $1 - \alpha$.

Example: Genetically Linked Bleeding in Doberman Pinschers

Investigators planned a prospective study to determine differences in coagulation profiles between healthy Doberman Pinschers and those that are autosomally recessive for the mutation for von Willebrand disease (vWD), and therefore at high risk for severe bleeding. From the literature, they estimated a 25% prevalence of homozygous affected Dobermans. The investigators needed to estimate the number of dogs that would have to be screened (N) and the number of homozygous affected dogs n they could expect to obtain in that sample with a probability of 95% and a precision of 5%.

Normal approximation. For a probability of 95%, $\alpha = 0.05$ and $z_{1-\alpha/2} = 1.96$, prevalence $p = 0.25$, and precision $\delta = 0.05$. Then the total number of animals to be screened is:

$$N \cong (1.96)^2 \cdot \left[0.25\,(1 - 0.25)/(0.05)^2 \right] = 288.1$$
$$\cong 290$$

and the expected number of homozygous affected is $n = N \cdot p \cong 290 \times 0.25 \cong 73$.

When the Agresti-Coull adjustment (Agresti and Coull 1998) is applied (Chapter 11):

$$\tilde{p} = (25 + 2)/(100 + 4) = 0.2596,$$
$$N \cong (1.96)^2 \cdot \left[0.2596\,(1 - 0.2596)/(0.05)^2 \right]$$
$$- 4 = 291.4 \cong 292$$

and the expected number of homozygous affected is $n = 76$.

With the asymptotic normal approximation, approximately 290–292 dogs need to be screened to have a 95% probability of capturing 73–76 homozygous affected dogs within the general Doberman population.

Exact binomial calculation. Investigators wished to determine the total number of dogs that would have to be screened to have 95% confidence of obtaining at least 75 or more homozygous affected dogs from a population with an expected prevalence of 25%. The cumulative binomial probability is $\Pr(X \geq 75) = 0.95$. Sample size is then approximated by iteration over a range of potential sample sizes to find the sample size that is closest to $\Pr(X \geq 75) = 0.95$. For this example, a potential sample size range of 10–400 was chosen. SAS code is:

```
%let numSamples = 400; *set maximum N subjects;
data prob;
do n = 10 to &numSamples by 1;
    y = PDF('BINOMIAL',75,0.25,n) + 1 - CDF
('BINOMIAL',75,0.25,n);
 output;
   end;
run;
```

With the exact method, it is estimated that 351 dogs need to be screened to have a 95% probability of obtaining at least 75 homozygous affected dogs. This revised estimate suggests that sample size based on the asymptotic normal approximation may underestimate the number of subjects that will have to be screened.

8.3.1   Rare or Non-Existent Events

Non-occurrence of an event in a series does not mean that it cannot happen at all, especially if the series is relatively short. Assessing the probability that an event could yet occur is especially important if the study is assessing the risk of rare but potentially catastrophic adverse events (for example, serious injury or death in a series of surgeries). Therefore, the question becomes one of determining the worst-case scenario based on these data. This is done by estimating the one-sided upper bound of the 95% confidence interval for p. For a one-sided confidence limit, all of α must apply to that limit, not half of it ($\alpha/2$). Therefore, for a one-sided 95% confidence limit, we need to solve the equation for $\alpha = 0.05$ (not $\alpha/2 = 0.025$ as is the case for a two-sided limit).

When no events have occurred ('Rule of Threes'). If there are no observed events in n trials, then $x = 0$ and the observed proportion p is $0/n$. The approximation for the upper confidence limit is obtained from

$$\Pr(X = 0) = \binom{n}{0} p^0 (1-p)^{n-0}$$

which reduces to

$$(1-p)^n = \alpha$$

Setting α to 0.05 and solving for p, the upper bound of the 95% upper confidence limit for p is $\cong 3/n$ (Louis 1981, Hanley and Lippman-Hand 1983, Jovanovic and Levy 1997).

Alternatively, the exact upper confidence limit can be calculated from the cumulative probability distribution of the binomial distribution as follows:

```
%let p = 1; *maximum proportion;
data prob;
*iterate over range of proportions from 0 to 1;
   do p = 0 to &p by 0.001;
*Pr(X <= x)describes the upper confidence
limit;
y = CDF('BINOMIAL', 0, p, n);
output;
end;
run;
```

When events are rare but not zero. The number of observations to capture one adverse event in n trials with a given probability is approximated by rearranging $(1-p)^n = \alpha$ to solve for n:

$$n = \ln(\alpha) / \ln(1-p)$$

Example: Occurrence of Adverse Events

Suppose one adverse event is expected for every 100 surgeries ($p = 1\%$, or 0.1). Then the number of observations needed to have 95% probability of seeing one event,

$$(1-p)^n = \alpha$$

$$(1-0.01)^n = 0.05$$

Solving for n:

$$n \cdot \ln(0.99) = \ln(0.05)$$

$$n = \ln(0.05) / \ln(0.99) = 298.1$$

Given an expected prevalence of adverse events of 0.1% and rounding up, we will need to observe approximately 300 surgeries to capture one adverse event with 95% probability.

Exact binomial calculation. Exact confidence intervals for nonzero rare events can be calculated from binomial simulations. Two confidence limits must be calculated: the upper confidence limit at $p_U = \alpha/2$ and the lower confidence limit at $p_L = \alpha/2$.

Example: Computing an Expected Range of Rare Events

We observe 3 events out of 100 trials. Then $p = 3/100 = 0.03$. The lower limit for the 95% confidence interval is the value of p for which $Pr(X \geq 3) = 0.025$, and the upper limit is the value of p for which $Pr(X \leq 3)$.

```
%let p = 1; *maximum proportion p;
data prob;
do p = 0 to &p by .001; *iterate over a
range of p from 0 to 1 with desired step
size;
  yL = PDF('BINOMIAL',3,p,100) + 1 - CDF
('BINOMIAL',3,p,100); *lower CL;
  yU = CDF('BINOMIAL', 3, p, 100); *upper
CL;
  output;
  end;
  run;
```

From the output, select the values of p for which y_L and $y_U = \alpha/2 = 0.025$. From the output $y_L = 0.025$ corresponds to $p = 0.006$, and $y_U = 0.025$ corresponds to 0.085. Therefore the 95% confidence interval for $p = 0.03$ is (0.006, 0.085), or 0.6–8.5%.

The large confidence intervals in this example show that there can be considerable loss of precision with small sample sizes, and therefore estimates are much less reliable.

Example: Predicting the Probabilities of Adverse Events

Previous data suggest 5% of animals administered a certain drug will show serious adverse effects. What is the probability that more than five will show adverse effects? More than 10?

The probability for a range of sample sizes ($n = 1$–50) is calculated by simulation on the binomial distribution for a prevalence of 5% (or an expected prevalence p of 5 subjects out of 100), so $p = 0.05$. Choose the probability y for the target sample size ($s = 5$ and $n = 10$). Sample SAS code is as follows:

```
%let numSamples = 50; *maximum of s subjects;

data prob;
do s = 1 to &numSamples by 1; *iterate over
a range of s from 1 to 50;
  y = 1 - CDF('BINOMIAL',s,0.05,100); *Pr(X > s)
more than s;
  output;
  end;
  run;
```

The probability that more than $s = 5$ animals will show an adverse effect is 0.384 (38.4%) and more than 10 is 0.0115 (1.2%).

Example: Risk of Adverse Events in Future Surgeries

(Data from Eypasch et al. 1995). A 'serious adverse event' during laparoscopic appendectomy is defined as an intraoperative vascular injury resulting in limb loss or death. A series of 25 laparoscopic surgeries had no observed serious adverse events. What is the risk that a serious adverse event could still occur with 95% probability?

The approximate 95% upper bound on the rate of occurrence is $3/N = 3/25 = 0.12$, or 12%. The exact bound for $a = 0.05$ is $p = 0.113$, or 11.3%. In this case, the rule of thumb approximation is very close to the exact calculation. The 95% confidence interval for zero events is (0, 0.113). Even though no adverse events had been observed in the preceding series of 25 procedures, there is still the risk that a severe adverse event could occur in 11–12 out of every 100 procedures.

8.4 Batch Testing for Disease Detection

A special case of binomial sampling is estimating sample sizes for batch testing of pooled samples (Box 8.3). Pooled samples are used when the prevalence of disease-positive individuals must be estimated, and resources are limited relative to

BOX 8.3
Applications of Batch Testing

- Number of available test kits are limited relative to the size of the population pool that needs to be tested.
- Diagnostic screening for disease prevalence in livestock and poultry operations.
- Rabies surveillance testing in bat colonies.

the size of the population that needs to be tested. Batch testing can result in considerable savings per unit of information. It is especially useful if diseases are rare (Hou et al. 2017), reagents and other resources are scarce, and the number of available tests is limited relative to the number of potential test subjects that require them (Litvak et al. 1994, Hughes-Oliver 2006), or if sample volumes are too small for processing singly and must be pooled over several subjects for analysis (Giles et al. 2021).

The concept of batch testing was first developed during World War II for detecting syphilis in US army conscripts (Dorfman, 1943). It has been adapted for use in large-scale testing of diseases, such as HIV and other sexually transmitted diseases (Hughes-Oliver 2006, Shipitsyna et al. 2007), bacteriological screening of livestock herds and poultry flocks (Arnold et al. 2005, Arnold et al. 2009), screening for environmental contaminants, such as lead, and assessment of large number of molecular targets for drug discovery (Hughes-Oliver 2006). The SARS-COV2 pandemic of 2020 renewed interest in further application to the problems of an emergent disease pandemic (Hitt et al. 2020, Zhou and O'Leary 2021, FDA `https://www.fda.gov/medical-devices/coronavirus-covid-19-and-medical-devices/pooled-sample-testing-and-screening-testing-covid-19`). Batch testing is also used when individual animals are too small to allow the collection of sufficient tissue or blood for single-sample analyses, such as surveillance testing of bat colonies (Giles et al. 2021).

In batch testing, samples from m subjects are pooled and tested for the presence of the disease (yes = positive; no = negative). The simplest batch-testing design tests samples in two stages. The first stage tests the pooled samples, where each individual sample is part of one batch. If a batch tests negative, no further testing is required. If a batch tests positive, then each sample in the batch is retested separately.

The sample size problem is determining the optimum batch size m. If the population is large and expected proportion is small, then the probability that a batched sample B is positive is derived from the binomial distribution as

$$P(B = 1) = 1 - (1-p)^m$$

where p is the disease prevalence and m is the batch size or number of samples in each batch.

The 'most efficient' group size m is that which minimises $E(T)/m$, where $E(T)$ is *expected number of tests*:

$$E(T) = 1 + m \cdot P(B = 1)$$

As a first-pass approximation, m can be estimated by

$$m = 1/\sqrt{p}$$

with rounding up to the nearest integer. Then the expected number of tests needed is approximately:

$$E(T) \cong 2 \cdot N\sqrt{p}$$

for a population of N subjects (Finucan 1964).

For feasibility and planning purposes, it is useful to estimate the *percentage reduction in the number of tests* by batch testing compared to testing individual samples:

$$R_T\,(\%) = 100 \cdot [1 - E(T)/m]$$

The *expected percentage increase in testing capacity* is approximated by the following:

$$I_{TC}\,(\%) = 100 \cdot \left[\frac{1}{E(T)/m} - 1\right]$$

Exact binomial determination. The above formulations assume that both the *sensitivity* (the probability of detecting a true positive) and *specificity* (the probability of detecting a true negative) of the tests are equal to 1.0. Exact binomial determinations that incorporate more realistic test sensitivity and specificity provide more rigorous estimates of m and $E(T)$. An open-access shiny app (Hitt et al. 2020) numerically iterates over a range of batch sizes to determine optimal batch size m, $E(T)$, $E(T)/m$, percent reductions in the expected number of tests with batch testing, and the expected increase in testing efficiency.

An analogous batch size determination algorithm has been developed for microarray or microplate studies (Bilder et al. 2019). In microarray studies, samples are arranged in a two-dimensional grid. In the first stage of testing, samples are pooled by row and by column. In the second stage, only samples located at the intersections of positive rows and columns are retested. See Hitt et al. (2019, 2020) and `https://bilder.shinyapps.io/PooledTesting/` for more details.

Example: Screening for SARS CoV-2 (COVID-19)

Abdalhamid et al. (2020) report an evaluation of batch-testing strategies for SARS-CoV-2 from specimens collected by nasopharyngeal swabs (the 'gold standard' for SARS-CoV-2. diagnosis). They used the Shiny application for pooled testing (available at `https://www.chrisbilder.com/shiny`).

For an expected disease prevalence of 5% ($p = 0.05$), test sensitivity of 0.95 and test specificity of 1.0, with testing in two stages, they obtained an optimal batch size m of 5, $E(T) = 2.07$, and expected percentage reduction in the number of tests of $R_T = 100[1 - 0.41] = 59\%$. However, if prevalence is only 0.1% ($p = 0.001$) then $m = 33$, $E(T) = 2.02$, and $R_T = 94\%$.

Zhou and O'Leary (2021) report lower sensitivity (82–88%) for anterior nares or mid-turbinate (nasal) swabs. If sensitivity is assumed to be 0.85 with prevalence 5% and test specificity of 1, then $m = 6$, $E(T) = 2.35$, and $R_T = 61\%$.

8.5 Negative Binomial Distribution

The negative binomial distribution can be used to estimate the total number of samples required to obtain a pre-specified number of subjects with the given condition. For example, suppose an investigator wishes to pilot a new drug therapy on five cats with primary brain tumours. However, the rarity of the condition means that determining the total number of subjects to be screened will be potentially very large. The investigator will need to know how many potentially eligible subjects will need to be screened until five subjects are obtained, when screening will stop. The number N of repeated trials to produce s successes is a negative binomial random variable. It is assumed that the trials are independent – the outcome of one trial does not affect the outcome of other trials.

The negative binomial distribution is the probability distribution of a negative binomial random variable. It describes the probability that $N = s + f$ trials will be required to obtain s successes. Because sampling continues until a predetermined number of s is observed, this means there will be many failures in the sequence (because f subjects will be selected that do not have the condition). When the pre-specified number of successes has been achieved, the sampling stops, so s defines the stopping criterion. The exact probability is given by:

$$\Pr[X = f] = \binom{f + s - 1}{s - 1} p^{s \cdot} (1 - p)^f$$

where the number of successes s is fixed, and the number of failures is f. Because N is the number of 'trials' or draws, then $f = N - s$. The expected proportion of 'successes' is p, and $(1 - p)$ is the expected proportion of 'failures'. Then the probability to be determined is the number of m failures given r successes.

The geometric distribution is a special case of the negative binomial distribution, when the number of successes to be obtained stops at the first subject with the target condition: $s = 1$.

Example: Osteosarcoma Pilot Study: Determining Screening Numbers

---

Investigators wish to enrol 10 client-owned dogs with osteosarcoma for a pilot clinical trial with a new drug treatment. They expect only 40% of clients will consent to participate. How many clients of eligible dogs will need to be screened to have a 95% probability of obtaining 10 that agree to participate?

The expected proportion of successes p is 0.4, and the number of desired successes is $r = 10$. Enrolment of 10 dogs means that the total number of dogs screened is therefore $N = m + 10$. To find N, the number of expected 'failures' f needs to be estimated. The probability of obtaining f or fewer failures is $\Pr(X \leq f)$. Then the probabilities over a range of n are computed using the cumulative distribution function of the negative binomial distribution. The target sample size is selected for which the computed probability is equal to or exceeds $(1 - \alpha) = 0.95$:

```
    *maximum number of n eligible subjects
to be sampled;
    %let numSamples = 50;
    data prob;
    do m = 1 to &numSamples by 1; *set
increment to desired step size;
    y = CDF('NEGBINOMIAL',m,0.4,10); *y is
the set of probabilities;
    output;
    end;
    run;
```

The number of failures that associated with a target probability ≥ 0.95 is $f = 26$ (with a realised probability of 0.955). Then the total number of dogs to be screened is approximately $N = 26 + 10 = 36$.

8.6 Hypergeometric Distribution

The hypergeometric distribution is used to determine a sample size when the total population is small (Box 8.4). Subsequent draws without replacement of each item during the selection process reduce the total population size. As a result, the

probability of selection of each item will change, and sampling events are not independent. (In contrast, for the binomial 'experiments', the size of the total population is extremely large relative to the size of the sample to be drawn (>95%), so the pool will not be depleted by repeated sampling).

The hypergeometric distribution describes the probability of drawing a predefined number of subjects with the 'condition' (x) out of a random sample n drawn without replacement from a finite population of size N, with K members having the condition:

$$\Pr(X = x) = \frac{\binom{K}{x}\binom{N-K}{n-x}}{\binom{N}{n}}$$

Each subject in a sample of size n drawn from a population pool of size N will fall into one of four categories: subject has the condition and is selected (x), subject has the condition and is not selected $(K - x)$, subject does not have the condition and is selected $(n - x)$, or the subject does not have the condition and is not selected $(N - K) - (n - x)$. Sampling is performed without replacement; that is, once each subject is selected, it is not returned to the pool.

BOX 8.4

Examples of Small Defined Target Populations for Hypergeometric Sampling Problems

Rosters listing all clients visiting a clinic within a given time period.
All individuals listed in a breed or disease registry.
All animals surveyed in a defined location.
Pollinators visiting flowers in a given location.
All blood samples in a given batch.

Example: Maine Coon Cats and Hypertrophic Cardiomyopathy Markers

Approximately 30% of Maine Coon cats have an MYBPC3 gene mutation associated with hypertrophic cardiomyopathy (HCM). There were 320 cats listed in the Maine Coon cat registry. An investigator wished to determine how many cats would need to be screened to have 95% probability that at least five cats in the sample would have the mutation.

The total number of subjects in the registry population is $N = 320$. The prevalence of the condition is approximately 30%; therefore $R = 0.3 \cdot N = 96$. The number of mutation-positive cats required is at least 5, so $\Pr(x \geq 5)$. SAS code for computations is given in Appendix 8.B.

Results are shown in Figure 8.1. A random selection of 28 cats from the registry list must

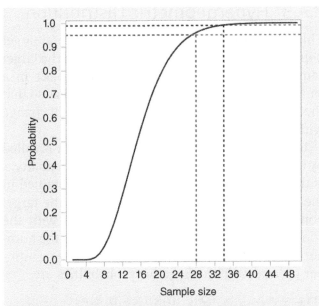

Figure 8.1: Determining the number of Maine Coon cats that must be selected at random from a small population to have a pre-specified probability of obtaining at least five with the MYBPC3 gene mutation. Twenty-eight cats must be selected at random to have a probability of 95% of obtaining at least five with the mutation and 34 cats for a probability of 99%.

be made to have a greater than 95% probability of obtaining at least 5 cats with the mutation and 34 cats for a probability of 99%.

8.6.1 Estimating the Proportion of Subjects with the Target Effect

The extended hypergeometric function can be used to estimate the anticipated proportion of subjects that will show a target response change from the expected mean value $d - \bar{x}$ (Zelterman 1999). The method requires three steps:

1. Calculate the z-score from values for the target change d, the mean $\bar{x}$, the sample standard error SE, and sample size n:

$$z = \frac{d - \bar{x}}{SE\sqrt{n}}$$

2. Convert z to the probability determined by the pdf of the standard normal distribution:

$$Pr = \varphi(z)$$

The probability is the proportion of the sample expected to show the desired change in the response. In SAS, this is obtained by

$$Pr = probnorm(z)$$

Example: Weight Loss in Overweight and Obese Cats

(Data from Flanagan et al. 2018). Cats on a weight reduction diet were fed dry commercially available weight loss food and monitored for three months. Weight loss of 10–15% of initial body weight is necessary for a one-unit change in body condition score. Female cats ($n = 155$) lost an average of 11.34% body weight (SE 0.556) and males ($n = 180$) lost 8.95% (SE 0.467). What percentage of cats in a new study could be expected to show a 15% decrease in body weight?

In this example, $d = -15$, and the mean observed weight loss $\bar{x}$ is −11.4 for females and −8.95 for males.

The z-score for females is $[-15 - (-11.34)]/(0.556 \cdot \sqrt{155}) = -0.5289$, and the proportion is *probnorm*(−0.5289) = 0.298, or 30%.

The z-score for males is $[-15 - (-8.95)]/(0.467 \cdot \sqrt{180}) = -0.8067$, and the proportion is *probnorm*(−0.8067) = 0.210, or 21%.

Therefore in a new sample of obese and overweight cats, approximately 30% of females and 21% males in the sample are anticipated to show the desired 15% decrease in weight for the target one-unit change in body condition score.

8.A Determining Cumulative Probabilities for Binomial, Negative Binomial, and Hypergeometric Distributions with SAS

Calculation of probabilities requires the specification of the cumulative distribution function (cdf) and probability density function (pdf). The *cumulative distribution function* (cdf) gives the probabilities of a random variable X being smaller than or equal to some value x, $\Pr(X \leq x) = F(x)$. The inverse of the cdf gives the value x that would make $F(x)$ return a particular probability p: $F-1$ $(p) = x$. The *probability density function* (pdf) returns the relative frequency of the value of a statistic or the probability that a random variable X takes on a certain value. The pdf is the derivative of the cdf.

8.A.1 Binomial Distribution

This distribution is applicable when the population is very large relative to the sample so that the outcome on one trial does not affect the outcome on other trials, the probability of success p is the same on every trial (that is, $p = 1/2$ or 0.5), and the trials are independent. In SAS, the binomial distribution is coded as ('BINOMIAL', s, p, n) where s is the number of subjects with the condition ('successes'), p is the prevalence (probability of success, or proportion of subjects with the condition), and n is the number of 'trials' (sample size).

The exact and cumulative probabilities are:

```
Pr(X = x)  y = PDF('BINOMIAL',s,p,n);
Pr(X < x)  y = CDF('BINOMIAL',s,p,n) - PDF
('BINOMIAL',s,p,n);
Pr(X ≤ x)  y = CDF('BINOMIAL',s,p,n);
Pr(X > x)  y = 1 - CDF('BINOMIAL',s,p,n);
Pr(X ≥ x)  y = PDF('BINOMIAL',s,p,n) + 1 - CDF
('BINOMIAL',s,p,n);
```

8.A.2 Negative Binomial Distribution

The negative binomial distribution is applicable for estimating the total sample size N necessary to obtain a pre-specified number of subjects with the condition. The 'experiment' continues until a predetermined number of successes s is observed, and s defines the stopping criterion (there is a 'success' on the last trial).

The total number of samples required to obtain s successes is $N = s + f$ trials. In SAS, the syntax is set up as s is the number of successes, $f =$ the number of failures ($= N - s$), and p is the probability of success.

```
Pr(X = f) y = PDF('NEGBINOMIAL', f, p, s);
Pr(X ≤ f) y = CDF('NEGBINOMIAL', f, p, s);
```

8.A.3 Hypergeometric Distribution

This distribution is applicable when the total target population N is small and finite relative to the target sample. Successive selection of each subject in a small population reduces the population size so that the probability of selection of each item will change with each sampling event.

In SAS, the hypergeometric distribution is called by ('HYPER', s, N, R, n) where y is the probability, s is the target number of successes to be obtained in the sample, N is the total sample size of the target population, R is the number of subjects with the condition in the study population, and n is the number of subjects to be screened (sample size or number of 'trials'). The respective probabilities are:

```
Pr(X = s)  y = PDF('HYPER',s,N,R,n);
Pr(X < s)  y = CDF('HYPER',s,N,R,n) - PDF
('HYPER',s,N,R,n);
Pr(X ≤ s)  y = CDF('HYPER',s,N,R,n);
Pr(X > s)  y = 1 - CDF('HYPER',s,N,R,n);
Pr(X ≥ s)  y = PDF('HYPER',s,N,R,n) + 1 - CDF
('HYPER',s,N,R,n);
```

8.B SAS Programme for Maine Coon Cat Example

```
%let numSamples = 50; *maximum of n subjects
to be sampled;
data prob;
do n = 1 to &numSamples by 1; *set start n and
increment to desired values;

*iterate over a range of potential sample
sizes to estimate the probability of a capture
of at least 4 events;

Y = PDF('HYPER', 5, 320, 96, n) + 1 - CDF
('HYPER', 5, 320, 96, n);
    output;
    end;
 run;
proc print;
    run;

/*plot change in probability with sample size
n*/
    proc sgplot data=prob;
    series x=n y=y;
        xaxis label="Sample size" grid
values=(0 to 50 by 1);
        yaxis label="Probability" grid
values=(0 to 1 by 0.1);
            refline 0.9 0.95 0.99;
    run;
```

References

Abdalhamid, B., Bilder, C.R., McCutchen, E.L. et al. (2020). Assessment of specimen pooling to conserve SARS COV-2 testing resources. *American Journal of Clinical Pathology* 153 (6): 715–718. `https://doi.org/10.1093/ajcp/aqaa064`.

Agresti, A. and Coull, B.A. (1998). Approximate is better than 'exact' for interval estimation of binomial proportions. *The American Statistician* 52: 119–126.

Arnold, M.E., Cook, A., and Davies, R. (2005). A modelling approach to estimate the sensitivity of pooled faecal samples for isolation of *Salmonella* in pigs. *Journal of the Royal Society Interface* 2: 365–372. `https://doi.org/10.1098/rsif.2005.0057`.

Arnold, M.E., Mueller-Doblies, D., Carrique-Mas, J.J., and Davies, R.H. (2009). The estimation of pooled-sample sensitivity for detection of *Salmonella* in turkey flocks. *Journal of Applied Microbiology* 107: 936–943. `https://doi.org/10.1111/j.1365-2672.2008.04273.x`.

Bilder, C., Tebbs, J., and McMahan, C. (2019). Informative group testing for multiplex assays. *Biometrics* 75 (1): 278–288.

Dorfman, R. (1943). The detection of defective members of large populations. *Annals of Mathematical Statistics* 14: 436–440.

Eypasch, E., Lefering, R., Kum, C.K., and Troidl, H. (1995). Probability of adverse events that have not yet occurred: a statistical reminder. *BMJ* 311 (7005): 619–620.

Finucan, H.M. (1964). The blood testing problem. *Journal of the Royal Statistical Society: Series C: Applied Statistics* 13 (1): 43–50. `https://doi.org/10.2307/2985222`.

Flanagan, J., Bissot, T., Hours, M.A. et al. (2018). An international multi-centre cohort study of weight loss in overweight cats: differences in outcome in different geographical locations. *PLoS ONE* 13 (7): e0200414. `https://doi.org/10.1371/journal.pone.0200414`.

Giles, J.R., Peel, A.J., Wells, K. et al. (2021). Optimizing noninvasive sampling of a zoonotic bat virus. *Ecology and Evolution* 11: 12307–12321. `https://doi.org/10.1002/ece3.7830`.

Hanley, J.A. and Lippman-Hand, A. (1983). If nothing goes wrong, is everything all right? Interpreting zero numerators. *JAMA* 249 (13): 1743–1745.

Hitt, B., Bilder, C., Tebbs, J., and McMahan, C. (2019). The objective function controversy for group testing: much ado about nothing? *Statistics in Medicine* 38 (24): 4912–4923.

Hitt, B., Bilder, C., Tebbs, J., McMahan, C. (2020). A shiny app for pooled testing. `https://bilder.shinyapps.io/PooledTesting/` (accessed 2022).

Hou, P., Tebbs, J., Bilder, C., and McMahan, C. (2017). Hierarchical group testing for multiple infections. *Biometrics* 73 (2): 656–665.

Hughes-Oliver, J.M. (2006). Chapter 3: Pooling experiments for blood screening and drug discovery. In: *Screening Methods for Experimentation in Industry, Drug Discovery, and Genetics* (ed. A. Dean and S. Lewis), 48–68. Springer 330 pp.

Jovanovic, B.D. and Levy, P.S. (1997). A look at the rule of three. *The American Statistician* 51 (2): 137–139.

Litvak, E., Tu, X.M., and Pagano, M. (1994). Screening for the presence of a disease by pooling sera samples. *Journal of the American Statistical Association* 89: 424–434.

Louis, T.A. (1981). Confidence intervals for a binomial parameter after observing no successes. *The American Statistician* 35 (3): 154.

Newcombe, R.G. (1998). Two-sided confidence intervals for the single proportion: comparison of seven methods. *Statistics in Medicine* 17: 857–872.

Newman, S.C. (2001). *Biostatistical Methods in Epidemiology*. New York: Wiley.

Shipitsyna, E., Shalepo, K., Savicheva, A. et al. (2007). Pooling samples: the key to sensitive, specific and cost-effective genetic diagnosis of *Chlamydia trachomatis* in low-resource countries. *Acta Dermato-Venereologica* 87: 140–143.

Zelterman, D. (1999). *Models for Discrete Data*. Oxford: Clarendon Press.

Zhou, Y. and O'Leary, T.J. (2021). Relative sensitivity of anterior nares and nasopharyngeal swabs for initial detection of SARS-CoV-2 in ambulatory patients: rapid review and meta-analysis. *PLoS ONE* 16 (7): e0254558. `https://doi.org/10.1371/journal.pone.0254559`.

III Sample Size for Description

9 Descriptions and Summaries

9.1 Introduction

Descriptive statistics describe the properties of sample data and summarise the results of statistical inference testing. Many studies are inherently descriptive rather than evaluative and inferential. These include surveys, species inventories, demand–supply assessments, feasibility studies, documentation of novel methodologies, and protocols. For these studies, the goal is to provide empirical descriptions of specific events, outcomes, or categories in the sample, not to test hypotheses (Gerring 2012). However, hypothesis-testing studies require descriptive statistics to summarise results and synthesise information across multiple groups or studies (Box 9.1).

Descriptive statistics are a condensation of large amounts of data down to two or a few metrics: *measures of central tendency* and *measures of variation*. The central tendency is a single value identifying the central point of the data distribution (point estimate). It is a measure of the magnitude or size of the variable. Measures of central tendency include the mean and median for continuous sample data. Measures of variation describe the spread or dispersion of the data around the central point. Variability between observations within the sample is described by the range, standard deviation, interquartile range,

> **BOX 9.1**
> *Descriptive Statistics*
>
> **1.** Describing sample data
>   Measures of central tendency
>   Measures of variation
>   Five-number summary
>   Counts and percentages
>
> **2.** Describing results of inference tests
>
> Range of estimates containing the unknown population parameter with a specified probability
>
>   Confidence intervals
>   Prediction intervals
>   Tolerance intervals
>   Reference intervals

and variance. The five-number summary (minimum, lower quartile, median, upper quartile, maximum) is appropriate for non-normally distributed continuous data. Categorical data are summarised as counts and percentages. Descriptive statistics require reporting of total sample size, sample sizes for each experimental group, and the number of any animals lost through attrition. The results of statistical hypothesis tests are used to make inferences

A Guide to Sample Size for Animal-based Studies, First Edition. Penny S. Reynolds.
© 2024 John Wiley & Sons Ltd. Published 2024 by John Wiley & Sons Ltd.

about the mean for the whole population. The uncertainty in the estimates is quantified by *confidence intervals*. Several other types of intervals (prediction, tolerance, or reference) can be computed, depending on study objectives.

The best-practice reporting standard is to present descriptive statistics in at least two summary tables. So-called 'Table 1' is for reporting study animal characteristics ('Who was studied?'). 'Table 2' summarises the results ('What was found?'). Total sample size, sample sizes per group, and numbers lost through attrition must always be reported. Summary descriptive data and meticulous sample sizes accounting enable the assessment of internal and external validity and are required for systematic reviews and meta-analyses (Macleod et al. 2015). Reporting of both sets of descriptive and summary statistics is recommended by all international consensus reporting guidelines, including ARRIVE 2.0 for animal-based research (Percie du Sert et al. 2020a, 2020b), and STROBE-VET for observational clinical studies (O'Connor et al. 2016; Sargeant et al. 2016).

9.2 Describing Sample Data

'Table 9.1' presents summary data for basic descriptors: animal signalment (source, species, strain, age, sex, body weight) and baseline (or pre-intervention) traits (e.g. baseline biochemistry, haematology, hemodynamics, scores, behaviour metrics, etc.) (Box 9.2). Table 9.1 information allows assessment of the representativeness of the sample and any potential biases (Roberts and Torgerson 1999). The more the study animals are representative of the target population, the more likely the results will be generalisable. The descriptive statistics in

these tables provide information about the animals in the study sample and are analogous to patient demographic data reported for human clinical trials. Information for variables recorded at baseline is important for providing context for subsequent interpretation of results. Effects of the test intervention and potential benefit are assessed by the magnitude of post-intervention changes compared to baseline (Fishel et al. 2007). Between-study differences in animal characteristics and baseline or 'starting' conditions may also contribute to substantial differences in results (Mahajan et al. 2016). Standard deviations are used to describe the sampling distribution of the mean (Altman and Bland 2005). Performing significance tests for differences on Table 1 characteristics is usually illogical and not recommended (Altman 1985; Senn 1994; Roberts and Torgerson 1999; de Boer et al. 2015; Pijls 2022).

Example: Sample 'Table 1' Data for a Two-Arm Crossover Study of Stress in Dogs

'Table 9.1' shows group sample sizes and descriptive statistics for signalment, baseline heart rate, and baseline fear, anxiety, and stress (FAS) scores for dogs in a crossover study of clinical exam location effects.

Table 9.1: Example of Descriptive Statistics Reporting for Animal Signalment and Baseline Characteristics.

Variable	Group 1	Group 2
Number of dogs, n	24	20
Sex, n males (%)	12 (50%)	12 (70%)
Age, years; median (IQR; minimum, maximum)	5.0 (2.5, 9.0; 1,15)	5.7 (4.7, 5.0; 2, 11)
Body weight, kg; median (IQR; minimum, maximum)	19 (9, 26; 3, 65)	22 (10, 30; 3, 39)
Heart rate, bpm; mean (SD)	103 (21)	107 (25)

Baseline FAS scores	Number of dogs	
0–1	9	11
2–3	7	7
>3	2	3

Source: Data from Mandese et al. (2021).

BOX 9.2
Describing Descriptive Data: 'Who Was Studied?'

'Table 1' information:

What: Summary of major signalment, clinical, and pre-intervention characteristics for each study group.

Descriptors: Sample statistics: sample size per group n, mean (SD), median (interquartile range), and counts (per cent).

9.3 Describing Results

'Table 2' information consists of the summary data for results per group and/or differences between intervention and control. These are reported as effect sizes (such as mean differences) and confidence (or other) intervals (Box 9.3). Intervals measure the uncertainty around the point estimate and are the best estimate of how far the estimate is likely to deviate from the true value (Altman and Bland 2014a). Intervals should be used to present the results of statistical inference so that the practical or biological importance of differences can be properly evaluated (Gardner and Altman 1986; Cumming 2012; Altman and Bland 2014a, b).

Table 9.2: Example of Descriptive Statistics Reporting for Results of Hypothesis Tests.

Variable	Adjusted mean difference (95% confidence interval)	*P*-value
Sample size, *n* breeding pairs	59	
Number of pups extra born per pair, tunnel versus tail-lift	+1.04 (0.95, 1.14)	0.41
Number of pups weaned per pair, tunnel versus tail-lift	+1.07 (0.94, 1.20)	0.33

Source: Data from Hull et al. (2022).

BOX 9.3
Describing Descriptive Data: 'What Was Found?'

'Table 2' information:

What: Summary of major results for each study group, and results of hypothesis tests for groups differences in primary and secondary outcomes, population-based measures of precision.

Descriptors: Sample size per group *n*, means, confidence intervals (or other relevant interval measure).

Example: Sample 'Table 2' Results from a Two-Arm Randomised Controlled Study of Mouse Productivity

'Table 9.2' provides descriptive statistics expressed as adjusted means and 95% confidence limits for differences in productivity for two methods of mouse handling. Means were adjusted for parental and parity effects. The differences in favour of tunnel handling were operationally important but not statistically significant.

9.4 Confidence and Other Intervals

There are four main types of intervals: confidence, prediction, tolerance, and reference. Choice of interval will depend on study objectives. Confidence intervals are related to the use of significance tests but provide more useful, actionable, and interpretable information on the size, direction, and uncertainty associated with the observed effect compared to a *P*-value (Gardner and Altman 1986; Steidl et al. 1997; Cumming 2012; Rothman and Greenland 2018).

The interval is an estimate based on the sampling distribution of the statistic. The general form of the interval is:

Sample point estimate $\pm$ *width*(SD, n, α)

Therefore the width of the interval is determined by the variation in the sample (the standard deviation SD), the inverse square root of the sample size $1/\sqrt{N}$, and the pre-specified significance level α.

There are four main types of intervals: *confidence, prediction, tolerance,* and *reference* (Box 9.4). Choice of interval will depend on study objectives.

BOX 9.4
Four Types of Descriptive Intervals

Confidence intervals describe the range within which *a given population parameter* is expected to be located with a specified probability. *Precision* is one-half the confidence interval or margin of error. Interval width is determined by sampling error.

Prediction intervals describe the range within which *one or more future observations* are expected to be located with a specified probability. Interval width is determined by sampling error and the expected variation in the predicted values.

Tolerance intervals describe the range within which *a specified proportion of the population* is expected to be located with a *specified probability*. Interval width is determined by sampling error and expected variation in the population.

Reference intervals describe the range of clinical read-out values between which typical results for a healthy subject might be expected to occur. Interval width is determined by sampling error variation and variation around the cut-off limits separating healthy from abnormal values.

Table 9.3: Coverage Probabilities for 95% and 99% Confidence Intervals for the Mean of a Normally Distributed Continuous Variable with Sample Size n of 10, 20, 50, and 100. Coverage probabilities were obtained by $N = 1000$ simulations for each sample size and confidence interval in R coversim.

Sample size	Confidence intervals	
	95%	**99%**
$n = 10$	0.928	0.988
$n = 20$	0.934	0.984
$n = 50$	0.944	0.993
$n = 100$	0.948	0.994

Confidence intervals define the range of values from a sample that is likely to contain the population parameter (such as a mean or variance). The confidence interval puts bounds on the variation around a *single, or point, estimate of population parameter* with a given significance α. The endpoints of the confidence interval are the *confidence limits*. The width of the confidence interval is determined by the variation in the observed responses or sampling error. As sample size increases, sampling error becomes smaller as the sample converges on the true population parameter, and the confidence interval approaches zero. Because of sampling variation, the confidence interval for any given sample may not contain the parameter. The *coverage probability* means that for a large number of samples of the sample size n, the proportion of intervals that will contain the population parameter will be $1 - \alpha$. For example, for significance level of $\alpha = 0.05$, the 95% confidence interval ($1 - 0.05 = 0.95$) means that approximately 95% of confidence intervals calculated for a large number of samples of size n will contain the population parameter (Table 9.3). *Precision* is defined as one-half of the confidence interval or *margin of error*.

When comparing two methods or devices that measure the same thing on the same scale of measurement, the *limits of agreement* describe the interval within which a proportion of the differences between measurements occur with a given probability (usually 95%) for a sample of n paired measurements (Bland and Altman 1986, 2003).

Prediction intervals describe the bounds on variation for one or more *future observations*, given the distribution of a sample of previous observations. Prediction interval width is determined by two variance components, the variance in the observed responses and the variance due to the prediction. Therefore prediction intervals are wider than confidence intervals. Prediction intervals are most frequently calculated for regression-based models where the magnitude of a response depends on one or more covariates (Bland and Altman 2003).

Tolerance intervals are constructed to estimate a range that contains a specified *proportion of the population* with a specified *confidence* or *coverage*. Tolerance intervals are most appropriate if the interval is intended to capture coverage for many future samples. The width of the tolerance interval is determined by both sampling error and the variance in the population (Meeker et al. 2017).

Reference intervals describe the range of clinical read-out values between which typical results for a healthy subject might be expected to occur. The width of the reference interval is defined by the lower and upper bounds on the central 95% of the reference sample measurements. These cut-points also have an associated precision defined as a proportion of the reference interval width (Jennen-Steinmetz and Wellek 2005).

9.5 Relationship Between Interval Width, Power, and Significance

Intervals are related to statistical tests of significance but provide more practical information about the true size of the specified difference or *effect*.

Figure 9.1 shows the relationship between interval width, statistical significance, and power (Clayton and Hills 1993). The power of the study is the probability that the lower confidence limit is greater than the value specified by the null hypothesis, the hypothesis of no difference between means. The probability of type I error is specified by the *significance threshold* α, the probability of obtaining a false positive (type I error), or incorrectly rejecting the null hypothesis when it is true. The *confidence interval* $(1 - \alpha)$ is a range that is expected to contain the population parameter with a specified probability.

For sufficient power to detect a specified effect, the lower limit of the confidence interval must exceed the upper limit for null hypothesis value. The bounds for the null value are defined by the range of d standard deviations from the null value. The confidence interval covers a range of c standard deviations on either side of the mean, so the limits of the confidence interval are defined by $\pm c \cdot$ SD.

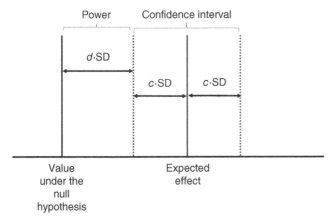

Figure 9.1: Relationship between interval width, statistical significance, and power. The significance level α defines the bounds for the confidence interval as a multiple of c standard deviations. Power defines how far the lower confidence limit is from the null value and is a multiple of d standard deviations.

Source: Adapted from Clayton and Hills (1993).

The values of c and d are obtained from the standard normal z-scores for the desired level of significance and power, respectively. The z-score is the number of standard deviations between an observation and the mean:

$$z = \frac{observed\ value - mean}{SD} = \frac{x - \mu}{\sigma}$$

The z-score is therefore a standardised normal value for the observation. The maximum possible z-score is $(n - 1)/\sqrt{n}$. Probability and z-scores are obtained from the standard normal distribution (Box 9.5). For the normal distribution, the probability for the value in that interval is the area under the normal curve. For example, a 95% confidence interval is defined by the lower and upper confidence limits z_l and z_u. The value of z_l is -1.96 and for z_u is 1.96. The area under the curve to the left of z_l is 0.025, and the area to the right of z_u is 0.025. The area under the curve between the two limits is 0.95. For power of 90%, $z = d = 1.282$.

The difference between the null value and the lower limit of the observed value is expressed as a multiple of the number of standard deviations. To be statistically significant, the distance $(c + d) \cdot$ SD must be greater than the effect so that the lower limit of the confidence interval (effect $- c \cdot$ SD) is larger than the upper limit for power (null value $+ d \cdot$ SD). Because confidence and power are pre-specified, sample size is chosen so that SD $\geq$ effect/$(c + d)$, and $(c + d)$ is approximately three. That is, the effect to be detected must be equal to or greater than three standard deviations (Clayton and Hills 1993; Table 9.4).

BOX 9.5

Calculating z Scores for Confidence and Power (SAS Code)

```
*significance alpha;
*0.1 for 90% power, 0.2 for 80% power;

alpha = 0.05;
beta = 0.2;

zleft = quantile('normal',alpha/2);
Zright = quantile('normal', 1-alpha/2);

zbeta = quantile('normal', 1-beta);
```

Table 9.4: Confidence, Power, and Corresponding z-Values for a Standard Normal Distribution.

α	Confidence 100(1 − α)%	$z_{1-\alpha/2}$	β	Power 100(1 − β)%	$z_{1-\beta}$
0.01	99	2.576	0.25	75	0.675
0.05	95	1.960	0.20	80	0.842
0.10	90	1.645	0.10	90	1.282
0.20	80	1.282	0.05	95	1.645

Source: Adapted from Clayton and Hills (1993).

Interval plots with different confidence levels are especially useful for evaluating empirical and translational pilot studies (Chapter 6). Pilot studies are small by design and therefore underpowered to detect any but very large effect sizes. As a result, non-significant p-values are often interpreted as evidence of 'no effect', and potentially meaningful results are rejected (Altman and Bland 1995; Lee et al. 2014). However, the data may still provide sufficient evidence of promise (always assuming the pilot study was designed with sufficient internal validity), although the study may be too small to statistically detect an effect size with a specified power. Lee et al. (2014) recommend that the strength of preliminary evidence can be assessed by comparing the effect relative to the null with confidence intervals of different widths (e.g. 95%, 90%, 80%, 75%).

References

Altman, D.G. (1985). Comparability of randomised groups. *The Statistician* 34: 125–136. https://doi.org/10.2307/2987510.

Altman, D.G. and Bland, J.M. (1995). Absence of evidence is not evidence of absence. *BMJ* 311 (7003): 485. https://doi.org/10.1136/bmj.311.7003.485.

Altman, D.G. and Bland, J.M. (2005). Standard deviations and standard errors. *BMJ* 331: 903. https://doi.org/10.1136/bmj.331.7521.903.

Altman, D.G. and Bland, J.M. (2014a). Uncertainty and sampling error. *BMJ* 349: g7064. https://doi.org/10.1136/bmj.g7064.

Altman, D.G. and Bland, J.M. (2014b). Uncertainty beyond sampling error. *BMJ* 349: g7065. https://doi.org/10.1136/bmj.g7065.

Bland, J.M. and Altman, D.G. (1986). Statistical methods for assessing agreement between two methods of clinical measurement. *Lancet* 1: 307–310.

Bland, J.M. and Altman, D.G. (2003). Applying the right statistics: analyses of measurement studies. *Ultrasound in Obstetrics & Gynecology* 22 (1): 85–93. https://doi.org/10.1002/uog.122.

de Boer, M.R., Waterlander, W.E., Kuijper, L.D. et al. (2015). Testing for baseline differences in randomized controlled trials: an unhealthy research behavior that is hard to eradicate. *International Journal of Behavioral Nutrition and Physical Activity* 12: 4. https://doi.org/10.1186/s12966-015-0162-z.

Clayton, D. and Hills, M. (1993). *Statistical Models in Epidemiology*. Oxford: Oxford University Press 375 pp.

Cumming, G. (2012). *Understanding the New Statistics: Effect sizes, Confidence Intervals, and Meta-Analysis*. New York: Routledge.

Fishel, S.R., Muth, E.R., and Hoover, A.W. (2007). Establishing appropriate physiological baseline procedures for real-time physiological measurement. *J Cogn Engin Decision Making* 1: 286–308.

Gardner, M.J. and Altman, D.G. (1986). Confidence intervals rather than P values: estimation rather than hypothesis testing. *British Medical Journal* 292 (6522): 746–750. https://doi.org/10.1136/bmj.292.6522.746.

Gerring, J. (2012). Mere description. *British Journal of Political Science* 42 (4): 721–746. https://www.jstor.org/stable/23274165.

Hull, M.A., Reynolds, P.S., and Nunamaker, E.A. (2022). Effects of non-aversive versus tail-lift handling on breeding productivity in a C57BL/6J mouse colony. *PLoS ONE* 17 (1): e0263192. https://doi.org/10.1371/journal.pone.0263192.

Jennen-Steinmetz, C. and Wellek, S. (2005). A new approach to sample size calculation for reference interval studies. *Statistics in Medicine* 24 (20): 3199–3212. https://doi.org/10.1002/sim.2177.

Lee, E.C., Whitehead, A.L., Jacques, R.M., and Julious, S.A. (2014). The statistical interpretation of pilot trials: should significance thresholds be reconsidered? *BMC Medical Research Methodology* 14: 41. https://doi.org/10.1186/1471-2288-14-41.

Macleod, M.R., Lawson McLean, A., Kyriakopoulou, A. et al. (2015). Risk of bias in reports of *in vivo* research: a focus for improvement. *PLoS Biology* 13 (11): e1002301.

Mahajan, V.S., Demissie, E., Mattoo, H. et al. (2016). Striking immune phenotypes in gene-targeted mice are driven by a copy-number variant originating from a commercially available C57BL/6 strain. *Cell Reports* 15 (9): 1901–1909. https://doi.org/10.1016/j.celrep.2016.04.080.

Mandese, W.W., Griffin, F.C., Reynolds, P.S. et al. (2021). Stress in client-owned dogs related to clinical exam location: a randomised crossover trial. *Journal of Small Animal Practice* 62 (2): 82–88. https://doi.org/10.1111/jsap.13248.

Meeker, W.Q., Hahn, G.J., and Escobar, L.A. (2017). *Statistical Intervals: A Guide for Practitioners and Researchers*, 2e. New York: Wiley.

O'Connor, A.M., Sargeant, J.M., Dohoo, I.R. et al. (2016). Explanation and elaboration document for the STROBE-Vet statement: strengthening the Reporting of Observational Studies in Epidemiology – Veterinary extension. *Journal of Veterinary Internal Medicine* 30 (6): 1896–1928. https://doi.org/10.1111/jvim.14592.

Percie du Sert, N., Hurst, V., Ahluwalia, A. et al. (2020a). The ARRIVE guidelines 2.0: updated guidelines for reporting animal research. *PLoS Biology* 18 (7): e300049. https://doi.org/10.1371/journal.pbio.3000410.

Percie du Sert, N., Ahluwalia, A., Alam, S. et al. (2020b). Reporting animal research: explanation and elaboration for the ARRIVE guidelines 2.0. *PLoS Biology* 18 (7): e3000411. https://doi.org/10.1371/journal.pbio.3000411.

Pijls, B.G. (2022). The Table I fallacy: P values in baseline tables of randomized controlled trials. *Journal of Bone and Joint Surgery* 104 (16): e71. https://doi.org/10.2106/JBJS.21.01166.

Roberts, C. and Torgerson, D.J. (1999). Understanding controlled trials: baseline imbalance in randomised controlled trials. *BMJ* 319 (7203): 185. https://doi.org/10.1136/bmj.319.7203.185.

Rothman, K.J. and Greenland, S. (2018). Planning study size based on precision rather than power. *Epidemiology* 29: 599–603.

Sargeant JM, O'Connor AM, Dohoo IR, Erb HN, Cevallos M, Egger M, Ersboll AK, Martin SW, Neilsen LR, Pearl DL, Pfeiffer DU, Sanchez J, Torrence ME, Vigre H, Waldner C, Ward MP (2016). Methods and processes of developing the Strengthening the Reporting of Observational Studies in Epidemiology-Veterinary (STROBE-Vet) statement. *Preventative Veterinary Medicine*, 134:188–196. https://doi.org/10.1016/j.prevetmed.2016.09.005

Senn, S. (1994). Testing for baseline balance in clinical trials. *Statistics in Medicine* 13 (17): 1715–1726. https://doi.org/10.1002/sim.4780131703.

Steidl, R.J., Hayes, J.P., and Schauber, E. (1997). Statistical power analysis in wildlife research. *Journal of Wildlife Management* 61 (2): 270–279.

10 Confidence Intervals and Precision

CHAPTER OUTLINE HEAD

10.1 Introduction

The confidence interval defines the limits within which a given population parameter (such as the mean) is expected to lie with a specified degree of confidence. The confidence interval enables an estimate of the precision of the sample estimate relative to the true population value (Box 10.1). The width of the confidence interval is entirely determined by the amount of variation in the sample due to sampling error. Therefore a 'precise' estimate indicates that measurements are stable and consistent, with a relatively narrow range of variation around the point estimate of the sample. Because a precise estimate is a reliable estimate, Cumming (2012) refers to precision as a measure of the 'informative-ness' of an experiment.

BOX 10.1
Definitions

- *Confidence interval* is the range within which *a given population parameter* is expected to be located with a specified probability.
- *Precision* is one-half the confidence interval. Sometimes called the *margin of error*.

A Guide to Sample Size for Animal-based Studies, First Edition. Penny S. Reynolds.
© 2024 John Wiley & Sons Ltd. Published 2024 by John Wiley & Sons Ltd.

10.2 Definitions

The confidence interval puts bounds on the variation around a *single estimate of population parameter* (for example, the mean or variance) with a given confidence level. The *confidence level* defines how close the confidence limits are to the point estimate for the effect size. The significance level α is used to compute the confidence level $100(1 - \alpha)\%$. For example, 95% confidence level is set by $\alpha = 0.05$. The endpoints of the *confidence interval* are the *confidence limits*. The upper and lower limits, or bounds of the confidence interval, are defined by the effect size $\pm$ precision d.

The amount of precision is determined entirely by sampling error. Therefore as sample size increases, the interval converges to a single point (the true population parameter), and the width converges to zero. There is a trade-off between precision, confidence levels, and sample size. A very precise estimate (high precision) has narrow confidence intervals. Narrower confidence intervals require larger sample sizes. As the confidence level increases, the confidence interval becomes wider and precision is accordingly reduced. A confidence level of 99% requires a larger sample size than confidence level of 95%. Therefore, increasing the information content of a study can be accomplished by reducing sample variation, as well as simply increasing sample size.

Precision is one-half of the confidence interval: (upper limit – lower limit)/2. It can also be computed from the product of the critical z-score for the confidence level and the standard error of the estimate. Estimating sample size for precision depends on:

Type of primary outcome variable. Variable type (continuous, proportion, time to event) determines the underlying distribution of the population parameter of interest.

Effect size. The biologically or clinically important difference to be detected.

Expected variability. Larger variation in the sample requires larger sample sizes.

The desired *confidence level* $100(1 - \alpha)\%$.

Precision required on either side of the parameter estimate.

Spatial or temporal correlation. Larger sample sizes are required if observations are not independent.

Confidence intervals can be obtained by bootstrapping, as an alternative to direct computation. Bootstrapping is a computationally-intense method for estimating the sampling distribution of most statistics. It uses random sampling with replacement to generate an approximating distribution for observed data (Efron and Tibshirani 1993; Davison and Hinkley 1997).

Bootstrapping involves for steps:

1. Generate new data by drawing a random sample with replacement multiple times from the observed data set. It is usually recommended that 5000–10,000 draws are performed.
2. Fit the model and calculate the mean (or mean difference) and SE for each bootstrap sample.
3. Calculate the confidence intervals for each bootstrap sample.
4. Estimate confidence intervals from the empirical quantiles. For example, for a 95% confidence interval, the quantiles for the limits are 2.5% and 97.5%.

10.3 Sample Size Calculations

The basis for sample size selection is the level of desired precision for the estimate of the true difference in the outcome. For example, a study may be designed to estimate the true range of biomarker expression within $\pm 5\%$ of the population value. The confidence interval will be bounded by 5% on either size of the sample estimate for a total width of 10%.

Sample size for confidence interval width is determined in four steps:

1. *Decide target precision d.* Precision is one-half the width of the desired confidence interval.

It can also be expressed as a proportional measure of the deviation from the mean. For example, if the desired confidence interval width is 10% of the mean (or 0.1), then precision d will be $\pm(0.1/2)$ or $\pm5\%$ (±0.05).

2. *Calculate the standard error of the estimate.* The formula for the standard error (SE) depends on the type of variable and the associated distribution. For example, the SE for the mean of a continuous normal variable $\mathrm{SE}(\overline{Y})$ is $s/\sqrt{n}$; the standard error for a proportion is $\mathrm{SE}(p) = \sqrt{p(1-p)/n}$.

3. *Specify confidence.* The most common α is 0.05 (5%), corresponding to a 95% confidence interval. Other commonly used settings for α are 0.01 (1%; 99% confidence) and 0.1 (10%; 90% confidence). A one-sided confidence interval is constructed by using α rather than $\alpha/2$ in the expression for the lower or upper limit. For most practical purposes and for ease of calculation, the z-score is used. For example, for a two-tailed 95% confidence interval $z_{1-\alpha/2} = 1.96$. If sample size is small, the t statistic can be substituted. However, sample size approximations will require iterations over a range of sample sizes to determine the value that is closest to the target power for a given effect size.

4. Divide the confidence interval by the square of the precision, d^2.

10.3.1 Absolute Versus Relative Precision

'Precision' can be either absolute or relative (Box 10.2). Research objectives will determine which measure of precision to use (Lwandga and Lemeshow 1991).

Absolute precision is a fixed difference of percentage points on either side of the parameter

---

BOX 10.2

Absolute Versus Relative Precision

Absolute precision is a fixed difference of percentage points on either side of the parameter estimate.
Relative precision is the fraction of the true value of the parameter estimate.

---

estimate and refers to the actual uncertainty in the metric itself. Absolute precision is used when the goal is to estimate the population parameter to within d percentage points of the true value (*estimate $\pm d$*), where *percentage point* is the unit for the arithmetic difference of two percentages. For example, if the prevalence of a given disease is 25% with a precision of 10%, the absolute precision of the estimate is 10%.

Relative precision is used when the goal is to estimate the population parameter to within a given percentage of the value itself. For example, if prevalence is $25\% \pm 10\%$, the relative uncertainty is 10% of 25%, or 2.5%. Relative precision scales the desired amount of precision to the parameter estimate; that is, the variation in the sample is a fraction or proportion of the true value of the parameter: *estimate$\pm d\cdot$ estimate* (Lwandga and Lemeshow 1991)

10.4 Continuous (Normal) Outcome Data

The confidence interval is computed from the product of the standard error for the estimate and the critical value of the test statistic. For a mean $\overline{Y}$ and standard error $\mathrm{SE}(\overline{Y}) = s/\sqrt{n}$, the confidence interval for the mean is

$$\overline{Y} \pm z_{1-\alpha/2} \cdot s/\sqrt{n}.$$

Re-arranging, sample size is therefore:

$$n = z_{1-\alpha/2}^2 \cdot s^2.$$

For *absolute* precision, the standard error is divided by the square of the precision:

$$n = z_{1-\alpha/2}^2 \cdot s^2/d^2$$

For *relative* precision, the standard error is divided by both the precision and the mean.

$$n = z_{1-\alpha/2}^2 \cdot s^2/d^2 \cdot \overline{Y}$$

The *coefficient of variation* (CV) is the ratio of the sample standard deviation to the sample mean $s/\overline{Y}$. It is a form of relative precision.

Example: Lizard Body Temperatures

An investigator studying lizard body temperatures wished to estimate the mean difference between preferred temperature and critical thermal maximum temperature (T_{diff}) for a population of lizards with 95% confidence. Mean T_{diff} ($\overline{T}$) for x species was estimated as 10°C, with variance $s^2 = 6.3$ (°C)^2. The investigator considered that a precision of 2.5°C around $\overline{T}_{diff}$ was biologically relevant and operationally feasible (that is, within the measurement error of the device).

Assuming the data are normally distributed, then $d = 2.5°C/2 = 1.25°C$ and the required sample size is

$$N = (1.96)^2 \cdot (6.3)/(1.25)^2 = 15.5 \cong 16$$

If the investigator wished to determine $\overline{T}$ with a precision of 1°C, then $d = 0.5°C$ and

$$N = (1.96)^2 \cdot (6.3)/(0.5)^2 \cong 97$$

10.4.1 Simultaneous Confidence Intervals

Simultaneous confidence intervals (SCI) are constructed for assessing multiple outcome variables. Simultaneous assessment is inherently a multiple comparisons problem. Therefore, SCI must be computed to reduce the probability of Type 1 error (false positives) and adjust for family wise error rate. For a given sample size, SCI are much wider than conventional confidence intervals computed for each variable separately. Confidence intervals that do not intersect the pre-specified null value suggest evidence against the null hypothesis and for the existence of an effect. No evidence is provided against the null hypothesis if confidence intervals contain the null value.

There are four steps in SCI construction:

1. If units of measurement for the variables are not the same, the variables must be standardised before analysis. For each of the p variables, compute its ratio r as the observed value divided by the corresponding reference value for that variable

$$r = y_i/(\text{reference value}).$$

The reference ratio across all standardised variables will equal 1.

2. For each variable, compute its mean ratio $\overline{r}_j$, standard deviation s_j, and sample size n_j.
3. Calculate the pooled standard error $s_{zj}/\sqrt{n}$ across all variables.
4. Compute the simultaneous confidence bands $100(1 - \alpha)\%$ as:

$$\overline{r}_j \pm \frac{s_{zj}}{\sqrt{n}} \cdot \sqrt{\frac{p_j(n_j - 1)}{n_j - p_j} \cdot F_{\alpha,p,n-p}}$$

where $F_{\alpha,p,n-p}$ is the critical F value. Appendix 10.A provides sample SAS code for calculating simultaneous confidence bands and generating profile plots.

Example: Nutrient Screening of Obese Dogs in a Weight Loss Study

(Data from German et al. 2015). Daily intake of 20 essential nutrients was measured for 27 obese dogs on an energy-restricted weight loss diet. The goal was to establish if average daily nutrient intakes complied with recommended dietary levels.

In this example, observations were first standardised to the specific minimum recommended daily allowance value (RDA) for each nutrient, so the null value is 1 for all variables. For example, crude protein intake is standardised by dividing each observation by 3.28, and the NRC recommended daily intake amount. Evidence for nutrient levels exceeding RDA is suggested if the lower confidence limit exceeds 1, below RDA if the upper confidence limit is less than 1, and no evidence against correspondence with RDA recommendations if confidence intervals contain 1.

A comparison of conventional and SCI for five nutrients are given in Table 10.1. Note that the SCI are much wider than conventional confidence intervals.

Table 10.1: Comparison of Conventional and Simultaneous Confidence Intervals for Standardised Nutrient Values in a Feeding Study of Obese Dogs.

Nutrient	Standardised value	95% confidence interval		95% simultaneous confidence interval	
Potassium	1.22	1.17	1.28	0.81	1.63
Selenium	0.43	0.41	0.45	0.29	0.56
Crude protein	1.93	1.85	2.01	1.32	2.55
Total fat	1.12	1.07	1.16	0.76	1.47
Vitamin D_3	32.7	31.4	34.1	22.3	43.2

10.5 Proportions

The proportion $p = x/N$, where x is the number of subjects responding or showing the condition of interest, and N is the total number of subjects. Choosing 'correct' confidence intervals for proportions need some care (Box 10.3).

The classic *Wald method* uses the z-score from the large-sample normal approximation. The confidence interval is $p \pm z_{1-\alpha/2} \cdot SE(p)$, where the standard error is

$$SE(p) = \sqrt{p(1-p)/N}$$

Sample size is:

$$N = z_{1-\alpha/2}^2 \, p(1-p)$$

This method may be useful as a first approximation for sample size estimate, but it is not recommended as a descriptor. It is highly unstable and is reliable only for intermediate values of p, and only if N is large (>50). It may result in inappropriate negative values for the lower limit of a proportion, which by definition can only occur between 0 and 1.

The *Clopper-Pearson method* is based on the exact binomial distribution and is usually estimated from quantiles from the beta distribution. It is accurate when $np > 5$, or $n(1-p) > 5$, and unlike the Wald method, it can be calculated when $p = 0$ or 1. However, the coverage probability can be very much larger than $(1 - \alpha)$, so it is not recommended for most practical applications (Brown et al. 2001).

The *Wilson method* is calculated by solving for the lower and upper levels of p separately, with $z_{\alpha/2}$ to solve for the lower limit of p and $z_{1-\alpha/2}$ to solve for the upper limit. It performs better than either the Wald or Clopper-Pearson intervals.

The *Agresti-Couill* (or *adjusted Wald*) method is recommended when assessing if p is statistically significantly different than some specified benchmark. The method involves 'add two successes and two failures' to the numerator and denominator of the proportion. This fix stabilises the confidence interval and improves coverage probability (that is, confidence intervals more closely match the designated confidence level) compared to the conventional Wald interval. A further advantage is that the lower limit cannot be negative (Agresti and Coull 1998; Agresti and Caffo 2000).

The *Jeffreys method*. The previous four methods are based on frequentist statistics; that is, the parameter is a fixed quantity that is expected to occur with a desired confidence (usually 95%) within the confidence interval. The Jeffreys method is Bayesian, in that it assumes the parameter is a random variable that lies within a credible interval that incorporates both prior information for the

---

BOX 10.3

Confidence Intervals for a Proportion

The classic *Wald* method for computing the standard error is best known but not recommended.

Clopper-Pearson method is based on the exact binomial distribution.

Wilson method solves for the lower and upper confidence levels of p separately.

Agresti-Couill (or *adjusted Wald*) 'add two successes and two failures'.

Jeffreys method is based on the beta distribution and is Bayesian, incorporating prior information for the proportion and the probability of the observed data.

binomial proportion, as well as the probability of the observed data. It is based on the beta distribution, with confidence limits:

$$\beta(\alpha/2, n_1 + 0.5, n - n_1 + 0.5),$$
$$\beta(1 - \alpha/2, n_1 + 0.5, n - n_1 + 0.5)$$

where $\beta(\alpha, b, c)$ is the αth percentile of the beta distribution with shape parameters b and c, and priors set to 0.5.

Brown et al. (2001) and Newcombe (1998) provide comprehensive review of all methods. They recommend either the Wilson or Jeffreys interval for small N, and Agresti-Coull interval for larger N. Calculations for all methods can be performed in SAS, either in *proc freq* (Wald, Clopper-Pearson, Wilson, Agresti-Coull, Jeffreys) with option *CL = ALL*, or customised (Newcombe 1998; Hu 2015; Appendix 10.B). R code is available with 'descTools' (https://cran.r-project.org/package=DescTools/). Hartnack and Roos (2021) have an excellent tutorial on binary variables.

Example: Simulated Equine Leg Bone Fractures

(Data from Hartnack and Roos 2021). Investigators used an impact device to simulate kicks on cadaveric leg bones from $N = 16$ horses to test the effect of horseshoe material on fracture rates. Fractures from a horn impactor were observed in 2 of 16 leg bones, $p = 2/16 = 0.125$. What is the 95% confidence interval?

The following 95% confidence intervals were estimated using SAS *proc freq* (Appendix 10.B), based on Newcombe (1998):

	95% confidence interval	
Wald	−0.037	0.287
Wilson	0.035	0.360
Clopper-Wilson (exact)	0.016	0.384
Agresti-Coull	0.022	0.373
Jeffrey's Bayesian	0.027	0.344

Note that the large-sample Wald approximation gives inappropriate negative values for the lower confidence limits, although the interval

width is approximately the same as that for the Wilson method. The Clopper-Wilson (exact) method gives the widest interval, and the Jeffreys the narrowest.

Sample size determinations. For absolute precision d, sample size is

$$N = z^2_{1-\alpha/2} \cdot p(1-p)/d^2$$

Sample size for relative precision is estimated as

$$N = z^2_{1-\alpha/2} \cdot (1-p)/(d^2 \cdot p)$$

When little or no preliminary information on precision is available, d can be approximated using simple rules of thumb:

When events are 'rare' ($p < 10\%$)	then d is ~ $p/2$
For most studies, p will be between 10% and 90%,	then d is ~5–10%.
When events are 'very common' ($p > 90\%$),	then d is ~ $(1-p)/2$

For an absolute precision, maximum sample size is attained with 50% prevalence; that is, N is maximum when $p_1 = p_2 = 0.5$. Therefore, when there is no information on p, sample size can be approximated with the so-called 'worst case' value $p = 0.5$ (Figure 10.1). Because it is a fixed value, absolute

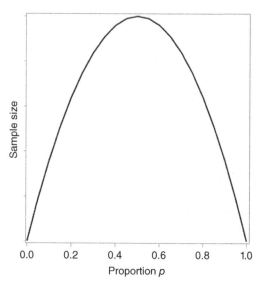

Figure 10.1: Sample size requirements for absolute precision of a proportion or prevalence p. Note that with absolute precision, maximum sample size N occurs at $p = 0.5$.

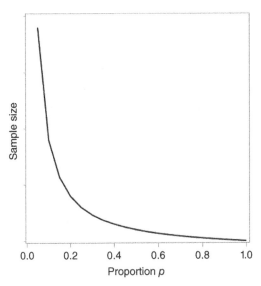

Figure 10.2: Sample size requirements for relative precision. Sample size N declines with increasing p.

precision can produce impossibly wide confidence intervals for small values or when events are rare.

If sample size is based on relative precision, sample size increases as prevalence decreases (Figure 10.2). Relative precision can estimate the interval without a lot of bias but may require substantial and often impossibly large sample sizes. Sample size can be reduced by increasing α, and more drastically, by increasing d_i. However, these make estimates less precise.

Example: Planning Trial Recruitment Numbers: von Willebrand Disease in Dobermans

Investigators planned a prospective observational study to determine differences in coagulation profiles between healthy Doberman Pinschers and those autosomally recessive for the mutation for von Willebrand disease (vWD) and, therefore, at high risk for associated bleeding. They determined from the literature that they could expect that 25% of Dobermans are homozygous affected (p_{HOM}).

How many dogs need to be recruited and screened (N) to ensure the proportion of homozygous affected (n_{HOM}) dogs in the sample is within five percentage points of the true value with 95% confidence?

This is an absolute precision problem. The absolute precision range is $\pm 5\%$ of the population prevalence or 20–30%. The expected number needed to screen (N) is

$$N \cong (1.96)^2 \left[0.25\,(1-0.25)/(0.05)^2 \right]$$
$$\cong 288.12 = 289$$

The expected number of homozygous affected $n_{HOM} = N \cdot p_{HOM}$ is therefore

$$n_{HOM} \cong 289(0.25) \cong 72$$

Therefore, the investigators need to recruit approximately 300 subjects to capture at least 72 homozygous affected dogs within the general Doberman population.

How many dogs must be sampled to capture the expected proportion of affected dogs within 10% of the true population prevalence with 95% confidence?

This is the relative precision question. The relative precision range is 10% of the true prevalence 25% (0.1×0.25), or 22.5–27.5%:

$$N = z_{1-\alpha/2}^2 \cdot \frac{(1-p)}{d^2 \cdot p} = 1.96^2 \cdot \frac{(0.75)}{0.1^2 \cdot 0.25}$$
$$= 1152.48 = 1153$$

Example: Estimating Prevalence of Feline Calicivirus (FCV) in Shelter Cats

A researcher wished to estimate the prevalence of feline calicivirus (FCV) in shelter cats. Prior data established that true prevalence was unlikely to exceed 30%.

How many cats need to be sampled to obtain prevalence of FCV within 5 percentage points of the true value with 95% confidence?

This is an absolute precision problem, so:

$$N = z_{1-\alpha/2}^2 \cdot p(1-p)/d^2$$

then $\quad N = 1.96^2 \cdot \dfrac{0.3(0.7)}{0.05^2} = 322.7 = 323$ cats

How many cats need to be sampled if the estimate of FCV prevalence is to occur within 10% of the true value with 95% confidence?

This is a relative precision problem:

$$N = z_{1-\alpha/2}^2 \cdot (1-p)/d^2 \cdot p$$

$$n = 1.96^2 \cdot (0.7)/0.1^2 \cdot 0.3 = 896.4 = 897 \text{ cats}$$

With 5% relative precision, sample size increases to 3586 cats.

10.6 Multinomial samples

Multinomial outcomes describe the proportion of observations occurring in each of k categories. For some applications, the most appropriate sample size for a specified precision will be based on estimates of the SCI for all categories rather than confidence intervals calculated for separate category cells (Tortora 1978; Thompson 1987). Sample size is found numerically by selecting the largest p (or the p closest to 0.5), then calculating N for that p_i. The probability threshold is adjusted by dividing the probability by the number of groups, k (See Appendix 10.C for SAS computation code).

Sample size for absolute precision is

$$N_{abs} = max \; z^2_{(\alpha/2)/k} \cdot p_i(1-p_i)/d^2$$

where d is the target precision, $z^2_{(\alpha/2)/k}$ is the two-tailed critical z-value with probability level $\alpha/2/k$ for k categories, and max is the maximum value across all k cells.

Sample size for relative precision is calculated as

$$N_{rel} = max \; z^2_{(\alpha/2)/k} \cdot (1-p_i)/(d^2/p_i)$$

'Worst case' sample size. If there is little or no information as to what the expected proportions should be, set $p = 0.5$. Then

$$N_{worst} = z^2_{(\alpha/2)/k} \cdot 0.5(1-0.5)/d^2$$

Older papers may present the multinomial sample size calculation as a product of the χ^2 distribution: $= max \; \chi^2_{\alpha/k,1} \cdot [p(1-p)/d_i^2]$, with the critical value of χ^2 at probability level α/k and 1 degree of freedom. Under certain conditions, χ^2 and z^2 are mathematically identical to t^2 and F-statistics:

$$z^2_{(\alpha/2)/k} = \chi^2_{\alpha/k,1} = t^2_{(\alpha/2)/k} = F_{\alpha(1),1}$$

Example: Categories of Bladder Disease in Dogs

Data from Jones et al. (2020) Histological diagnoses of five categories of bladder disease were made for 267 dogs. Percentages were cystitis 30%; neoplasia 28%, normal 15%, urolithiasis 14%, other 13%.

What sample size for a new study would enable determination of the observed sample proportion with a precision of 5% with 95% confidence? With a precision of 10%?

There are $k = 5$ categories, $\alpha = 0.05$ for 100 $(1 - \alpha) = 95\%$; $d = 0.05$. The critical value for $z^2_{0.025/5}$ is 3.291. The largest p is for the cystitis category, so $p = 0.30$. Then

$$N_{abs} = (3.291)^2 \cdot 0.3(1-0.3)/(0.05)^2 = 909.5 \cong 910$$

For a precision of 10%, $d = 0.10$, and

$$N_{abs} = (3.291)^2 \cdot 0.3(1-0.3)/(0.10)^2 = 227.4 \cong 228$$

10.7 Skewed Count Data

The *Poisson* and the *negative binomial* distributions are applicable to discrete count data (Box 10.4). Simulation results show that likelihood-based methods for calculating sample size and power for negative binomial or Poisson distributed data are preferred to nonparametric methods, such as Wilcoxon test (Aban et al. 2008). Count data are not normally distributed, usually highly skewed, so it is inappropriate to describe these data with normal distribution parameters such as the mean and standard deviation. Assuming a normal distribution to estimate confidence intervals will result in confidence intervals that are too narrow unless the sample size is very large.

Poisson distributions are used for modelling independent items distributed randomly in time or space. The Poisson parameter estimates a 'rate' of events occurring per unit time or per some linear

BOX 10.4
Distributions for Count Data

Poisson. Count data expressed as a rate (events/unit time, length, area, volume)

Variance = mean

Negative binomial. Count data showing aggregation, usually with over-representation of zeros

Variance > > mean

BOX 10.5
Examples of Poisson applications

- Number of birds visiting a feeder over a day
- Cells per square in a haemocytometer
- Number of foalings per year
- Number of adverse events in a clinical trial
- Bacterial cell counts per mm^2
- Arrival rate of dogs at a clinic or shelter
- Number of cases of bubonic plague per year
- Biological dosimetry, mutation detection & mutation rates
- Spiking activity rates of individual neurons; inter-spike intervals (sometimes)
- Lifetime encounter rate between parasitoid and host
- Visits of pollinators to a flower
- Predator attack rates
- Accident rates

metric such as distance, area, or volume. A key feature of the Poisson model is that the mean and variance are equal (*equidispersion*).

Aggregated data are more appropriately modelled using *a negative binomial distribution*. Many biological count distributions are characterised by a highly skewed and uneven frequency distribution of subjects for a given characteristic, such that only a few subjects have the characteristic, and many or most subjects do not. This is called aggregation or clumping. Aggregated distributions are characterised by over-representation of zeros and relatively few observations in the remaining categories. The negative binomial distribution differs from the Poisson in that the variance exceeds the mean, a property referred to as *over-dispersion*. If over-dispersion is present and a Poisson distribution is incorrectly assumed, then the confidence intervals will be too wide. The negative binomial distribution is widely used for ecological applications (White and Bennetts 1996; Stoklosa et al. 2022). The best-known examples are patterns of macroparasite distribution among hosts (Pennycuick 1971; Shaw et al. 1998; McVinish and Lester 2020).

10.7.1 Poisson Distribution

The Poisson distribution is the limiting case of the binomial distribution where the binomial probability p for each 'trial' or event occurrence is unknown, but the average $\bar{p}$ is known. Examples are given in Box 10.5.

The probability of x number of events during an interval t is

$$\Pr(X = x) = \frac{e^{-\lambda}\lambda^x}{x!}$$

where λ is the mean occurrence rate or number of events x per unit time and e is a constant (2.718282). Because the mean and variance are equal, they are both estimated as

$$\lambda = x/N$$

The number of events x that occur during a time interval t is estimated by substituting $\lambda = at$, so that the number of events x expected during any time interval t is a.

Model fit is assessed by evaluation of the scaled deviance goodness of fit statistic. If the Poisson model is not a good fit, the deviance statistic will be >1.

Sample size. The large-scale approximation for sample size N for a target precision d is

$$N = \left(\frac{z_{1-\alpha/2}}{d}\right)^2 \cdot \lambda$$

Confidence intervals. The Poisson distribution converges to the binomial distribution at large sample sizes (or as the interval between events approaches zero). Therefore, for $x > 20$ and for first-pass approximations, the two-sided confidence interval using the critical z-value can be used. Confidence intervals estimated by the large-sample normal approximations are given by:

$$\lambda \pm z_{1-\alpha/2} \sqrt{\lambda/N}$$

When the data are x events in a specified time period t, the confidence interval is computed by substituting t for N in the denominator.

When $x < 20$, exact confidence intervals are recommended. The lower and upper confidence limits are obtained from the χ^2 distribution as

$$\left[\frac{0.5 \cdot \chi^2_{\alpha/2,2x}}{t}, \frac{0.5 \cdot \chi^2_{1-\alpha/2,2(x+1)}}{t} \right]$$

where x is the number of events and t is the time period (Hahn and Meeker 1991; Meeker et al. 2017). (Appendix 10.D shows SAS code for calculating the critical χ^2 values and confidence intervals).

Example: Visitation Rates of Sunbirds to Flowers: Confidence Intervals

---

(Data from Burd 1994) In a study of avian pollinators, sunbirds made 114 visits to giant lobelia flowers over approximately 90 hours of observation, averaging 1.05 visits/plant/h. It is assumed arrivals occur independently of each other at a constant rate. What is the 95% confidence interval for the visitation rate?

The visitation rate $\lambda = 1.05$. The conservative 95%confidence interval is $1.05 \pm 1.96 \sqrt{1.05/90}$ $= (0.84, 1.26)$ visits/plant/h.

Example: Zebrafish Mortality and Experiment Requirements

Current recommended stocking densities for laboratory zebrafish (*Danio rerio*) welfare are 5/L, or 50 per 10 L tank, although density requirements may vary according to fish life-stage and research

objectives (Andersson and Kettunen 2021). A new study on fish lifespan and senescence is planned, with stocking density as an explanatory variable. However, it is necessary to establish preliminary baseline estimates for the maximum number of deaths that might be expected and the amount of time until all fish in a tank might be expected to die.

Outbred zebrafish have an average lifespan of approximately 42 months (Gerhard et al. 2002). Suppose routine husbandry data indicated that zebrafish mortality per tank averaged approximately three fish per week after 42 months.

Given this attrition rate, what is the *maximum number* of weekly deaths per tank that could be expected 95% of the time?

λ = Mean number of deaths per tank per week = 3.

x = Total number of deaths

Solve for $\dfrac{e^{-3}3^x}{x!} = 0.95$ by iterating over the candidate values of x and selecting the value of x that gives a cumulative probability closest to 0.95. If no calculated cumulative probabilities are exactly equal to 0.95, round up to the next value. The SAS code to approximate the target 95% confidence level is given in Appendix 10.D. The output is as follows:

Number of deaths	Exact probability	Cumulative probability
0	0.050	0.050
1	0.149	0.199
2	0.224	0.423
3	0.224	0.647
4	0.168	0.815
5	0.101	0.916
6	0.050	0.966
7	0.022	0.988
8	0.008	0.996
9	0.003	0.999
10	0.001	1.000

The cumulative Poisson probability indicates that a maximum number of six deaths per week can be expected at least 95% of the time.

If the experiment starts with 50 zebrafish in each tank, at the expected rate of attrition of 3 fish per week per tank after 42 months (168 weeks), how long can the experiment last?

Since $\lambda = a \cdot t$, the time to 50 deaths

$$t = a/\lambda = 50/3 \cong 17 \text{ weeks}$$

The expected duration is $17 + 168 = 185$ weeks ≈ 46 months.

10.7.2 Negative Binomial Distribution

Counts of individuals in nature frequently follow a negative binomial distribution, where individuals are clumped rather than being randomly dispersed. Examples are given in Box 10.6. The negative binomial is described by two parameters, the mean μ and the dispersion parameter k. When k is small, the population becomes more clumped, and therefore shows much more skew in the distribution than a Poisson with the same mean. As the variance of the negative binomial approaches the mean, k becomes very large and the distribution converges to the Poisson distribution. Unfortunately, it is usually difficult to distinguish between competing distributional models by the usual inference tests

---

BOX 10.6
Examples of Negative Binomial Applications

- Parasite distribution among hosts
- Hospital length of stay
- Number of clinic readmissions
- Brain lesion counts
- Frequency of disease exacerbations
- Field quadrat counts of organisms
- Bacterial counts per microscope field of view
- Cells per haemocytometer square
- Ear posture in sheep
- Incidence rate of mastitis in cattle.
- Incidence of infectious diseases
- Fish net catch data
- Species richness and diversity
- Tag counts across multiple gene libraries

---

unless the sample is large and represents a random sample from the population (Gardner et al. 1995).

The mean of the negative binomial is

$$m = x/p$$

where m is the mean, x is 'success', and p is the probability of a 'success' on any given trial. The mean can be calculated from total number of 'events' ($E = \sum x_i$) divided by the number of experimental units N:

$$m = E/N$$

The variance is $s^2 = m + (m^2/k)$ and

$$k = m^2/(s^2 - m)$$

The parameter k is estimated by maximum likelihood (Cox 1983; Ross and Preece 1985; Shaw et al. 1998; Saha and Paul 2005; Lloyd-Smith 2007; Shilane et al. 2010). In SAS, it is most easily obtained by fitting a generalised linear model in *proc genmod, proc countreg, proc logistic,* or *proc probit*, and obtaining k. If k is known, it can be incorporated into the model as a constant.

Sample size is approximated as

$$N = \frac{(z_\alpha)^2}{d^2}\left[\frac{1}{m} + \frac{1}{k}\right]$$

(Cundill and Alexander 2015). Methods for sample size determination for two-group comparisons are described in Chapter 17.

Confidence intervals. Methods for estimating confidence intervals depend on the size of the sample N and the magnitude of k (Shilane et al. 2010). Confidence intervals estimated from the conventional large-sample approximations ($\pm z_{1-\alpha} \cdot \text{SE}$) can result in under-coverage (confidence intervals that are too narrow). This is primarily because k is over-estimated when N is too small, or bias due to under-counting zero class events (Lloyd-Smith 2007). If k is large, the Wald approximation or bootstrapping will provide adequate coverage, regardless of the size of n. When the number of successes r is known, the confidence interval for p may be calculated from the quantiles of the χ^2 distribution, but these may result in over-coverage. Methods for testing differences in means and k between subgroups are described in Shaw et al. (1998).

Example: Parasitic worm burdens in Soay sheep: Estimates from prior data

(Data from Gulland 1992). Lungworm counts were obtained for 67 Soay sheep that died during the 1989 population crash on St Kilda. Worm burdens were described by a negative binomial distribution, with mean of 47.5 worms/sheep and $k = 0.841$. The variance of the worm count is

$$s^2 = 47.5 + (47.5^2/0.841) = 2730.3$$

What sample size for a new study would enable the determination of the expected mean worm count with a precision of 10% and 95% confidence?

$$N = \frac{(1.96)^2}{(0.1)^2} \left[\frac{1}{47.5} + \frac{1}{0.841} \right] = 317.5 \cong 318 \text{ sheep}$$

Example: Counts of Red Mites on Apple Leaves: Estimating μ and k from Raw Data

(Data from Bliss and Fisher 1953). A total of 172 female mites were counted on 150 apple leaves (Figure 10.3). Appendix 10.E provides data and SAS code.

Simple descriptive statistics (mean, variance) were calculated from the raw count data. The mean number of mites per leaf is $172/150 = 1.147$ and the variance is $s^2 = 2.274$. The data are clearly over-dispersed with an over-representation of zeros. The Poisson distribution is not a good fit, as the variance is nearly twice as large as the mean, and the scaled deviance statistic is 1.91 (much greater than 1.0).

The negative binomial provided a better fit to the data. The maximum likelihood estimate for k is 0.976. The standard error is, therefore

$$\text{SE} = \sqrt{1.147 + 0.967/150} = 1.074.$$

The asymptotic 95% confidence interval is $1.147 \pm 1.96(1.074) = [-0.95, 3.18]$, or approximately $(-1, 3.2)$ mites per leaf.

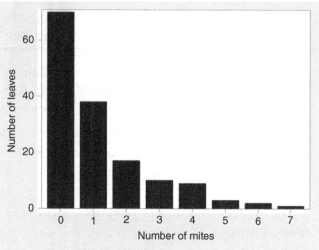

Figure 10.3: Counts of female mites on apple leaves follow a negative binomial distribution.
Source: Data from Bliss and Fisher (1953).

10.A SAS Code for Computing Simultaneous Confidence Intervals (Data from German et al. 2015)

```
%let p=20;
DATA DOG;
INPUT ID $ PROTEIN ARGIN HIST ISOLEUC METCYS
LEUC LYSINE PHETYR THRE TRYPT VALINE TFAT
LINOL CALCIUM PHOS MG SODIUM K CHLORIDE IRON
COPPER ZN MANG SEL IODINE VITA VITD3 VITE
THIAMIN RIBO PYRID NIACIN PANTO COBAL FOLIC;

*calculate standardised variables;
data new;
set DOG;
var="PROTEIN";   ratio=PROTEIN/3.28;  output;
var="METCYs";    ratio=METCYS/0.21;   output;
var="CALCIUM";   ratio=CALCIUM/0.13;  output;
var="PHOS";      ratio=PHOS/0.1;      output;
var="MG";        ratio=MG/19.7;       output;
keep var         ratio; run;
:

proc sort; by var; run;

*calculate summary statistics means SD and n
for each standardised variable;
proc means;
 by var;
 var ratio;
 output out=a n=n mean=xbar var=s2;
 run;

*calculate simultaneous confidence intervals;
data b;
 set a;
```

```
f=finv(0.95,&p,n-&p); *set confidence level,
here (1-alpha) = 0.95;

 ratio=xbar; output;

*calculate lower limit;
 ratio =xbar-sqrt(&p*(n-1)*f*s2/(n-&p)/n);
output;
```

```
*calculate upper limit;
 ratio =xbar+sqrt(&p*(n-1)*f*s2/(n-&p)/n);
output;
 run;

proc print; run;
quit;
```

10.B Sample SAS Code for Computing Confidence Intervals for a Single Sample Proportion where *x* is the Number of Events, *N* is the Sample Size, and Proportion *p* = *x/N* (Adapted from Newcombe 1998; Hu 2015)

SAS proc freq

```
proc freq data=test;
tables outcome / binomialc (CL=ALL);
weight Count;
run;
```

Wald method (large-sample normal)

```
data wald;
x = 2;
n = 16;
alpha = 0.05;
p = x / n;
z = probit (1-alpha/2);
*calculate standard error;
se = (sqrt(n*p*(1-p)))/n;
L = p - z * se;        *Lower confidence
limit;
 U = p + z * se;       *Upper confidence
                        limit;

put L= U= ;
run;
```

Clopper –Pearson method

```
data CP;
x = 2;
n = 16;
alpha = 0.05;
L = 1 - betainv(1 - alpha/2,n-x+1,r);
U = betainv(1 - alpha/2,r+1 ,n-x);
put L= U=;  *Lower & upper confidence
limit;
run;
```

Wilson method

```
data Wilson;
x = 2;
n = 16;
alpha = 0.05;
p = x / n;
q = 1-p;
z = probit (1-alpha/2);
L = ( 2*x+z**2 - (z*sqrt(z**2+4*x*q))) /
(2*(n+z**2));
U = ( 2*x+z**2 +(z*sqrt(z**2+4*x*q)) ) /
(2*(n+z**2));
put L= U=;
run;
```

Agresti-Coull method

```
data AC;
x= 2;
n = 16;
alpha = 0.05;
z = probit (1-alpha/2);
do psi = z**2/2, 2, 1, 3;
p2=(x+psi)/(n+2*psi);
L = p2 - z*(sqrt(p2*(1-p2)/(n+2*psi)));
U = p2 + z*(sqrt(p2*(1-p2)/(n+2*psi)));
put L= U= ;
output;
end;
run;
```

Jeffreys Bayesian method

```
data jeffreys;
x = 2;
n = 16;
alpha = 0.05;
*priors are set to 0.5;
L = betainv( alpha/2, x+0.5, n-x+0.5);
U = betainv(1-alpha/2, x+0.5, n-x+0.5);
put L= U=;
run;
```

10.C SAS Code for Calculating the Critical Values for $z_{(\alpha/2)/k}$ and $\chi^2_{\alpha/k,1}$

10.C.1 Calculating the critical values for $z_{(\alpha/2)/k}$

```
data zval;
   alpha=0.05;
   k = 5; *number of categories - in this
   example k equals 5;
   Pr = (alpha/2)/k; *for a two-sided
   probability;
   z = quantile('normal',Pr);
   z2 = z*z;
 run;
 proc print; *output;
 run;
```

10.C.2 Calculating the critical values for $\chi^2_{\alpha/k,1}$

```
data chisq;
   alpha = 0.05; *desired type I probability;
   k = 5; *number of categories - in this
   example k equals 5;
   df = 1;
   Pr = 1- alpha/k;
   chi_crit=cinv(Pr,1);
   put chi_crit = ;
run;
proc print; *output;
run;
```

10.D SAS Code for Calculating Confidence Limits for Poisson Data

```
*large scale normal;
data large;
alpha = 0.05; *define desired type I
probability;
x = [...];     *define desired number of
               events x;
lambda= [event rate];
lower = lambda - probit(alpha/2)*sqrt(lambda/
x);
upper = lambda + probit(1- alpha/2,)*sqrt
(lambda/x);
run;
proc print; *output;
run;
```

```
*defining exact confidence intervals when
number of events
data exact;
alpha = 0.05; *define desired type
I probability;
x = [..];     *define desired number of
               events x;
lower = quantile('CHISQ', alpha/2,2*x)/2;
upper = quantile('CHISQ',1-alpha/2,2*(x+1))/2;
run;
proc print; *output;
run;
```

Code to approximate the target 95% confidence level by iterating over a range of sample sizes n and selecting the value that gives a cumulative probability closest to 0.95.

```
%let nmax = 10; *define a reasonable estimate
of the maximum;
data event;
    lambda = 3; * the expected mean event
    rate λ;

*calculate n! factorial;
    do n = 0 to &nmax by 1;
    f=fact(n);

*calculate probabilities for number of events;
prob = exp(-lambda)*(lambda**n)/f;
    if n=0 then cumprob=prob;
       else cumprob=cumprob+prob;
   output;
 end;
    run;
proc print;
run;
```

10.E Evaluating Poisson and negative binomial distributions for fitting counts of red mites on apple leaves (Data from Bliss and Fisher 1953)

```
data mite;
input n @@;
datalines;
0 0 0 0 0 0 0 0 0 0 0 0 0 0 0 0 0 0 0 0 0 0 0 0
0 0
0 0 0 0 0 0 0 0 0 0 0 0 0 0 0 0 0 0 0 0 0 0 0 0
0 0
```

```
0 0 0 0 0 0 0 0 0 0 0 0 0 0 0 0 0 0 1 1 1 1 1 1
1 1
1 1 1 1 1 1 1 1 1 1 1 1 1 1 1 1 1 1 1 1 1 1 1 1
1 1
1 1 1 1 2 2 2 2 2 2 2 2 2 2 2 2 2 2 2 2 2 2 3 3 3
3 3
3 3 3 3 3 4 4 4 4 4 4 4 4 4 5 5 5 6 6 7;

run;
proc means n sum mean var maxdec=2;
    run;

*Poisson;
proc genmod data=mite;
      model n = / dist=poisson scale=pearson
      CL;
    run;

*negative binomial;
proc genmod data=mite;
      model n = / dist=negbin CL;
    run;
```

References

Aban, I.B., Cutter, G.R., and Mavinga, N. (2008). Inferences and power analysis concerning two negative binomial distributions with an application to MRI lesion counts data. *Computational Statistics and Data Analysis* 53 (3): 820–833. https://doi.org/10.1016/j.csda.2008.07.034.

Agresti, A. and Caffo, B. (2000). Simple and effective confidence intervals for proportions and differences of proportions result from adding two successes and two failures. *The American Statistician* 54: 280–288. https://doi.org/10.2307/2685779.

Agresti, A. and Coull, B.A. (1998). Approximate is better than 'exact' for interval estimation of binomial proportions. *The American Statistician* 52: 119–126. https://doi.org/10.2307/2685469.

Andersson, M. and Kettunen, P. (2021). Effects of holding density on the welfare of zebrafish: a systematic review. *Zebrafish* 2021: 297–306. https://doi.org/10.1089/zeb.2021.0018.

Bliss, C.I. and Fisher, R.A. (1953). Fitting the negative binomial distribution to biological data. *Biometrics* 9 (2): 176–200. https://doi.org/10.2307/3001850.

Brown, L.D., Cai, T.T., and Das Gupta, A. (2001). Interval estimation for a binomial proportion. *Statistical Science* 16 (2): 101–133. https://doi.org/10.1214/ss/1009213286.

Burd, M. (1994). A probabilistic analysis of pollinator behaviour and seed production in *Lobelia deckenii*. *Ecology* 75: 1635–1646.

Cox, D.R. (1983). Some remarks on overdispersion. *Biometrika* 70: 269–274.

Cumming, G. (2012). *Understanding the New Statistics: Effect Sizes, Confidence Intervals, and Meta-Analysis*. New York: Routledge.

Cundill, B. and Alexander, N.D. (2015). Sample size calculations for skewed distributions. *BMC Medical Research Methodology* 5: 28. https://doi.org/10.1186/s12874-015-0023-0.

Davison, A.C. and Hinkley, D.V. (1997). *Bootstrap Methods and Their Application*. Cambridge: Cambridge University Press.

Efron, B. and Tibshirani, R.J. (1993). *An Introduction to the Bootstrap*. Boca Raton: Chapman & Hall/CRC.

Gardner, W., Mulvey, E.P., and Shaw, E.C. (1995). Regression analyses of counts and rates: Poisson, overdispersed Poisson, and negative binomial models. *Psychological Bulletin* 118: 392–404.

Gerhard, G.S., Kauffman, E.J., Wang, X. et al. (2002). Life spans and senescent phenotypes in two strains of Zebrafish (*Danio rerio*). *Experimental Gerontology* 37 (8-9): 1055–1068. https://doi.org/10.1016/s0531-5565(02)00088-8.

German, A.J., Holden, S.L., Serisier, S. et al. (2015). Assessing the adequacy of essential nutrient intake in obese dogs undergoing energy restriction for weight loss: a cohort study. *BMC Veterinary Research* 11: 253. https://doi.org/10.1186/s12917-015-0570-y.

Gulland, F.M.D. (1992). The role of nematode parasites in Soay sheep (*Ovis aries* L.) mortality during a population crash. *Parasitology* 105 (Pt 3): 493–503. https://doi.org/10.1017/s0031182000074679.

Hahn, G.J. and Meeker, W.Q. (1991). *Statistical Intervals: A Guide for Practitioners*. New York: Wiley.

Hartnack, S. and Roos, M. (2021). Teaching: confidence, prediction and tolerance intervals in scientific practice: a tutorial on binary variables. *Emerging Themes in Epidemiology* 18 (1): 17. https://doi.org/10.1186/s12982-021-00108-1.

Hu, J. (2015). Confidence intervals for binomial proportion using SAS®: the all you need to know and no more. Paper SD103. https://www.lexjansen.com/sesug/2015/103_Final_PDF.pdf (accessed 2022).

Jones, E., Alawneh, J., Thompson, M. et al. (2020). Predicting diagnosis of Australian canine and feline urinary bladder disease based on histologic features. *Veterinary Sciences* 7 (4): 190. https://doi.org/10.3390/vetsci7040190.

Lloyd-Smith, J.O. (2007). Maximum likelihood estimation of the negative binomial dispersion parameter for highly overdispersed data, with applications to infectious diseases. *PLoS ONE* 2 (2): e180. https://doi.org/10.1371/journal.pone.0000180.

Lwandga, S.K. and Lemeshow, S. (1991). *Sample Size Determination in Health Studies: A Practical Manual*. Geneva: World Health Organization.

McVinish, R. and Lester, R.J.G. (2020). Measuring aggregation in parasite populations. *Journal of the Royal Society Interface* 17: 20190886. https://doi.org/10.1098/rsif.2019.0886.

Meeker, W.Q., Hahn, G.J., and Escobar, L.A. (2017). *Statistical Intervals: A Guide for Practitioners and Researchers*, 2e. New York: Wiley.

Newcombe, R.G. (1998). Two-sided confidence intervals for the single proportion: comparison of seven methods. *Statistics in Medicine* 17: 857–872.

Pennycuick, L. (1971). Frequency distributions of parasites in a population of three-spined sticklebacks, *Gasterosteus aculeatus* L, with particular reference to the negative binomial distribution. *Parasitology* 63 (3): 389–406. https://doi.org/10.1017/S0031182000079920.

Ross, G.J.S. and Preece, D.A. (1985). The negative binomial distribution. *Statistician* 34: 323–336.

Saha, K. and Paul, S. (2005). Bias-corrected maximum likelihood estimator of the negative binomial dispersion parameter. *Biometrics* 61: 179–185.

Shaw, D.J., Grenfell, B.T., and Dobson, P. (1998). Patterns of macroparasite aggregation in wildlife host populations. *Parasitology* 117: 597–610. https://doi.org/10.1017/S0031182098003448.

Shilane, D., Evans, S.N., and Hubbard, A.E. (2010). Confidence intervals for negative binomial random variables of high dispersion. *The International Journal of Biostatistics* 6 (1): Article 10. https://doi.org/10.2202/1557-4679.1164.

Stoklosa, J., Blakey, R.V., and Hui, F.K.C. (2022). An overview of modern applications of negative binomial modelling in ecology and biodiversity. *Diversity* 2022 (14): 320. https://doi.org/10.3390/d14050320.

Thompson, S.K. (1987). Sample size for estimating multinomial proportions. *The American Statistician* 41 (1): 42–46.

Tortora, R. (1978). A note on sample size estimation for multinomial populations. *The American Statistician* 32 (3): 100–102.

White, G.C. and Bennetts, R.E. (1996). Analysis of frequency count data using the negative binomial distribution. *Ecology* 77 (8): 2549–2557.

11 Prediction Intervals

CHAPTER OUTLINE HEAD

11.1 Introduction

A *prediction interval* is an interval within which one or more future observations from a population will fall with a specified degree of confidence $(1 - \alpha)$. Prediction intervals are useful for a variety of applications (Box 11.1), although they are most often used for regression when the variable of interest is expected to vary with a second explanatory variable or covariate. Other applications include reliability and survival analysis (Landon and Singpurwalla 2008), 'personalised' reference intervals in laboratory medicine (Coşkun et al. 2021), assessing feasibility based on pilot data, replication studies (Spence and Stanley 2016), and meta-analyses (Riley et al. 2011; Inthout et al. 2016; Deeks et al. 2022). Karl Pearson called prediction 'the fundamental problem of practical statistics' (Pearson 1920).

Prediction intervals differ from confidence intervals in that a prediction interval accounts for both the random variation in the observations and the uncertainty in the estimation of the population mean, rather than random variation alone. As a result, prediction intervals are always wider than the corresponding confidence interval and, unlike confidence intervals, will not converge to zero as N increases (Hahn and Meeker 1991; NIST/SEMATECH 2012; Meeker et al. 2017).

BOX 11.1
Applications of Prediction Intervals

- 'Personalised' reference intervals Regression models
- Forecasting models
- Relationship between two traits for an unmeasured species in phylogenetic analyses
- Replication studies
- Meta-analyses.

11.2 Prediction Intervals: Continuous Data

The prediction interval is calculated as

$$\hat{y}_{new} \pm (t - \text{value}) \cdot (SE_{new})$$

where $\hat{y}_{new}$ is the sample estimate for the new predicted observation and SE_{new} is the standard error of the new predicted value.

A Guide to Sample Size for Animal-based Studies, First Edition. Penny S. Reynolds.
© 2024 John Wiley & Sons Ltd. Published 2024 by John Wiley & Sons Ltd.

11.2.1 Continuous Data, Single Observation

The standard error SE for a prediction interval for a single new observation $\hat{y}_{new}$ is

$$SE = s\sqrt{\frac{1}{m} + \frac{1}{N}}$$

where s is the standard deviation and N is the number of observations in the original sample, and m is the number of new observations. Then the prediction interval is

$$\hat{y}_{new} \pm t_{1-\alpha/2,\ N-1} \cdot s\sqrt{\left(\frac{1}{m} + \frac{1}{N}\right)}$$

where $t_{1-\alpha/2,N-1}$ is the $100(1 - \alpha/2)$ percentile of the t-distribution with $N - 1$ degrees of freedom. For a single future observation, $m = 1$ and the equation reduces to

$$\hat{y}_{new} \pm t_{1-\alpha/2,N-1} \cdot s\sqrt{\left(1 + \frac{1}{N}\right)}$$

The *two-sided prediction interval* $100(1 - \alpha)\%$ for the difference between means $(\bar{y}_1 - \bar{y}_2)$ is

$$(\bar{y}_1 - \bar{y}_2) \pm t_{1-\alpha/2,n_1 + n_2 - 2} \cdot SE$$

The *one-sided lower prediction limit* of the difference between future observations is

$$(\bar{y}_1 - \bar{y}_2) - t_{\alpha,n_1 + n_2 - 2} \cdot SE$$

Example: Predicting a Single Future Value: Creatinine Levels in Obese Dogs

(Data from German et al. 2015). Serum creatinine was measured in 27 obese dogs before entering a weight loss programme, averaging 81.5 (SD 19.6) µmol/L. What is the predicted value of creatinine for a single future obese dog at $\alpha = 0.05$?

$$81.5 \pm 2.052 \cdot 19.6$$
$$\sqrt{\left(1 + \frac{1}{27}\right)} = (40.5, 122.5)\ \mu mol/L$$

Therefore creatinine of a single future randomly selected obese dog will occur between (40.5, 122.5) µmol/L in 95 out of 100 cases.

11.2.2 Continuous Data, Comparing Two Means

The standard error for the prediction interval for the difference between two means is

$$SE = s_p\sqrt{\frac{1}{m_1} + \frac{1}{m_2} + \frac{1}{n_1} + \frac{1}{n_2}}$$

where the pooled standard deviation is

$$s_p = \sqrt{\frac{(n_1 - 1)s_1^2 + (n_2 - 1)s_2^2}{(n_1 + n_2 - 2)}}$$

Here, n_1 and n_2 are the sample sizes for each group, s_1^2 and s_2^2 are the variances for each group, and m_1 and m_2 are the number of new observations.

Example: Predictions Based on a Fixed Sample Size: Anaesthetic Immobilisation Times in Mice

(Data from Dholakia et al. 2017). Time of complete immobilisation following anaesthesia was measured in two groups of 16 CD-1 male mice randomly allocated to receive intraperitoneal injections of either ketamine-xylazine (KX) or ketamine-xylazine combined with lidocaine (KXL). Immobilisation time averaged 38.8 (SD 7.9) min for KX mice and 33.3 (SD 3.9) min for KXL mice for a difference of 5.5 minutes.

If a new study is planned using only five mice per group, what is the expected difference in future immobilisation times for KX mice compared to KXL with 95% confidence?

Summary data for the original study are $\bar{y}_1 = 38.8$, $\bar{y}_2 = 33.3$, $n_1 = n_2 = 8$, and $t_{1-0.05/2,n_1+n_2-2} = 2.145$. For the new study, $m_1 = m_2 = 5$

Then

$$(38.8 - 33.3) \pm 2.145\,(3.9)\sqrt{\frac{1}{8} + \frac{1}{8} + \frac{1}{5} + \frac{1}{5}}$$

$$= 5.5 \pm 6.74 = (-1.2,\ 12.2)\ \text{min}$$

Therefore, immobilisation times for KXL could be as much as 12 min longer than KX or 1.2 min shorter.

Example: Predicting a Range of Future Observations: Creatinine Levels in Obese Dogs

(Data from German et al. 2015). Serum creatinine was measured in 27 obese dogs before entering a weight loss programme, averaging 81.5 (SD 19.6) µmol/L. Suppose a new study was planned using a sample size of 100 dogs. What range of mean creatinine values can be expected with 95% confidence?

For this query, the prediction interval on the difference between two means must be computed, and both the predicted mean and sample standard deviation are hypothetical but assumed to be equal to those in the original sample. The prediction interval for the new study is then

$$\textit{original mean} \pm t_{df, N_1 - 1}\sqrt{\frac{s_1^2}{N_1} + \frac{s_1^2}{N_2}}$$

$$= 81.5 \pm 2.052\sqrt{\frac{19.6^2}{27} + \frac{19.6^2}{100}}$$

$$= 81.5 \pm 8.7 = 72.8,\ 90.2$$

Therefore the range of average creatinine values that can be expected in a future study with a sample size of 100 is between 72.8 and 90.2 µmol/L.

11.2.3 Continuous Data, Linear Regression

Prediction intervals are most often used for regression when one or more future observations are predicted from the relations between the response and one or more explanatory (or independent) variables. Therefore the prediction interval on the new observation will be conditional on the predictors in the regression model.

The two-sided prediction interval is

$$\hat{y}_{new} \pm t_{\alpha/2, N - (K + 1)}\sqrt{MSE + \left(SE(\hat{y}_{new})\right)^2}$$

where MSE is the mean square error of the regression with associated degrees of freedom $N - (K + 1)$ and the term $\sqrt{MSE + \left(SE(\hat{y}_{new})\right)^2}$ is the standard error of the prediction. In SAS, the 95% prediction interval for a new observation at a given value of X is called by the command cli in the model option statement.

The ordinary least squares regression is fitted under the assumption that the independent variable is measured without error, or at least the error is negligible compared with that for the response variable (Draper and Smith 1998).

Prediction intervals must be interpreted with caution if there is a substantial error for values of the independent variable. del Río et al. (2001) describe methods for constructing prediction intervals for linear regression that account for errors on both axes. Gelman et al. (2021) provide an excellent guide to modern methods of regression, including Bayesian inference methods (incorporation of prior information into inferences) and R commands for computation.

Example: Rodent p_{50} in Relation to Body Mass.

Oxygen affinity for haemoglobin is quantified by P_{50}, the partial pressure of oxygen at which haemoglobin is 50% saturated. In mammals, P_{50} scales negatively with body mass, such that small mammals have larger P_{50} than large mammals (Schmidt-Nielsen and Larimer 1958). Body mass (kg) and P_{50} (mm Hg) data for 23 rodent species are given in Table 11.1.

Table 11.1: Mean body mass (kg) and P_{50} (mm Hg) data for 23 rodent species.

Species	1	2	3	4	5
Body mass	0.018	0.018	0.020	0.025	0.030
P_{50}	52.0	40.0	28.8	33.5	33.2
	6	**7**	**8**	**9**	**10**
Body mass	0.045	0.047	0.049	0.068	0.088
P_{50}	33.8	53.0	32.0	41.0	29.0
	11	**12**	**13**	**14**	**15**
Body mass	0.100	0.157	0.162	0.193	0.196
P_{50}	38.4	25.0	39.0	23.0	29.5
	16	**17**	**18**	**19**	**20**
Body mass	0.196	0.226	0.454	0.500	0.517
P_{50}	29.5	39.0	26.8	27.8	36.0
	21	**22**	**23**		
Body mass	1.000	1.200	3.500		
P_{50}	24.0	22.0	27.0		

Source: Compiled from Schmidt-Nielsen and Larimer (1958).

The regression on $\log_{10}$-transformed data was

$$\log_{10} P_{50} = 1.43 - 0.1(\log_{10} \text{ mass})$$
$$\left(R^2 = 0.34, \text{MSE} = 0.00736\right).$$

The 95% confidence and prediction intervals for P_{50} were estimated for body mass of 0.1 kg (values were back-transformed to original units). The predicted mean P_{50} is 33.2. The 95% confidence intervals are (30.4, 36.2) mmHg and the 95% prediction intervals are (21.8, 50.5) mmHg.

11.3 Prediction Intervals: Binary Data

Binary data are expressed as a proportion p. The standard error SE for a prediction interval for a new observation from a binomial distribution is

$$SE = z_{1-\alpha/2}\sqrt{m \cdot p \cdot (1-p) \cdot \frac{m+n}{mn}}$$

The prediction interval is

$$m \cdot p \pm z_{1-\alpha/2}\sqrt{m \cdot p \cdot (1-p) \cdot \frac{m+n}{n}}$$

Hartnack and Roos (2021) recommend the Bayesian Jeffreys prediction interval. This is based on the posterior predictive distribution combining the binomial distribution with a conjugate beta prior. Prediction intervals can be computed in R using the package DescTools; see Hartnack and Roos (2021) for details.

Example: Predicting Number of Future Events: Simulated Equine Orbital Fractures

The major cause of long bone and facial fractures in horses is kicks from other horses. The type of shoe can greatly affect the severity of the injury and probability of bone fracture (Sprick et al. 2017).

Investigators used a drop-weight impact device to simulate kicks on cadaveric skulls from 17 horses to test the effect of shod versus unshod hoof on fracture rates. Orbital fractures from steel shoes were observed in 12 of 17 skulls (Joss et al. 2019). How many orbital fractures can be expected in a future sample of $m = 10$ skulls at $\alpha = 0.05$?

The observed proportion of fractures is $12/17 = 0.706$, $n = 17$, and $m = 10$. The approximate normal 95% prediction interval is

$$10(0.706) \pm 1.96\sqrt{10 \cdot (0.706) \cdot (0.294) \cdot \frac{37}{27}}$$
$$= (3.75, 10.37)$$

Therefore, for a future sample of 10 skulls, we can expect to find that between 4 and 11 fractures will occur with 95% confidence.

11.4 Prediction Intervals: Meta-Analyses

For a random-effects meta-analysis, the mean describes systematically different effects for the compiled studies. Confidence intervals calculated for mean effects in a meta-analysis will usually be too narrow for an adequate description of the range of possible study effects. If the number of studies in the meta-analysis is >10, then the $100(1 - \alpha)\%$ prediction interval can be estimated as

$$\overline{Y} \pm t_{1 - \alpha, k - 2} \sqrt{\tau^2 + SE(\overline{Y})^2}$$

where $\overline{Y}$ is the summary mean of absolute measures of effect (e.g. risk difference, mean difference, standardised mean difference), $SE(\overline{Y})^2$ is the variance, $t_{1 - \alpha, k - 2}$ is the critical t-value for $1 - \alpha$ and $k - 2$ degrees of freedom, and τ is the estimate of the variation of the true effects (heterogeneity). For relative measures, such as RR and OR, the interval must be calculated from the logarithm (ln) of the summary estimate, as is the case for confidence intervals (Higgins et al. 2009).

These prediction intervals may have very poor coverage when k is small, resulting in intervals that are too narrow (Partlett and Riley 2017). Nagashima et al. (2019a, b) have developed an R package *pimeta* that compiles several alternative methods for constructing prediction intervals based on bootstrapping. The package also includes methods for estimating prediction intervals when the number of studies is very small ($k < 5$).

References

Coşkun, A., Sandberg, S., Unsal, I. et al. (2021). Personalized reference intervals in laboratory medicine: a new model based on within-subject biological variation. *Clinical Chemistry* 67 (2): 374–384. https://doi.org/10.1093/clinchem/hvaa233.

Deeks, J.J., Higgins, J.P.T., and Altman, D.G. (2022). Chapter 10: Analysing data and undertaking meta-analyses. In: *Cochrane Handbook for Systematic Reviews of Interventions version 6.3 (updated February 2022)* (ed. H. JPT, J. Thomas, J. Chandler, et al.). Cochrane Available *from* www.training.cochrane.org/handbook.

del Río, F.J., Riu, J., and Rius, F.X. (2001). Prediction intervals in linear regression taking into account errors on both axes. *Journal of Chemometrics* 15: 773–788. https://doi.org/10.1002/cem.663.

Dholakia, U., Clark-Price, S.C., Keating, S.C.J., and Stern, A.W. (2017). Anesthetic effects and body weight changes associated with ketamine-xylazine-lidocaine administered to CD-1 mice. *PLoS ONE* 12 (9): e0184911. https://doi.org/10.1371/journal.pone.0184911.

Draper, N.R., and Smith, H. (1998). *Applied Regression Analysis*, 3e. New York: Wiley.

Gelman, A., Hill, J., and Vehtari, A. (2021). *Regression and Other Stories*. Cambridge University Press.

German, A.J., Holden, S.L., Serisier, S. et al. (2015). Assessing the adequacy of essential nutrient intake in obese dogs undergoing energy restriction for weight loss: a cohort study. *BMC Veterinary Research* 11: 253. https://doi.org/10.1186/s12917-015-0570-y.

Hahn, G.J. and Meeker, W.Q. (1991). *Statistical Intervals: A Guide for Practitioners*. New York: Wiley.

Hartnack, S. and Roos, M. (2021). Teaching: confidence, prediction, and tolerance intervals in scientific practice: a tutorial on binary variables. *Emerging Themes in Epidemiology* 18: 17. https://doi.org/10.1186/s12982-021-00108-1.

Higgins, J.P.T., Thompson, S.G., and Spiegelhalter, D.J. (2009). A re-evaluation of random-effects meta-analysis. *Journal of the Royal Statistical Society: Series A (Statistics in Society)* 172: 137–159.

IntHout J, Ioannidis JPA, Rovers MM, Goeman JJ (2016) Plea for routinely presenting prediction intervals in meta-analysis. *BMJ Open*, 6: e010247. https://doi.org/10.1136/bmjopen-2015-010247

Joss, R., Baschnagel, F., Ohlerth, S. et al. (2019). The risk of a shod and unshod horse kick to create orbital fractures in equine cadaveric skulls. *Veterinary and Comparative Orthopaedics and Traumatology* 32 (4): 282–288.

Landon, J. and Singpurwalla, N.D. (2008). Choosing a coverage probability for prediction intervals. *The American Statistician* 62 (2): 120–124. https://doi.org/10.1198/000313008x304062.

Meeker, W.Q., Hahn, G.J., and Escobar, L.A. (2017). *Statistical Intervals: A Guide for Practitioners and Researchers*, 2e. New York: Wiley.

Nagashima, K., Noma, H., and Furukawa, T.A. (2019a). Prediction intervals for random-effects meta-analysis: a confidence distribution approach. *Statistical Methods in Medical Research* 28 (6): 1689–1702. https://doi.org/10.1177/0962280218773520.

Nagashima, K., Noma, H., and Furukawa, T.A. (2019b) *pimeta:* prediction intervals for random-effects meta-analysis. R package version 1.1.2. Available from: https://CRAN.R-project.org/package=pimeta (accessed 2022).

NIST/SEMATECH (2012) *e-Handbook of Statistical Methods* www.itl.nist.gov/div898/handbook, https://doi.org/10.18434/M32189

Partlett, C. and Riley, R.D. (2017). Random effects meta-analysis: coverage performance of 95% confidence and prediction intervals following REML estimation. *Statistics in Medicine* 36 (2): 301–317.

Pearson, K. (1920). The fundamental problem of practical statistics. *Biometrika* 13 (1): 1–16. https://doi.org/10.2307/2331720.

Riley, R.D., Higgins, J.P.T., and Deeks, J.J. (2011). Interpretation of random effects meta-analyses. *BMJ* 342: d549.

Schmidt-Nielsen, K. and Larimer, J.L. (1958). Oxygen dissociation curves of mammalian blood in relation to body size. *American Journal of Physiology* 195 (2): 424–428. https://doi.org/10.1152/ajplegacy.1958.195.2.424.

Spence, J.R. and Stanley, D.J. (2016). Prediction interval: what to expect when you're expecting … a replication. *PLoS ONE* 11 (9): e0162874. https://doi.org/10.1371/journal.pone.0162874.

Sprick, M., Fürst, A., Baschnagel, F. et al. (2017). The influence of aluminium, steel and polyurethane shoeing systems and of the unshod hoof on the injury risk of a horse kick: an *ex vivo* experimental study. *Veterinary and Comparative Orthopaedics and Traumatology* 30: 339–345. https://doi.org/10.3415/VCOT-17-01-0003.

12 Tolerance Intervals

CHAPTER OUTLINE HEAD

12.1 Introduction

A *tolerance interval* is defined as the interval (coverage) between which a pre-specified *proportion (p)* of observations fall with a pre-specified level of confidence. That is, statistical tolerance limits are limits within which a given proportion of the population is expected to lie. Unlike prediction intervals for which coverage accounts for a pre-specified number of future observations, coverage for tolerance limits accounts for any number of future observations (Hahn and Meeker 1991; Vardeman 1992; NIST/SEMATECH 2012; Meeker et al. 2017; Hartnack and Roos 2021).

Tolerance intervals have numerous biological and clinical applications for the quantification of 'acceptable' or 'reference' performance limits (Box 12.1). Examples include drug and device quality control, environmental and toxicology

BOX 12.1
Applications of Tolerance Intervals

- Diagnostic and clinical reference ranges
- Regulatory thresholds for performance horse medication
- Medical device performance
- Method comparison tests for medical devices
- Quality control applications
- Safety limits
- Medication residue limits
- Environmental regulatory limits for pesticide concentrations

Potential applications

- Humane endpoints
- Physiological 'tolerance polygons'
- Critical thermal maxima and minima for ectotherms.

A Guide to Sample Size for Animal-based Studies, First Edition. Penny S. Reynolds.
© 2024 John Wiley & Sons Ltd. Published 2024 by John Wiley & Sons Ltd.

monitoring (Smith 2002; Gibbons et al. 2009; Komaroff 2018), safety (Chen and Kishino 2015), medication levels in performance horses (RMTC 2016) and dairy animals (CVMP 2000), comparative performance of biosimilar drugs (Chiang et al. 2021), and method comparison assessments for medical devices (Francq et al. 2020). Tolerance intervals have been recommended overconfidence and prediction intervals for constructing clinical reference ranges (Liu et al. 2021).

For process validation and quality control applications, 'confidence' is defined as the amount of certainty that the tolerance interval contains a specified percentage of each individual measurement in the population. 'Reliability' is the proportion of the population sample contained by the interval. For clinical purposes, such as medical device testing, a defined 'risk' component can be incorporated that combines the probability of occurrence and potential severity of harm resulting from product failure or defect (Durivage 2016 a, b).

12.2 Tolerance Interval Width and Bounds

The width of a tolerance interval results from both sampling error and population variation. As sample size increases, sampling error is reduced so that the percentiles estimated from the sample approach are the true population percentiles. Therefore the bounds of a tolerance interval are the upper or lower confidence interval bounds of a quantile of the underlying data distribution. For example, a common upper threshold for regulatory purposes is the 95th quantile (95% of the sampled population should fall at or below the threshold). Then the tolerance interval limit is the upper bound of the $(1 - \alpha)$ confidence interval for the 95th quantile.

Tolerance intervals may be either two-sided or one-sided (Box 12.2). A *two-sided tolerance* interval is bounded by both an upper and lower limit. Examples include quality control applications in pharmaceutical development, device performance in comparison to a reference standard, and comparative performance of biosimilar drugs (Chiang and Hsiao 2021; Chiang et al. 2021). *A one-sided tolerance interval* is calculated when the objective is to determine if a given proportion of observations fall outside some upper or lower threshold. Examples

> **BOX 12.2**
> *One-Sided or Two-Sided Tolerance Intervals?*
>
> > *Two-sided tolerance interval:* When it is important to define an accurate and reliable *range* for a given proportion of observations.
>
> What length of the interval will contain *p* observations with a specified level of confidence?
>
> > *One-sided tolerance interval:* When it is important to define an accurate and reliable performance *threshold*.
>
> What length of interval will ensure that *p* observations do not fall below a lower threshold limit *L* (or exceed an upper threshold limit *U*) with a specified level of confidence?

included the determination of upper threshold for environmental contamination or toxicology responses, or for determining a specific cut-point in biomarker expression when evaluating response to a drug (Pan 2015).

12.3 Parametric Formulations

For normally distributed observations obtained from a random sample, two-sided tolerance intervals are bounded by the corresponding lower (L) and upper (U) tolerance *limits,* calculated as differences from the mean $\overline{Y}$. Therefore the lower limit Y_L is $\overline{Y} - k \cdot s$ and the upper limit Y_U is $\overline{Y} + k \cdot s$. The value for k is a function of the coverage or proportion of the population p to be covered with confidence α. If the data are not normally distributed, log-transformation may be sufficient to ensure normalisation. Values for k are obtained from the appropriate non-central distributions, with critical values generated from the respective inverse functions (Appendix 12.A).

12.3.1 Two-Sided Limits

For normally distributed continuous data, the two-sided tolerance interval is $\overline{Y} \pm k_{(2)} \cdot s$, where $k_{(2)}$ is based on the non-central chi-squared distribution:

$$k = z_{(1+p)/2} \sqrt{\frac{df\left(1 + \frac{1}{N}\right)}{\chi^2_{1-\alpha,df}}}$$

with coverage p (the proportion of observations that need to lie within the interval), $\chi^2_{1-\alpha,df}$ is the critical value distribution, with $df = N-1$ degrees of freedom, and $z_{(1+p)/2}$ is the critical value of the normal distribution with cumulative probability $(1+p)/2$. For example, suppose the required coverage is 99% ($p = 0.99$). Then $(1+p)/2 = (1+0.99)/2 = 0.995$, and $z_{0.995}$ is 2.576. For 95% coverage, $(1+p)/2 = 0.975$ and $z_{0.975} = 1.96$.

For small samples ($N < 30$), Guenther (1977) suggests a weighted correction for $k_{(2)}$ as $k'_{(2)} = k_{(2)} \cdot w$, where w is

$$w = \sqrt{1 + \frac{N - 3 - \chi^2_{N-1,\alpha}}{2(N+1)^2}}$$

12.3.2 One-Sided Limits

For continuous normally distributed data with mean $\bar{y}$ and standard deviation s, the upper limit of a one-sided tolerance interval is $\bar{y} + k_{(1)} \cdot s$, and the lower limit is $\bar{y} - k_{(1)} \cdot s$. The one-sided normal tolerance intervals have an exact solution based on the non-central t-distribution. The z-distribution can be used for large-sample approximations ($N > 100$). In general, there is no difference between a one-sided tolerance bound and a one-sided confidence bound on a given quantile of the distribution. For example, a 95 per cent confidence limit on the upper 95th percentile and an upper tolerance limit on the 95th percentile at 95% confidence are the same (Meeker et al. 2017).

For small-sample one-sided tolerance intervals based on the non-central t-distribution

$$k_{(1)} = \frac{t_{\alpha,N-1,\lambda}}{\sqrt{N}}$$

where λ is the non-centrality parameter $z_p\sqrt{N}$. The sample size N is obtained by iteration to find the minimum N that satisfies $t_{\alpha,N-1,\lambda} \geq t_{1-\alpha,N-1,\lambda}$.

For one-sided tolerance intervals based on the large-sample normal distribution, $k_{(1)}$ is calculated as

$$k_{(1)} = \frac{z_p + \sqrt{z_p^2 - ab}}{a}$$

where $a = 1 - \frac{z_\alpha^2}{2(N-1)}$, and $b = z_p^2 - \frac{z_\alpha^2}{N}$

Example: Regulatory Threshold for Racehorse Medication Withdrawal

(Data from RMTC 2016.) The Racehorse Medication and Testing Consortium [RMTC] Scientific Advisory Committee has determined the regulatory threshold for specific medications to be the upper limit of the 95/95 tolerance interval; that is, the specified tolerance interval has 95% coverage with 95% confidence. Samples from 20 research horses were collected 24 hours after administration of a certain medication and assayed to determine medication residue. Observed values were:

$$y = 6.8, 3.4, 6.2, 5.4, 0.3, 0.5, 2.6, 0.1, 0.1, 4.5,$$
$$1.0, 2.3, 10.0, 3.5, 0.2, 1.2, 0.8, 1.0, 1.4, 20.0$$

The range is 0.1–20.0 ng/mL, with mean 3.565 ng/mL (SD = 4.596 ng/mL); median 1.85 (IQR 0.725, 4.725) ng/mL. Because the data were non-normal and right-skewed, they were ln-transformed for analysis. The transformed mean and variance are 0.43 and 1.50, respectively, with $p = 0.95$, and $\alpha = 0.95$.

Sample SAS and R (Young 2010) codes for calculating tolerance limits are provided in Appendix 12.A. The value for k_1 is 2.383. The one-sided upper tolerance limit is exp(0.43 + 2.283 1.50) = exp(4.00523) for a threshold value of 54.6 ng/mL.

Example: One-Sided Lower Tolerance Interval: Osprey Eggshell Thickness

(Data from Odsjö and Sondell 2014.) Poor reproductive success in ospreys is related to eggshell thinning, mostly resulting from bioaccumulation of environmental contaminants. Investigators measured shell thicknesses in 166 eggs; average shell thicknesses were 0.51 (SD 0.039) mm.

From historical data, the investigators determined that a reduction in shell thickness by approximately 20% was associated with markedly increased rates of reproductive failure. A 20% reduction in shell thickness from the mean corresponds to an absolute eggshell thickness of 0.4 mm.

Suppose the effectiveness of environmental remediation was defined as at least 95% of eggs in the population to be above the breakage threshold of 0.4 mm with 90% confidence. What is the lower one-sided tolerance limit?

Using the `normtol.int` option in R library `tolerance`:

```
set.seed(166)
x <- rnorm(166, 0.515, 0.039)
out <- normtol.int(x = x, alpha = 0.10, P =
0.95, side = 1, method = "HE", log.norm =
FALSE)
out
```

The lower one-sided tolerance limit is 0.448 mm.

Example: Reference Intervals: Normally Distributed Data

Reference intervals for common veterinary haematology and biochemistry variables are usually based on the mean ± 2 SD if it can be assumed the data are normally distributed (Klaassen 1999). How do reference intervals based on this approximation compare with prediction and tolerance intervals for the same data?

(Data from Liu et al. 2021.) Fasting blood glucose values were obtained from 210 subjects and averaged 95.54 (SD 7.42) mg/dL.

The 95% confidence interval is:

$$RI = \overline{Y} \pm z_{1-\alpha/2} \cdot s/\sqrt{n} = 95.54$$
$$\pm 1.96 \left(7.42/\sqrt{210} \right)$$
$$= (94.5, 96.5)$$

The reference interval RI approximated by ± 2 SD (Klaassen 1999) is

$$RI = 95.54 \pm 2(7.42) = (80.71, 110.37) \text{ mg/dL}$$

The prediction interval for a single future observation is

$$RI = \overline{Y} \pm t_{1-\alpha/2,N-1} \cdot s\sqrt{(1 + 1/N)}$$
$$= 95.54 \pm 1.97 \cdot 7.42\sqrt{(1 + 1/210)}$$
$$= (80.9, 110.18)$$

Using the `normtol.int` option in R package `tolerance` and the Howe method for estimating the two-sided k, the tolerance interval for 95% coverage and 95% confidence is (80.1, 110.9) mg/dL

12.4 Non-parametric Tolerance Limits

Non-parametric, or distribution-free, tolerance limits are used if the data are not normally distributed and cannot be readily transformed or if the investigator chooses not to make distributional assumptions about the data. The only assumption is that the underlying distribution function is a non-decreasing continuous probability distribution. Non-parametric tolerance intervals are approximated by rank order methods. A major disadvantage of non-parametric approximations is the requirement for large sample sizes.

For count or binary data, the tolerance interval specifies the upper and lower bounds on the number of observations that are expected to show the event of interest (yes/no) in a future sample of m observations with a specified confidence $100 (1 - \alpha)\%$. For two-sided tolerance intervals with an upper and a lower limit, a specified proportion P of the population is contained within the bounds with a specified level of confidence. For one-sided tolerance intervals, the upper (or lower) limit describes the specified proportion P that meets or exceeds (or falls below) a specified threshold value P'. For example, to provide evidence of safety, an investigator might wish to show with 90% confidence that $\leq 1\%$ of subjects exposed to a test substance demonstrated adverse effects (Meeker et al. 2017; Hartnack 2019; Hartnack and Roos 2021).

For continuous data, non-parametric tolerance intervals are calculated from quantiles of the ranked sample data using the largest and smallest values in the sample. The major disadvantage of this method is that very large sample sizes are required for reasonable precision. Alternatively, if too high a tolerance is specified, this method may result in estimates of impossibly large sample sizes. Determining non-parametric tolerance limits usually require sample size N of at least 60 to ensure 90% coverage with 95% confidence (Hahn and Meeker 1991; Meeker et al. 2017).

Example: Blood Ammonia Tolerance Intervals: Non-normal Continuous Data

---

(Data from Tivers et al. 2014.) Blood ammonia concentration is measured routinely for dogs and cats with hepatic encephalopathy (HE). A study of 90 dogs without clinical signs of HE reported blood ammonia concentration averaging $\overline{Y} = 152.6$ µmol/L with $s = 101.6$ µmol/L. These data are clearly non-normal and probably right-skewed. Therefore, the estimates for the mean and standard deviation are ln-transformed prior to analysis, and then back-transformed to obtain the tolerance limits in the original units.

The transformed SD is $s_t = \sqrt{\ln\left(s^2 + \overline{Y}^2\right) - 2\ln\left(\overline{Y}\right)}$

$$= \sqrt{\ln(101.6^2 + 152.6^2) - 2\ln(152.6)} = 0.605738$$

The transformed mean is $\mu = \ln\left(\overline{Y} - s^2/2\right) = 4.844361$. Using the `normtol.int` option in R package `tolerance` and the Howe method for estimating the two-sided k, the tolerance interval for 95% coverage and 95% confidence is (3.568984, 6.168753), which back-transform to (35.5, 477.6) µmol/L. However, the tolerance interval estimated by this method is probably too wide to be practical or informative and seems to include values that lie outside the established normal range.

Example: Equine Skull Fractures: Binary Data

(Data from Joss et al. 2019.) Twelve out of 17 cadaveric equine skulls sustained orbital fractures following simulated hoof impact. The observed proportion was $12/17 = 0.706$. How does the 95/95 tolerance interval compare to the 95% prediction interval for $m = 10$ future observations?

Tolerance limits describe the proportion of 'defective' items (skulls with fractures) that bound the number of fractures expected in future samples. The two-sided 95/95 tolerance interval for 95% coverage and 95% confidence can be calculated with the bintol.int option in R library `tolerance`. Either the Wilson score method or Jeffrey's Bayesian method can be used to calculate the lower and upper bounds.

The approximate two-sided 95% prediction interval for $m = 10$ future observations is (4, 11). The two-sided 95/95 tolerance interval for 95% coverage and 95% confidence is (2, 10).

12.5 Determining Sample Size for Tolerance Intervals

Depending on the research problem, either of two different approaches can be used to choose sample size N so that the probability of either (a) including more than a specified proportion p of the population in the tolerance interval is small; or (b) the *entire tolerance interval* being within the specification limits is large.

Two-sided interval distribution-free sample size. The sample size required for a coverage p is estimated by iterating over candidate values for N to solve for confidence $100(1 - \alpha)\%$, where

$$1 - \alpha = 1 - Np^{N-1} + (N-1)p^N$$

Alternatively, N can be obtained by sorting the data from lowest to highest, then iterating over a range of coverage p from 0.5 to 0.999, which represent the proportion 'interval' between the smallest and largest observations in the sample. The two-sided non-parametric tolerance interval will be between the kth largest and $(n - k$th largest $+ 1)$

values for these data (Faulkenberry and Daly 1970; Meeker et al. 2017).

One-sided interval distribution-free sample size. For the one-sided tolerance interval, the above equality reduces to

$$1 - \alpha = 1 - p^N$$

Therefore, the smallest p to achieve a desired confidence interval is $(1 - \alpha)$ is

$$p = exp[(\ln \alpha)/N]$$

and the minimum sample size is

$$N = (\ln \alpha)/(\ln p)$$

Another approximation for N is

$$N \cong \frac{(1+p)}{4(1-p)} \cdot \chi^2_{\alpha,4} + \frac{1}{2}$$

where $\chi^2_{\alpha,4}$ is the critical value of the χ^2 distribution with 4 degrees of freedom that is exceeded with probability α (Hahn and Meeker 1991, NIST/SEMA-TECH 2012). Kirkpatrick (1977) provides tabulated values for determining sample size for one-sided and two-sided tolerance limits for both normally distributed and distribution-free data.

Example: Sample Size for Osprey Egg Study

How many eggs need to be randomly sampled to be able to claim that 95% of the osprey egg population will exceed the lower tolerance bound with 95% confidence?

Sample size is estimated from the approximation:

$$N \cong \frac{(1+p)}{4(1-p)} \cdot \chi^2_{\alpha,4} + \frac{1}{2}$$

For $p = 0.95$ and $\alpha = 0.95$ with $\chi^2_{0.95,4} = 9.488$, $N = 93$ and for $p = 0.99$ and $\alpha = 0.95$, $N = 473$.

Using the `norm.ss` option in R library `tolerance` (Young 2010, Young et al. 2016) for 95% coverage 95% confidence, and $P' = 0.97$, $N \cong 525$.

12.6 Sample Size for Tolerance Based on Permissible Number of Failures

When trialling novel medical procedures or devices, the consequences of error can be catastrophic. For sequential processes with risk of harm, a predetermined level of acceptable risk must be included with 'confidence' (percentage of occurrences) and 'reliability' (the proportion of the population sample) in sample size calculations. Durivage (2016 a, b) recommends setting confidence and reliability levels based on risk acceptance (Table 12.1). As risk increases, the predetermined level of reliability (or proportion of the sample population evaluated without failure) increases.

When number of failures is predetermined, sample size N is approximated as

$$N = \frac{0.5\left(\chi^2_{1-C,2(r+1)}\right)}{1-R}$$

where r is the number of failures, C is the confidence level, R is the 'reliability', and the value of χ^2 is determined for $(1 - C)$ confidence and $2(r + 1)$ degrees of freedom.

If zero failures are allowed:

$$N = \frac{\ln(1-C)}{\ln(R)}$$

Table 12.1: Confidence and reliability levels can be specified based on subjective levels of risk acceptance.

Risk	Defect, adverse event	Confidence	Reliability
Low	Minor	95%	90%
Medium	Major	95%	95%
High	Critical	95%	99%

Source: Adapted from Durivage (2016a).

Example: Surgical Procedure: No-Fail Performance

A new surgical procedure is being trialled where the consequences of error can be catastrophic. Therefore, the risk level is high. How many surgeries must be performed to ensure the investigators can have 95% confidence that the process is 99% reliable (99% of the sample is expected to have a successful procedure) when zero failures occur in the series? Two failures?

In this example, $C = 0.95$, $R = 0.99$ and $r = 0$ or 2.

Zero failures: $N = \dfrac{ln(1 - .95)}{ln(.99)} = 298.1 \cong 299$

Two failures: $N = \dfrac{0.5\left(\chi^2_{1 - 0.95, 2(2 + 1)}\right)}{1 - 0.99} = 629.6$
$\cong 630$

12.A SAS and R Code for Calculating Tolerance

12.A.1 Solving for k

The value k for the calculation of tolerance intervals is obtained from the non-central distributions for χ^2, and either z or t, with critical values generated from the respective quantile functions. The quantile function is the inverse of the CDF function.

For t, the format is $t = quantile$ ('[distribution]', [coverage/confidence], [degrees of freedom]).

For z, the format is $z = quantile$ ('[distribution]', [coverage/confidence]).

For χ^2, the format is chisq = $quantile$('[distribution]', alpha, [degrees of freedom]).

For the iterations, start at $N = 2$ and increase N by one until $k_1 \leq k_2$.

In SAS, for coverage p and confidence α

```
t_alpha = quantile("T",(1+p)/2,N - 1);
t_p = quantile("T",p,N - 1);

z_alpha = quantile("normal", (1+p)/2);
z_p = quantile("normal", p);

chi_crit = quantile("chisq",alpha, N-1);
```

Equivalent tolerance values can be calculated in *R* using the *tolerance* package (Young 2010).

12.A.2 SAS Code Racehorse Medication Threshold Limits for N = 20, Mean = 0.43 and Standard Deviation STD = 1.50

```
data tolerance;
set (keep=N mean STD);
     N = N;
     p=0.95; *coverage of 95%;
     alpha=0.05;
     alpha2=1-alpha; *confidence 95%;
     c=(1+p)/2;
     DF=N-1; *degrees of freedom;

*1. Small-sample method based on the non-
central t-distribution;
     lambda =zp*sqrt(N);
     t_crit = quantile("T",alpha,DF,lambda);
     k1_1 = t_crit/sqrt(N);
*calculate the upper tolerance limit;
*Because observations were ln-transformed,
they must be back-transformed to get the
original units;
     UL_t = exp(mean + k1_1*STD);

*2. Large-sample method;
     zc = quantile("normal", c);
     za = quantile("normal", alpha2);
     zp = quantile("normal", p);
     a = 1-((za*za)/(2*DF));
     b = zp*zp - (za*za/N);
k1 = (zp + sqrt(zp*zp - a*b))/a;

*Calculate the upper tolerance limit;
UL = exp(mean + k1*STD);
 run;

*print output;
 proc print; run;
```

12.A.3 R Code for Package Tolerance (Young 2010) for Racehorse Medication Threshold Limits

```
     set.seed(20)
     x <- rnorm(20, 0.43, 1.50)
     out <- normtol.int(x = x, alpha = 0.05,
P = 0.95, side = 1, method = "HE", log.norm =
FALSE)
     out

plottol(out, x, plot.type = "both", side =
"upper", 12.lab = "Normal Data")
```

References

Chen, H. and Kishino, H. (2015). Hypothesis testing of inclusion of the tolerance interval for the assessment of food safety. *PLoS ONE* 10 (10): e0141117. https://doi.org/10.1371/journal.pone.0141117.

Chiang, C., and Hsiao, C.-F. (2021). Tolerance interval testing for assessing accuracy and precision simultaneously. *PLoS ONE* 16 (2): e0246642. https://doi.org/10.1371/journal.pone.0246642.

Chiang, C., Chen, C.T., and Hsiao, C.F. (2021). Use of a two-sided tolerance interval in the design and evaluation of biosimilarity in clinical studies. *Pharmaceutical Satistics* 20 (1): 175–184. https://doi.org/10.1002/pst.2065.

Committee for Veterinary Medicinal Products. (2000). Note for guidance for the determination of withdrawal periods for milk. EMEA/CVMP/473/98-FINAL. https://www.ema.europa.eu/documents/scientific-guideline/note-guidance-determination-withdrawal-periods-milk_en.pdf (accessed 2022).

Durivage, M.A. (2016a). Risk-based approaches to establishing sample sizes for process validation https://www.meddeviceonline.com/doc/risk-based-approaches-to-establishing-sample-sizes-for-process-validation-0004 (accessed 2022).

Durivage, M.A. (2016b). How to establish sample sizes for process validation using the Success-Run theorem. Pharmaceutical Online https://qscompliance.com/wp-content/uploads/2017/10/How-To-Establish-Sample-Sizes-For-Process-Validation-Using-The-Success-Run-Theorem.pdf (accessed 2022).

Eberhardt, K.R., Mee, R.W., and Reeve, C.P. (1989). Computing factors for exact two-sided tolerance limits for a normal distribution. *Communications in Statistics Part B.* 1989: 397–413.

Faulkenberry, G.D., and Daly, J.C. (1970). Sample size for tolerance limits on a normal distribution. *Technometrics* 12 (4): 813–821.

Francq, B.G., Berger, M., and Boachie, C. (2020). To tolerate or to agree: a tutorial on tolerance intervals in method comparison studies with BivRegBLS R package. *Statistics in Medicine* 39: 4334–4349. https://doi.org/10.1002/sim.87092.

Gibbons, R.D., Bhaumik, D.K., and Aryal, S. (2009). *Statistical Methods for Groundwater Monitoring*, 2e. Hoboken: Wiley.

Guenther, W.C. (1977). *Sampling Inspection in Statistical Quality Control*. Griffin's Statistical Monographs, Number 37. London and High Wycombe: Griffin.

Hahn, G.J. and Meeker, W.Q. (1991). *Statistical Intervals: A Guide for Practitioners*. New York: Wiley.

Hartnack, S. (2019). *Confidence, Prediction and Tolerance Intervals in Classical and Bayesian Settings Master Thesis in Biostatistics*. Zurich: University of Zurich, Faculty of Science 82 pp. https://www.math.uzh.ch/li/index.php?file&key1=112446.

Hartnack, S., and Roos, M. (2021). Teaching: confidence, prediction and tolerance intervals in scientific practice: a tutorial on binary variables. *Emerging Themes in Epidemiology* 18: 17. https://doi.org/10.1186/s12982-021-00108-1.

Howe, W.G. (1969). Two-sided tolerance limits for normal populations – some improvements. *Journal of the American Statistical Association* 64 (326): 610–620.

Joss, R., Baschnagel, F., Ohlerth, S. et al. (2019). The risk of a shod and unshod horse kick to create orbital fractures in equine cadaveric skulls. *Veterinary and Comparative Orthopaedics and Traumatology* 32 (4): 282–288.

Kirkpatrick, R.L. (1977). Sample sizes to set tolerance limits. *Journal of Quality Technology* 9 (1): 6–12. https://doi.org/10.1080/00224065.1977.11980758.

Klaassen, J.K. (1999). Reference values in veterinary medicine. *Laboratory Medicine* 30 (3): 194–197. https://doi.org/10.1093/labmed/30.3.194.

Komaroff, M. (2018). The applications of tolerance intervals: make it easy. PharmaSUG 2018 – Paper AA-09. https://www.pharmasug.org/proceedings/2018/AA/PharmaSUG-2018-AA09.pdf (accessed 2022).

Liu, W., Bretz, F., and Cortina-Borja, M. (2021). Reference range: which statistical intervals to use? *Statistical Methods in Medical Research* 30 (2): 523–534. https://doi.org/10.1177/0962280220961793.

Meeker, W.Q., Hahn, G.J., and Escobar, L.A. (2017). *Statistical Intervals: A Guide for Practitioners and Researchers*, 2e. New York: Wiley.

NIST/SEMATECH. (2012). *e-Handbook of Statistical Methods*. http://www.itl.nist.gov/div898/handbook. http://doi.org/10.18434/M32189

Odeh, R.E., Chou, Y.-M., and Owen, D.B. (1987). The precision for coverages and sample size requirements for normal tolerance intervals. *Communications in Statistics: Simulation and Computation* 16: 969–985.

Odsjö, T., and Sondell, J. (2014). Eggshell thinning of osprey (Pandion haliaetus) breeding in Sweden and its significance for egg breakage and breeding outcome. *Science of the Total Environment* 470–471: 1023–1029. https://doi.org/10.1016/j.scitotenv.2013.10.051.

Pan, J. (2015). The application of tolerance interval in defining drug response for biomarker. PharmaSUG China 2015 – Paper 57 https://www.pharmasug.org/

Racehorse Medication and Testing Consortium [RMTC]. (2016). Explaining the 95/95 tolerance interval. https://rmtcnet.com/wp-content/uploads/2016-02-Explaining-the-95_95-Tolerance-Interval.pdf (accessed 2022).

Smith, R.W. (2002). The use of random-model tolerance intervals in environmental monitoring and regulation. *Journal of Agricultural, Biological, and Environmental Statistics* 7 (1): 74–94.

Tivers, M.S., Handel, I., Gow, A.G. et al. (2014). Hyper-ammonemia and systemic inflammatory response syndrome predicts presence of hepatic encephalopathy in dogs with congenital portosystemic shunts. *PLoS ONE* 9 (1): e82303. https://doi.org/10.1371/journal.pone.0082303.

Vardeman, S.B. (1992). What about the other intervals? *The American Statistician* 46 (3): 193–197. https://doi.org/10.2307/2685212.

Young, D.S. (2010). *tolerance*: an R package for estimating tolerance intervals. *Journal of Statistical Software* 36 (5): 1–39.

Young, D.S., Gordon, C.M., Zhu, S., and Olin, B.D. (2016). Sample size determination strategies for normal tolerance intervals using historical data. *Quality Engineering* 28 (3): 337–351. https://doi.org/10.1080/08982112.2015.1124279.

13 Reference Intervals

CHAPTER OUTLINE HEAD

13.1 Introduction

Reference intervals describe the range of clinical read-out values between which typical results for a healthy subject might be expected to occur. If the measurement falls outside these limits, it is flagged as potentially abnormal (Box 13.1). The *reference interval* is the proportion of measurements for healthy subjects contained between the upper and lower reference limits. The reference limits are defined by the percentiles of the reference population (Figure 13.1). Because the reference interval is the proportion of healthy subjects for which a correct negative test result is obtained, it corresponds to diagnostic *specificity*.

Typically, reference intervals are constructed for a single quantitative marker, usually a haematology or biochemistry variable or some other marker of disease. Markers are obtained from a sample of 'representative' reference subjects, usually healthy, in numbers 'sufficient' to provide reliable and consistent results. Defining what is 'representative' and 'sufficient' will be context-dependent. Reference intervals for the same marker may vary considerably between laboratories due to differences in subject populations, measurement devices, reagents, methods of analysis, and methods of verification. For a reference interval to be both clinically useful and generalisable, studies describing its construction must thoroughly document all methodology, including subject characteristics (signalment, health status, inclusion and exclusion criteria), methods of subject selection, sampling methods, and the type and number of biomarkers from which reference values are to be derived.

BOX 13.1
What Are Reference Intervals?

The *reference interval* is the range of clinical read-out values between which typical results for a healthy subject might be expected to occur.
Reference limits define potential abnormalities.
Reference intervals are descriptive; they do not test hypotheses.
Reference intervals are sample-based, not risk-based.

A Guide to Sample Size for Animal-based Studies, First Edition. Penny S. Reynolds.
© 2024 John Wiley & Sons Ltd. Published 2024 by John Wiley & Sons Ltd.

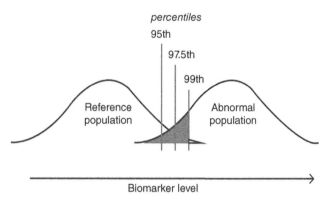

Figure 13.1: Read-out distributions for reference 'healthy' versus 'abnormal' subject populations.

Source: Adapted from Ekelund (2018).

BOX 13.2

Definitions for Selection and Spectrum Bias

Spectrum bias. Diagnostic performance differences resulting from differences in patient and subject composition between clinical settings. Differences in case mix affect diagnostic specificity (when the proportion of health subjects differs), sensitivity (when the proportion of subjects with the target condition differs), and estimates of prevalence.

Selection bias. Group differences in signalment, baseline characteristics, and/or outcome due to differences in subject characteristics resulting from poorly defined inclusion/exclusion criteria, sampling bias (different rates of selection), attrition bias (different rates of failure or dropout), and under-coverage bias (usually resulting from convenience sampling)

Selection bias and spectrum bias (Box 13.2) will limit the application of the reference interval because test subjects will not be representative of the desired target group (Ransohoff and Feinstein 1978; Goehring et al. 2004).

Although commonly used for clinical applications, *reference intervals do not measure risk.* They are sample-based and descriptive. Therefore, sample characteristics are the pre-eminent determinant of reliability and generalisability. Unless the reference interval is well-designed and verified, clinical interpretation must be performed with care. At worst, the reference interval will be neither applicable nor useful (Machin et al. 2018).

13.2 Constructing the Reference Interval

Construction of the reference interval and determination of an adequate sample size is a four-step process (Table 13.1). The process requires specification of the reference or target population (and health status), the subject pool, the marker variable, and coverage and confidence. Sample size can be justified by both rule-of-thumb methods and formal calculations, but in most real-life clinical settings it will be determined by convenience and the pool of available and accessible subjects. Convenience sampling is cheap and easy but is not a probability-based sampling method and is subject to sampling bias and spectrum bias (Ransohoff and Feinstein 1978, Mulherin and Miller 2002). Most formal sample size justification methods are based on the assumption that the sample is obtained by random probability-based selection. Reference interval determination studies must clearly describe methods of sampling and discuss the associated limitations.

Construction of the reference interval requires specification of three items: interval *coverage, reference limits,* and *cut-point precision* (Figure 13.2) Similar to 95% confidence intervals, bounds for a

Table 13.1: Four steps in construction of a reference interval.

1. Specify the reference population
   - Healthy or diseased?
   - If healthy, is subclinical disease an issue?
   - Inclusion/exclusion criteria
   - Consider laboratory analytics, methodology
2. Specify methods for subject selection
   - Prospective or retrospective
   - Random or cohort
   - Consider selection and spectrum bias
3. Specify marker variable
   - One or multiple?
   - Covariates? (age, sex, weight, other)
   - Distribution? (uniform, normal, non-normal)
4. Specify coverage and confidence settings
   - Target coverage for the interval
   - Confidence around the limits
5. Determine number of subjects
   - Rule of thumb = 120
   - Formal determinations
      - Parametric or non-parametric?
      - One- or two-tailed?

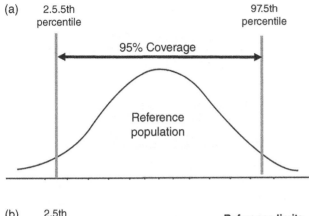

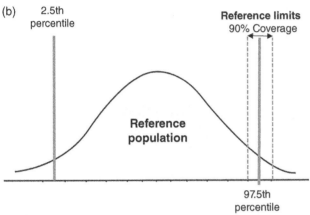

Figure 13.2: Construction of the reference interval requires specification of three items: interval *coverage*, *reference limits*, and *cut-point precision*. (a) By convention, *coverage p* is specified as 95%. Therefore the central 95% of the sample measurements for the reference distribution is bounded by the 2.5th and 97.5th percentiles, the *reference limits*. (b) *Cut-point precision* is defined by the coverage over the reference limits themselves. The desired coverage for the reference limit is usually specified as 90% ($\beta = 0.90$).

reference interval for continuous normally distributed data are determined from percentiles of the standard normal distribution. For non-normal data, non-parametric methods based on percentiles of rank-ordered values are used (Jennen-Steinmetz and Wellek 2005). The reference interval itself is defined as the central range of read-out values bounded by reference limit values. By convention, the central 95% of the sample measurements (*coverage*) fall between the lower 2.5th percentile and upper 97.5th percentile bounds (reference *limits* or *cut-points*). This means that 2.5% of observations are below the lower limit, and 2.5% are above the upper limit (Figure 13.2a). The reference interval requires an additional measure for the *precision of*

the cut-point. This is estimated as a designated proportion of the reference interval width, usually 1 to 5% (Figure 13.2b). The precision of the cut-points is the *confidence probability*. Coverage and confidence probability are analogous to the concepts of confidence and power used in statistical significance testing.

Frequently, reference ranges for common veterinary haematology and biochemistry markers are constructed on either the *95% confidence interval* or the *mean* ± 2 *SD* for normally distributed variables (Klaassen 1999). However, these approximations will be biased (especially if data are non-normal), resulting in under-estimation of true coverage. Regression-based reference intervals are used when clinical measurements vary with one or more covariates, such as age or body weight (Altman 1993; Wellek et al. 2014). Reference intervals may also be constructed using prediction intervals and tolerance intervals (Liu et al. 2021).

Example: Reference Range: Normally Distributed Data

(Data from Liu et al. 2021.) Fasting blood glucose values obtained from 210 subjects averaged 95.54 (SD 7.42) mg/dL. Reference ranges calculated from these summary data are summarised below.

The $\pm 2\,\mathrm{SD}$ approximation is RI_{2SD}
$$= 95.54 \pm 2(7.42) = (80.7, 110.4)\ \mathrm{mg/dL}$$

The reference range based on the 95% confidence interval is: $\mathrm{RI}_{CI} = \overline{Y} \pm z_{1-\alpha/2} \cdot s/\sqrt{n} = 95.54 \pm 1.96\ (7.42/\sqrt{210}) = (94.5, 96.5)$

The 95% prediction interval for a single future observation is

$$\mathrm{RI}_{pred} = \overline{Y} \pm t_{1-\alpha/2, N-1} \cdot s\sqrt{(1 + 1/N)}$$
$$= 95.54 \pm 1.97 \cdot 7.42\sqrt{(1 + 1/210)}$$
$$= (80.9, 110.2)$$

The 95/95 tolerance interval calculated with the `normtol.int` option in *R* library tolerance, with the Howe method for estimating the two-sided k (Young 2010, 2013) is

$$RI_{95/95} = (79.9, 111.8) \text{ mg/dL}$$

Example: Reference Range: Non-normal Continuous Data

(Data from Tivers et al. 2014.) Blood ammonia concentration is measured routinely for dogs and cats with suspected or overt hepatic encephalopathy (HE). A study of 90 dogs without clinical signs of HE reported blood ammonia concentrations averaging $\overline{Y} = 152.6$ μmol/L with SD 101.6 μmol/L. What is the reference interval based on (a) the ±2 SD approximation; (b) the 95/95 tolerance interval?

Because the mean is smaller than twice the standard deviation (Altman and Bland 1996), these data are clearly non-normal and right-skewed. Summary statistics were ln-transformed for analysis, as follows:

$$\sigma_{\ln} = \sqrt{\left[\ln(SD^2 + mean^2) - 2 \cdot \ln(mean)\right]} \text{ and } \mu_{\ln}$$
$$= \ln(mean - \sigma_{\ln}^2/2)$$

In this example, $\sigma_{\ln} = 0.605738$ and $\mu_{\ln} = 4.844361$

(a) The reference interval based on ±2SD approximation is $4.844361 \pm 2(0.605738)$, and back-transformed to give

$$RI_{2SD} = (37.8, 426.6) \text{ mg/dL}$$

The tolerance interval was estimated using `normtol.int` in the R library `tolerance`

```
        set.seed(20)
x <- rnorm(210, 4.844361, 0.605738)
out <- normtol.int(x = x, alpha = 0.05,
P = 0.95, side = 2, method = "HE", log.norm
= FALSE)
out
```

The output for the lower and upper limits are (3.568984, 6.168753). Back-transforming to obtain the original units gives:

$$RI_{95/95} = (35.5, 477.6) \text{ μmol/L}$$

Using the non-parametric method with `nptol.int` in the R library `tolerance`

$$RI_{95/95} = (-78.2, 414.9) \text{ μmol/L}$$

The 95/95 tolerance interval is wider than the ±2 SD approximation, but results are roughly comparable. In contrast, the interval calculated with non-parametric methods is extremely wide and includes impossible values. The sample size is not large enough to obtain realistic precision with non-parametric methods.

13.2.1 Regression-Based Reference Ranges

Regression-based reference ranges are constructed when clinical measurements vary with one or more covariates, such as age or body weight (Altman 1993; Jennen-Steinmetz 2013; Wellek et al. 2014). In the simplest case, with one covariate X, the linear regression model is $Y = b_0 + b_1 X + \varepsilon$, with intercept b_0, slope b_1 and error term ε, which is normally distributed with mean 0 and variance σ^2. The relative margin of error D is a pre-specified proportion of the width of the $100(1 - \alpha)\%$ confidence interval, say 10–30%. Sample sizes are calculated such that N is sufficiently large to obtain a pre-specified margin of error. Non-normal data may be log-transformed to approximate a normal distribution, then back-transformed to obtain values in the original units.

If data cannot be readily transformed or the data do not conform to the assumptions required for traditional least-squares regression, semi-parametric or non-parametric rank methods can be used (Wellek et al. 2014; Machin et al. 2018). Altman (1993) described a simple method using centiles on the absolute residuals (see also Wright and Royston 1997a, b). Quantile regression (Wei et al. 2019) may be indicated when clinical markers show uneven patterns of variation across the range of the covariate (Cade et al. 1999; Yu et al. 2003; Jennen-Steinmetz 2014) or when measurements falling below (or above) limits of detection may be an issue (Smith et al. 2006). Quantile regression can be performed in SAS *proc quantreg* or R *quantreg*

(Koenker 2022), and confidence intervals constructed by rank test statistics or bootstrap resampling. Bootstrap resampling is recommended for small samples (<200) and covariate skew (Koenker and Hallock 2001; Yu et al. 2003; Tarr 2012; Jennen-Steinmetz 2014). The Harrell-Davis quantile estimator (Harrell and Davis 1982) is another distribution-free method that constructs somewhat more efficient confidence intervals of quartiles than other non-parametric methods. The Harrell-Davis estimator routine can be found in the *R* Hmisc package (hdquantile function).

Methods for constructing regression-based tolerance intervals are available for several types of regression models, including linear regression with one or more predictors, polynomial regression, nonlinear regression, and non-parametric regression (Young 2013, Young et al. 2016; Liu et al. 2021). Regression tolerance intervals can be computed in *R* tolerance: regtol.int for linear regression, nlregtol.int for nonlinear regression, and npregtol.int for non-parametric regression (Young 2010, 2014). However, because tolerance interval estimation was developed for engineering quality control applications, it may not be generally applicable to clinical reference interval construction. Reference intervals based on tolerance estimation may be too wide, sacrificing sensitivity for specificity (Wellek and Jennen-Steinmetz 2022).

Example: Echocardiography Metrics in Cavalier King Charles Spaniels

(Data from Misbach et al. 2014.) Readings for 13 echocardiography and pulsed-wave Doppler variables were obtained for 134 clinically healthy adult Cavalier King Charles Spaniels, and effects of body weight, age, and sex were used to construct 90% confidence regression-based reference intervals. They reported the fitted linear regression of left ventricular end-diastolic diameter (LVDd, mm) on body weight (BW, kg) as LVDd = 20.57 + 0.987 · BW, with 90% reference range (23.4, 35.6) mm.

Computation of regression tolerance intervals for a 10 kg male dog (*R* tolerance: regtol. int; Young 2013) indicated that the predicted LVDd is approximately 30.4 mm, with the 90/95

tolerance interval (24.5, 36.2) mm. These values are very similar to the interval limits originally reported.

13.3 Sample Size Determination

13.3.1 Rules of Thumb

The Clinical and Laboratory Standards Institute (CLSI) recommends a minimum of 120 subjects as a rule of thumb for a reference interval study (Bautsch 2009; Hortowitz et al. 2010). At least 400 subjects in each homogeneous subgroup are recommended for multi-centre studies (Ichihara and Boyd 2010). For verification of a previously established reference interval, the minimum recommended sample size is 20 subjects (Henny 2009; Hortowitz et al. 2010; Katayev et al. 2010).

Rule of thumb approximations implicitly assume the data are normally distributed. This is frequently not the case for biological or biochemical data. Another disadvantage is that rules of thumb do not account for design and operational specifics or sample population characteristics. Rule of thumb estimates may considerably over- or under-estimate numbers actually required to achieve the necessary specificity for the new study.

13.3.2 Sample-Based Coverage Methods

Sample size estimation for a reference interval requires specification of three items: the coverage (*probability content p*) of the reference interval, the precision with which the reference limits or 'cut-points' are estimated (*confidence probability β*), and cut-point *precision* (*δ*) (Figure 13.2.) The cut-point precision is expressed as a proportion of the reference interval width, usually 1–5%.

Reference intervals may be determined by either parametric or non-parametric methods of estimation. Parametric methods are appropriate for normally distributed continuous data, or data that can be normalised with a log-transformation. For non-normal data, non-parametric methods are used. Non-parametric methods require much larger sample sizes than parametric approximations. Reference intervals may be either one- or two-sided.

13.3.3 Parametric Sample Size Estimates

Wellek et al. (2014) give the approximate sample size formula for a two-sided parametric reference interval based on the standard normal probability density function. Sample size is

$$N \geq 1 + (z_1^2/2) \cdot [\Phi \cdot z_2/(\delta/2)]^2$$

where z_1 is the $(1 + q)/2$ percentile for coverage, and $z_2 = (1 + \beta)/2$ percentile for the cut-point precision, δ. Typically, $q = 0.95$ and $\beta = 0.90$, so that $z_1 = z_2 = 1.96$ for the two-sided case, For the one-sided reference interval, z_1 is the q percentile and cut-point precision is δ (rather than $\delta/2$), and $z_1 = z_2 = 1.6440$. The quantity Φ is the pdf of the standard normal density function at $x = z_1$ and is calculated as

$$\Phi = \frac{1}{\sqrt{2\pi}} \cdot e^{-z_1^2/2}$$

If $z_1 = 1.96$, then $\Phi = 0.05845$ and if $z_1 = 1.6448$ then $\Phi = 0.10314$.

13.3.4 Non-parametric Sample Size Estimates

Non-parametric (distribution-free) formulations are based on weighted average of the rank orders of measurements. Reference intervals are determined by percentiles of the distribution of clinical biomarker values. The reference interval is defined by the qth and $[1 - (1 - q)/2]$th quantiles of the normal distribution. If a one-sided reference interval is required, the two-sided quantile $[1 - (1 - q)/2]$ is replaced by $[1 - (1 - q)] = q$.

The approximate sample size formula for a two-sided non-parametric reference interval (Wellek et al. 2014) is

$$N \geq [(1 + q)/2] \cdot [1 - (1 + q)/2] \cdot [z_2/(\delta/2)]^2$$

and for the one-sided case:

$$N \geq q(1 - q) \cdot (z_2/\delta)^2$$

Example: Determining Sample Size for a Non-Parametric Reference Interval

(From Wellek et al. 2014.) For two-sided reference coverage of 95%, cut-point precision δ of 0.01, and β of 90%, the parametric estimate of sample size is

$$N \approx \left(1 + \frac{1.96^2}{2}\right) \cdot [(0.05845 \cdot 1.96)/(0.01/2)]^2$$

$$\approx 1534$$

and the non-parametric sample size is

$$[[1+0.95]/2] \cdot [1-[1+0.95]/2] \cdot [1.96/[0.01/12]]^2$$
$$\approx (0.975 \cdot 0.025 \cdot (1.96/0.005)^2 \approx 3746$$

13.3.5 Covariate-Dependent Sample Size Estimates

When clinical measurements vary with one or more covariates, sample size estimates need to account for the structure of the regression model and how covariates are distributed, as well as the desired degree of precision. In general, sample sizes based on covariate-dependent samples will be very much larger than those that are independent of covariates.

Bellera and Hanley (2007) developed a simple regression-based method for estimating sample size when the value of the response depends on a single covariate, for example, age or body weight. Four variations of this method are available to account different distributions of the covariate in the study sample.

Uniform distribution. If the covariate is uniformly distributed across its range, minimum sample size N is:

$$N \geq \frac{z_{1-\alpha/2}^2 \cdot \left(4 + z_p^2/p\right)}{z_{1-\beta/2}^2 \cdot D^2}$$

for the $100(1 - \alpha)\%$ confidence interval, the $100 \cdot p\%$ reference limit compared to the $100(1 - \beta)\%$ reference range, and for margin of error D.

Normal distribution. If the covariate is approximately normally distributed, and its range is approximately four times that of its standard deviation:

$$N \geq \frac{z_{1-\alpha/2}^2 \cdot \left(5 + z_p^2/2\right)}{z_{1-\beta/2}^2 \cdot D^2}$$

Tertile method. If the covariate does not conform to either type of distribution, sampling for a new study could be approached by partitioning the covariate into thirds, or tertiles, and then obtaining samples of size *N/3* from each tertile:

$$N \geq \frac{z_{1-\alpha/2}^2 \cdot \left(5/2 + z_p^2/2\right)}{z_{1-\beta/2}^2 \cdot D^2}$$

Estimating at the covariate average. For estimating the reference range at the *average value* of the covariate, the sample size is:

$$N \geq \frac{z_{1-\alpha/2}^2 \cdot \left(1 + z_p^2/2\right)}{z_{1-\beta/2}^2 \cdot D^2}$$

Example: Canine Pulmonary Stenosis

(Data from Ackerman et al. 2022). Pulmonary stenosis severity in dogs showed a negative linear association with body weight, and body weights showed a right-skewed distribution. Median body weight for 90 dogs was 14 kg (IQR 7, 25 kg; range 1–46 kg). If a new study was planned based on these data, how many dogs would need to be sampled to obtain the 95% confidence interval for the 95% reference limit with a relative margin of error *D* of 10%?

Normal method. If weight is ln-transformed, the distribution of the covariate is approximately normal, with mean 2.617 (SD 0.74) and range 0.1–3.8. Then

$$N \geq \frac{1.96^2 \cdot \left(5 + 1.96^2/2\right)}{1.96^2 \cdot 0.10^2} = 692.1$$

or approximately 693 dogs.

Tertile method. One potential sampling scheme for a new study would be to divide the study sample into three weight classes, or tertiles (<10 kg, 10–20 kg, and >20 kg). Equal number of dogs would then be enrolled from each tertile.

The estimated minimum sample size for the new study based on the tertile method is

$$N \geq \frac{1.96^2 \cdot \left(5/2 + 1.645^2/2\right)}{1.96^2 \cdot 0.10^2} = 755.2$$

for a total of 756 dogs, with 252 dogs randomly sampled from each weight class.

References

Ackerman, L.H., Reynolds, P.S., Aherne, M., and Swift, S. T. (2022). Right axis deviation in the canine electrocardiogram for predicting severity of pulmonic stenosis: a retrospective cohort analysis. *American Journal of Veterinary Research* 83 (4): 312–316. https://doi.org/10.2460/ajvr.21.09.0138.

Altman, D.G. (1993). Construction of age-related reference centiles using absolute residuals. *Statistics in Medicine* 12: 917–924.

Altman, D.G. and Bland, J.M. (1996). Detecting skewness from summary information. *BMJ* 313 (7066): 1200. https://doi.org/10.1136/bmj.313.7066.1200.

Bautsch, W. (2009). Requirements and assessment of laboratory tests: part 5 of a series on evaluation of scientific publications. *Deutsches Ärzteblatt International* 106 (24): 403–406. https://doi.org/10.3238/arztebl.2009.0403.

Bellera, C.A. and Hanley, J.A. (2007). A method is presented to plan the required sample size when estimating regression-based reference limits. *Journal of Clinical Epidemiology* 60: 610–615.

Cade, B.S., Terrell, J.W., and Schroede, R.L. (1999). Estimating effects of limiting factors with regression quantiles. *Ecology* 80 (1): 311–323.

Ekelund, S. (2018). Reference intervals and percentiles – implications for the healthy patient. https://acute-caretesting.org/ (accessed 2022).

Goehring, C., Perrier, A., and Morabia, A. (2004). Spectrum bias: a quantitative and graphical analysis of the variability of medical diagnostic test performance. *Statistics in Medicine* 23 (1): 125–113.

Harrell, F.E. and Davis, C.E. (1982). A new distribution-free quantile estimator. *Biometrika* 69 (3): 635–640.

Henny, J. (2009). The IFCC recommendations for determining reference intervals: strengths and limitations. *Journal of Laboratory Medicine* 33 (2): 45–51. https://doi.org/10.1515/JLM.2009.0162.

Hortowitz GL, Altaie S, Boyd JC, Ceriotti F, Garg U, Horn P, Pesce A, Sine HE, Zakwski J. *Defining, establishing, and verifying reference intervals in the clinical laboratory; Approved guideline.* 3 Clinical and Laboratory Standards Institute (CLSI) Document EP28-A3C, Vol. 28, No. 30 Wayne, PA: Clinical and Laboratory Standards Institute; 2010.

Ichihara, K. and Boyd, J.C. (2010). An appraisal of statistical procedures used in derivation of reference intervals. *Clinical Chemistry and Laboratory Medicine* 48: 1537–1551.

Jennen-Steinmetz, C. (2013). Sample size determination for studies designed to estimate covariate-dependent reference quantile curves. *Statistics in Medicine* 33: 1336–1348.

Jennen-Steinmetz, C. (2014). Sample size determination for studies designed to estimate covariate-dependent reference quantile curves. *Statistics in Medicine*, 33 (8):1336–48. https://doi.org/10.1002/sim.6024.

Jennen-Steinmetz, C. and Wellek, S. (2005). A new approach to sample size calculation for reference interval studies. *Statistics in Medicine* 24 (20): 3199–3212. https://doi.org/10.1002/sim.2177.

Katayev, A., Balciza, C., and Seccombe, D.W. (2010). Establishing reference intervals for clinical laboratory test results: is there a better way? *American Journal of Clinical Pathology* 133 (2): 180–186. https://doi.org/10.1309/ajcpn5bmtsf1cdyp.

Klaassen, J.K. (1999). Reference values in veterinary medicine. *Laboratory Medicine* 30 (3): 194–197. https://doi.org/10.1093/labmed/30.3.194.

Koenker, R. (2022). R package *quantreg*. https://cran.r-project.org (accessed 2022).

Koenker, R. and Hallock, K.F. (2001). Quantile regression. *Journal of Economic Perspectives* 15 (4): 143–156.

Liu, W., Bretz, F., and Cortina-Borja, M. (2021). Reference range: which statistical intervals to use? *Statistical Methods in Medical Research* 30 (2): 523–534. https://doi.org/10.1177/0962280220961793.

Machin, D., Campbell, M.J., Tan, S.B., and Tan, S.H. (2018). *Sample Sizes for Clinical, Laboratory and Epidemiology Studies*, 4e. New York: Wiley.

Misbach, C., Lefebvre, H.P., Concordet, D. et al. (2014 Jun). Echocardiography and conventional Doppler examination in clinically healthy adult Cavalier King Charles Spaniels: effect of body weight, age, and gender, and establishment of reference intervals. *Journal of Veterinary Cardiology* 16 (2): 91–100. https://doi.org/10.1016/j.jvc.2014.03.001.

Mulherin, S.A. and Miller, W.C. (2002). Spectrum bias or spectrum effect? Subgroup variation in diagnostic test evaluation (PDF). *Annals of Internal Medicine* 137 (7): 598–602. https://doi.org/10.7326/0003-4819-137-7-200210010-00011.

Ransohoff, D.F, Feinstein, A.R. (1978). Problems of spectrum and bias in evaluating the efficacy of diagnostic tests. *New England Journal of Medicine (NEJM)*, 299 (17): 926–30. doi:https://doi.org/10.1056/NEJM197810262991705.

Smith, D., Silver, E., and Harnly, M. (2006). Environmental samples below the limits of detection – comparing regression methods to predict environmental concentrations. http://www.lexjansen.com/wuss/2006/Analytics/ANL-Smith.pdf (accessed 2022).

Tarr, G. (2012). Small sample performance of quantile regression confidence intervals. *Journal of Statistical Computation and Simulation* 82 (1): 81–94. https://doi.org/10.1080/00949655.2010.527844.

Tivers, M.S., Handel, I., Gow, A.G. et al. (2014). Hyperammonemia and systemic inflammatory response syndrome predicts presence of hepatic encephalopathy in dogs with congenital portosystemic shunts. *PLoS ONE* 9 (1): e82303. https://doi.org/10.1371/journal.pone.0082303.

Wei, Y., Kehm, R.D., Goldberg, M., and Terry, M.B. (2019). Applications for quantile regression in epidemiology. *Current Epidemiology Reports* 6: 191–199. https://doi.org/10.1007/s40471-019-00204-6.

Wellek, S. and Jennen-Steinmetz, C. (2022). Reference ranges: Why tolerance intervals should not be used. [Comment on Liu, Bretz and Cortina-Borja, Reference range: Which statistical intervals to use? SMMR, 2021,Vol. 30(2) 523–534]. *Statistical Methods in Medical Research* 31 (11): 2255–2256. https://doi.org/10.1177/09622802221114538.

Wellek, S., Lackner, K.J., Jennen-Steinmetz, C. et al. (2014). Determination of reference limits: statistical concepts and tools for sample size calculation. *Clinical Chemistry and Laboratory Medicine (CCLM)* 52 (12): 1685–1694. https://doi.org/10.1515/cclm-2014-0226.

Wright, E.M. and Royston, P. (1997a). Simplified estimation of age-specific reference intervals for skewed data. *Statistics in Medicine* 16: 2785–2803.

Wright, E.M. and Royston, P. (1997b). A comparison of statistical methods for age-related reference intervals. *Journal of the Royal Statistical Society, A* 160: 47–69.

Young, D.S. (2010). *tolerance*: an R package for estimating tolerance intervals. *Journal of Statistical Software* 36 (5): 1–39.

Young, D.S. (2013). Regression tolerance intervals. *Communications in Statistics: Simulation and Computation* 42 (9): 2040–2055. https://doi.org/10.1080/03610918.2012.689064.

Young, D.S. (2014). Computing tolerance intervals and regions using R. In: *Handbook of Statistics: Computational Statistics with R*, vol. 32 (ed. M.B. Rao and C.R. Rao), 309–338. Amsterdam: North Holland-Elsevier.

Young, D.S., Gordon, C.M., Zhu, S., and Olin, B.D. (2016). Sample size determination strategies for normal tolerance intervals using historical data. *Quality Engineering* 28 (3): 337–351. https://doi.org/10.1080/08982112.2015.1124279.

Yu, K., Lu, Z., and Stander, J. (2003). Quantile regression: applications and current research areas. *The Statistician* 52 (3): 331–350.

IV Sample Size for Comparison

14 Sample Size and Hypothesis Testing

CHAPTER OUTLINE HEAD

14.1 Introduction

Comparisons usually involve tests of statistical hypotheses. The purpose of hypothesis testing is to determine if there is sufficient evidence to support a claim (the statistical hypothesis) about a population parameter based on a sample of data. A hypothesis test can be designed to minimise the likelihood of making false claims and to maximise the ability of the test to detect a difference when one exists. Increasing sample size is one method of increasing the reliability of test interpretations and precision of estimates. However, increasing sample size increases costs and time to obtain the data and, of course, increases the number of animals. Right-sizing experiments involved trade-offs between statistical precision, balancing the probabilities of different kinds of false claims, and the constraints related to resource availability and ethical considerations.

14.2 Power and Significance

Power and significance are related but distinct statistical concepts (Box 14.1, Table 14.1). *Power* is

> **BOX 14.1**
> *Definitions*
>
> *Power:* probability of correctly rejecting the null hypothesis if the alternative hypothesis is true, *true positive* $(1 - \beta)$. The probability of a *false negative* is Type II error β.
>
> *Significance:* probability of observing the sample result if the null hypothesis is true. The probability of a *false positive* is Type I error α.
>
> *Non-centrality parameter:* a measure of the difference between the mean under the alternative hypothesis compared to that expected for the null.
>
> *P-value:* probability of observing the result or a result more extreme in the sample data if the null hypothesis is true.
>
> *Statistical significance* is declared when $P \leq \alpha$.

the conditional probability of correctly rejecting the null hypothesis H_0 when the alternate hypothesis H_A is true:

$$\text{Power} = \Pr(\text{Reject } H_0 \mid H_A \text{ is true})$$

A Guide to Sample Size for Animal-based Studies, First Edition. Penny S. Reynolds.
© 2024 John Wiley & Sons Ltd. Published 2024 by John Wiley & Sons Ltd.

Table 14.1: The relationship between significance, confidence, power, and Type I and Type II errors.

If the null hypothesis is	True	False
Rejected	α False positive Type I error	$1 - \beta$ (power) True positive
Not rejected	$1 - \alpha$ True negative	β False negative Type II error

BOX 14.2
Minimum Information Planning Sample Size for Comparative Studies

- Type I and type II error rates, α and β
- Primary outcome (response) variable
- Expected variance of the response
- The minimal biologically relevant difference to be detected
- Definition of the experimental unit
- Randomisation strategy for experimental unit selection, treatment allocation, and sequence allocation.

It is the probability of detecting a real effect or difference when one exists (the probability of a true positive). The Type II error β is the probability of obtaining a false negative, or not rejecting the null hypothesis when it is false. Therefore, power is $(1 - \beta)$. High power means that the test is less likely to miss a real effect – there is a low probability of making a type II error. Low power means that the test has a high risk of Type II error and is more likely to miss a real effect.

Recall that the Type I error is the probability of obtaining a false positive or incorrectly rejecting the null hypothesis when it is true. The probability of Type I error is specified by the *significance threshold* α. The *confidence level* $(1 - \alpha)$ is the probability of not rejecting the null hypothesis given it is true (true positive). When a statistical test is applied to the sample data, the resulting *P*-value is compared to the significance threshold value α. A *statistically significant* result occurs when $P \le \alpha$.

In addition to pre-specification of power and confidence, the minimum items of information for a comparative study will include definitions for the primary outcome variable, the biologically relevant difference between groups, and the measure of variance. The study is powered off the number of experimental units, so these must be defined, and a randomisation plan devised both for allocation of treatments to units, and the measurements sequence (Box 14.2).

Determining an appropriate sample size involves trade-offs between the risks of false positives and false negatives. Power is also determined by sample size, significance level, and effect size (Table 14.2). Increasing sample size can increase the power of a test and reduce standard error of the estimate, making it easier to detect an effect if it exists. However, 'wrong sample size bias' occurs when sample size is either too small or too large to meet study objectives (Catalogue of Bias Collaboration et al. 2017). A too-small study will

Table 14.2 Sample size, power, and significance.

Sample size is determined by significance (α), power $(1 - \beta)$, and the number of tails of the test (one-tailed or two-tailed).
For a given effect size, significance (α), and power $(1 - \beta)$, larger sample sizes are required for a two-tailed test compared to a one-tailed test.
Power increases as effect size increases.
Power increases with increasing significance (α) but smaller confidence ($1 - \alpha$). Larger values of α correspond to smaller confidence levels and therefore greater precision. A study with α at 5% (95% confidence) has greater power to detect a true effect than if α is at 1% (99% confidence).
Power is maximised as β (the probability of type II error) is minimised. However, if false positives (Type I errors) are considered more serious than false negatives (Type II errors), a rule of thumb to ensure a target power of 80% is to use a ratio of β to α as 4:1.

have both insufficient power to detect true effects if they exist (essentially wasting all the animals in the study) and a high probability of false positives, resulting in over-estimation of efficacy or benefit of a test intervention (Sena et al. 2010). Too-large a sample size wastes the excess animals, without contributing much additional information.

Statistical significance is frequently misinterpreted. The *P*-value is a statement of the statistical probability of observing the result or a result more extreme *if the null hypothesis is true*. Statistical significance does not mean the results are biologically meaningful or clinically actionable. Conversely, failure to reject the null hypothesis does not 'prove' the null or signify that results are unimportant

(Altman 1991; Altman and Bland 1995; Gardner and Altman 2012; Cumming 2012; Greenland et al. 2016; Amrhein et al. 2019). The threshold for significance (conventionally 0.05) is arbitrary and has no clinical or biological meaning (Altman 1991). Therefore, dichotomising results into 'significant' or 'non-significant' loses almost all quantitative information necessary for interpretation. In addition, P-values are unstable, especially when effect size is small and variation is large. P-values can vary considerably between studies simply due to random variation (Greenland et al. 2016). Statistically significant results from small underpowered studies are likely to be false positives and are unlikely to replicate (Button et al. 2013).

Performing sample size calculations to 'find a statistically significant result' should never be a study goal (Chapter 3). 'Chasing significance' (Ware and Munafò 2016; Szucs 2016; Marín-Franch 2018) leads to questionable research practices such as P-hacking and N-hacking. P-hacking is the selection of data and analysis methods to produce statistically significant results (Head et al. 2015). N-hacking is the selective manipulation of sample size to achieve statistical significance, usually by increasing sample size, cherry-picking observations, or excluding outliers without justification (Szucs 2016). Both practices increase the false positive rate and violate standards of ethical scientific and statistical practice (Wasserstein and Lazar 2016).

14.3 Non-centrality

The non-centrality parameter (Box 14.3) is a measure of the difference between the population means under the alternative hypothesis. It provides a method of evaluating the power of the test with respect to the alternative hypothesis (Carlberg 2014). A large non-centrality parameter represents a large effect size, providing more evidence against the null hypothesis and increasing the power of the test. Calculations based on non-central distributions of the relevant test statistic can be used to determine power for a given sample size, or alternatively, find the sample size necessary to achieve a target power ('Brien and Muller 1993; Harrison and Brady 2004). This is a useful property for animal-based studies that are frequently small, with data that often do not conform to a normal distribution, or when complex study designs are required. For these

> **BOX 14.3**
> *Centrality and Non-centrality*
>
> The *central probability distribution* describes the distribution of the test statistic under the null hypothesis.
>
> *The non-central distribution* describes the distribution of the test statistic under an alternate hypothesis.
>
> *The non-central parameter* λ is the difference by which the test statistic distribution estimated from sample data differs from that expected under the null hypothesis.

studies, sample sizes based on large-sample normal approximations will be highly inaccurate, and simple sample size formulae may not even exist.

The *central probability distribution* describes how the test statistic is distributed under the null hypothesis. Figure 14.1 shows an example of the central distribution for the F-statistic. The critical value of the test statistic defines the threshold of statistical significance. The critical F-value is defined by $F_{\alpha,\ \nu_1,\ \nu_2}$, with degrees of freedom υ_1 and υ_2, and probability of Type I error α. If the null hypothesis is true, the values of the test statistic calculated from observed data will occur outside the region defined by the cut-off with probability α, the probability of a false positive.

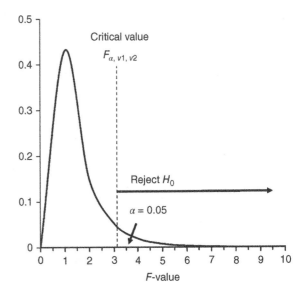

Figure 14.1: The central F distribution with degrees of freedom (3, 20) and $\alpha = 0.05$. The critical value is 14.098. The significance threshold is α. Values of the F statistic obtained from the sample data are statistically significant if they exceed the critical F-value.

The *non-central distribution* describes the distribution of the test statistic when the null hypothesis is false, and an alternative hypothesis is true. *The non-centrality parameter λ* is a measure of the difference by which the distribution of a test statistic estimated from sample data differs from the distribution expected under the null hypothesis of no difference (Kirk 2012). When there are no differences in group means, then λ is zero, and the distribution follows the central distribution. As λ gets larger, the distribution peak moves further away from zero and the distribution stretches out. The power of the test is the area to the right of the critical value under the non-central distribution (Figure 14.2). Therefore, as λ gets larger, power increases. Power is the probability that the non-central test statistic is equal to, or greater than, the critical value of the test statistic for a given level of significance:

Power = Pr[Non-central test statistic ≥ Critical (central) value of the test statistic]

The power of a test is not one value but is a range of values that depend on the parameters in the alternate hypothesis. Sample size is only one factor that determines the magnitude of the non-centrality parameter. It is also affected by the difference between the true population mean and the hypothesised mean, the variance, the distribution of the test statistic, the covariance structure of the statistical model, and the study design (Stroup 2012).

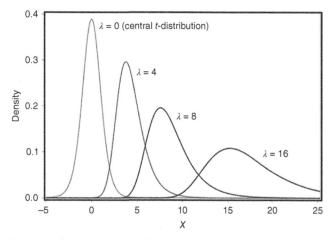

Figure 14.2: Central (λ = 0) and non-central (λ > 0) *t*-distributions with 10 degrees of freedom. As λ gets larger, the curve shifts to the right and becomes more asymmetrical.

BOX 14.4
Relationships Between Non-centrality Parameter λ, ϕ, and Cohen's Effect Size f

$$\lambda = f^2 / \sqrt{N}$$

$$\lambda = F_{crit} \cdot df_{numerator}$$

$$\lambda = SS_{tre} / MSE$$

$$\lambda = \phi^2 \cdot k$$

$$\phi = f \sqrt{N}$$

Definitions, terminology, and notation for the non-centrality parameter are inconsistent and can be confusing (Carlberg 2014). Older textbooks that rely on power charts and nomograms usually refer to the quantity phi (ϕ) for tabulated values of power (Pearson and Hartley 1958; Zar 2010); ϕ is sometimes identified as the non-centrality parameter itself. However, ϕ is not the same as λ. For example, for single-factor ANOVA, ϕ is equivalent to the square root of the non-centrality parameter and the number of groups. Conversion relationships for λ, ϕ, and f (Winer et al. 1991; Dollins 1995) are given in Box 14.4.

In G∗Power, the value of the non-centrality parameter is part of the automatic output for sample size calculations. Dudek (2022) provides a useful overview of power concepts with accompanying R code.

14.4 Estimating Sample Size

Sample size using non-centrality is computed by iterative numerical methods (Fenschel et al. 2011; Stroup 2011, 2012). Depending on the type of information available, estimates of the non-centrality parameter λ are determined from:

1. *Summary statistics* for each group. These can be means and standard deviations, or standard error and sample size obtained from prior data.

2. An estimate of *mean difference and the pooled variance of the difference.*

14. *Exemplary or raw data.* An exemplary or 'dummy' data set is artificial data that consist of values obtained from historical data or

simulated from statistical models used to model the expected responses. The λ is calculated from the mock data by fitting a prespecified analysis model (e.g. ANOVA, mixed models, generalised linear models) and extracting the necessary statistics from the output (O'Brien and Muller 1993; Castelloe and O'Brien 2001). The model output step provides F-values and degrees of freedom for calculating λ.

Calculations for λ and the critical value for the test statistic ($\lambda = 0$) are made by iterating over a range of candidate sample sizes (Littell et al. 2006; Fenschel et al. 2011; Stroup 2012). The sample size is chosen that results in the value of the computed power that equals or exceeds the prespecified target power. Total sample size N, sample size per group n_i, or maximum number of groups k for a specified total sample size can be calculated with this method.

14.4.1 Non-central t-Distribution

The one-sample t-statistic follows the Student's t-distribution with $n - 1$ degrees of freedom:

$$t = \frac{\bar{x} - \mu}{s/\sqrt{n}}$$

where $\bar{x}$ is the sample mean of a random sample from a normal population, s is the standard deviation, n is the sample size, and μ is the population mean. The effect size is

$$d = \frac{\bar{x} - \mu}{s}$$

The central t is distributed around zero. Suppose the true population mean is actually μ_1, then the t-statistic will follow a non-central t-distribution:

$$\lambda = \frac{\mu_1 - \mu}{\sigma/\sqrt{n}}$$

with values of t is distributed around the non-centrality parameter λ. Under the null hypothesis, the difference between μ and μ_1 is zero, and the test statistic has a standard t-distribution. The distributions move to the right as λ increases. Under the

alternative hypothesis, the test statistic now has a non-zero mean t-distribution.

The power to detect the difference d with significance α is approximately

$$1 - \beta = T_{n-1}\left[t_{\alpha/2, n-1} \frac{|d|\sqrt{n}}{\sigma} \right]$$

where T_{n-1} is the cumulative distribution function of the non-central t-distribution with degrees of freedom $n - 1$ and non-centrality parameter λ. (Box 14.5 gives definitions for cumulative and probability density functions).

The critical value of t is estimated by the inverse function of t, calculated from the total sample size n, degrees of freedom $df = n - 1$, and significance level α (Harrison and Brady 2004). In SAS and R, the *tinv* function returns the pth quantile from the Student's t-distribution:

```
t_crit = tinv(alpha/2,n-1)
```

For a one-sided test, the one-sided significance level α replaces the two-sided $\alpha/2$.

Power is then computed from the critical value of t, the associated degrees of freedom, and the non-centrality parameter (ncp):

```
in SAS as power = 1 - probt(t_crit, n-1, ncp)
and in R as
ncp <- d/(s/sqrt(n))
t <- qt(0.975,df=n-1)
pt(t,df=n-1,ncp=ncp)-pt(-t, df=n-1,ncp=ncp)
1-(pt(t,df=n-1,ncp=ncp)-pt(-t,df=n-1,ncp=ncp))
```

BOX 14.5

Cumulative Distribution Function (cdf) and Probability Density Function (pdf)

The *cumulative distribution function* (cdf) gives the probabilities of a random variable X being smaller than or equal to some value x, $\Pr(X \leq x) = F(x)$. The inverse of the cdf gives the value x that would make $F(x)$ return a particular probability p: $F^{-1}(p) = x$.

The *probability density function* (pdf) returns the relative frequency of the value of a statistic or the probability that a random variable X takes on a certain value. The pdf is the derivative of the cdf.

The two sample t-statistic follow the Student's t-distribution with $n - 2$ degrees of freedom. Here, the t statistic is

$$t = \frac{\bar{x}_1 - \bar{x}_2}{s_p \sqrt{\dfrac{1}{n_1} + \dfrac{1}{n_2}}}$$

where s_p is the pooled standard deviation of the two samples. The effect size is

$$d = \frac{\bar{x}_1 - \bar{x}_2}{s_p}$$

Example: Increasing Power By Reducing Variation: Rat Model of Bone Fracture

(Data from Prodinger et al. 2018.) Biomechanical evaluations of rodent long bone fractures frequently use a three-point bending apparatus, with support bars set to a fixed span. Femur failure loads in a fixed span model of femur fracture in $n = 20$ rats averaged 206 (SD 30) N. However, it was found that adjusting the span to accommodate individual bone length may reduce variation in response by up to 30%.

Suppose it was determined that a 20 N increase in a second experimental group was biologically important. What sample size would be required for this level of variation, and what is the effect on power and sample size if the variation could be reduced by changing from the fixed-span to the individualised-span protocol?

Simple simulations can be performed using G∗Power for a two-sample t-test for the difference between two independent means. Assuming standard deviation is similar to that for the original fixed-span protocol, effect size is $d/s = 20/30 = 0.67$. For a one-tailed test with $\alpha = 0.05$ and nominal power of 0.8, the critical t-value is 1.673 and the non-centrality parameter is 2.551. The total sample size is 58, with 29 rats in each group.

If the standard deviation can be reduced by 30%, then SD is approximately 20, and the effect size is now $20/20 = 1.0$. Then the critical t-value is 1.706 and the non-centrality parameter is 2.646. The total sample size is 28, with only 14 per group, with an increase in realised power to 0.82.

Suppose the target effect size is 2.5 for a non-centrality parameter of >14. The total sample size required for a power of 0.8 is now 6, or 3 per group. However, an effect size this large would require a standard deviation of approximately 8 N, which is unlikely to be practically or biologically feasible.

14.4.2 Non-central F-Distribution

The F-statistic is the ratio of the total variance to the error variance:

$$F = \frac{MST}{MSE} = \frac{\sigma_T^2 + \sigma_e^2}{\sigma_e^2}$$

where σ_T^2 is the estimate of the variation due to differences between group means ('treatments'), and σ_e^2 is the estimate of the variation due to error. The distribution of the F-ratio is determined by the degrees of freedom for the treatment group, the numerator degrees of freedom $\nu_1 = (k - 1)$ and the error, or denominator, degrees of freedom $\nu_2 = k(n - 1) = (N - k)$, where k is the number of 'groups', N is the total sample size, and n is the sample size per group (if group sizes are balanced).

If group sizes are equal, the non-centrality parameter for the non-central F is the weighted sum of squares of treatment means:

$$\lambda = \frac{SS_T}{MSE} = (k - 1)F$$

where $(k - 1)$ is the numerator degrees of freedom (Liu and Raudenbush 2004). Because $t^2 = F$, the methods used to calculate the non-central F-variate can also be used to calculate the non-central t-variate (Winer et al. 1991).

The shape of the non-central F-distribution is determined by the numerator and denominator degrees of freedom and the variance of the treatment group means, σ_T^2 (Carlberg 2014). If the null hypothesis cannot be rejected, there are no differences between group means, σ_T^2 is zero, and the F-value is equal to 1. Therefore the ratio of variances will follow the central F-distribution, and the non-

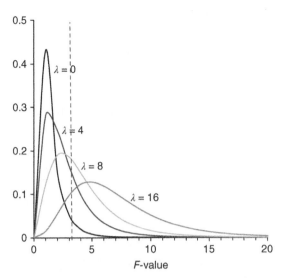

Figure 14.3: Central ($\lambda = 0$) and non-central ($\lambda > 0$) F-distributions with degrees of freedom (3, 20). The critical $F_{3,20} = 14.098$. Area under the curve to the right of the critical value (dotted line) for each non-central curve is the power of the test.

centrality parameter is zero ($\lambda = 0$). Under the alternate hypothesis, the sampling distribution of the F-ratio is the non-central F-distribution $F(\nu_1, \nu_2, \lambda)$, $\lambda > 0$ and the F-distribution stretches out to the right (Figure 14.3).

The critical F-value is obtained from the 100 $(1 - \alpha)$% quantile value of the central F distribution and corresponding degrees of freedom. For example, for $\alpha = 0.05$, the quantile is 100 $(1 - 0.05)$, or 95th quantile value of the central F-distribution with $\lambda = 0$. This is calculated from the inverse function of the corresponding cumulative density function cdf. The critical F-value is obtained by setting the non-centrality parameter to zero:

In SAS this is   `F_crit=Finv(1-alpha, numdf, dendf,0)`
and in R as   `F.crit = qf(alpha, numdf, dendf,lower.tail=F)`

where *numdf* is the numerator degrees of freedom $\nu_1 = (k - 1)$, and *dendf* is the denominator degrees of freedom $\nu_2 = k \cdot (n - 1) = (N - k)$.

The associated power for each N is then estimated as Power $= \Pr(F(\nu_1, \nu_{2,\lambda}) \geq F(\nu_1, \nu_2))$:

in SAS as   `Power = 1-probf(F_crit, NumDF, DenDF, NCP)`
and in R as   `pf(F.crit, NumDF, DenDF, ncp=delta, lower.tail=F)`

Example: Determining Sample Size from Exemplary Data: Mouse Breeding Productivity

(Data from Hull et al. 2022.) A study was designed to compare two handling methods on reproductive indices for laboratory mice. The study design was a two-arm randomised controlled trial, with breeding pair as random effect and handling method as the fixed effect.

The primary outcome was number of pups produced per pair. The outcome data were discrete counts, which are non-Gaussian. Analyses required a generalised linear mixed model with a Poisson distribution for counts. Therefore, conventional power calculation formulae for determining sample size were inappropriate, and power and sample size were estimated by simulation (Littell et al. 2006; Stroup 2011, 2012).

Sample size was estimated in three steps (Sample SAS code is in Appendix 14.A). First, an exemplary data set was created for the total number of pups expected in each group. Data were based on historical information for controls and the anticipated effect size. The expected total number of pups for control animals in this study was:

5 litters $\times$ 6 pups/litter = 30 pups per pair

The expected difference was an additional pup per litter, or 4–5 additional pups per pair, for a total of 34–35 pups in the test group. It was determined *a priori* that an increase of one extra pup per pair was operationally significant enough to justify the facility switching to the new handling method. In this example, exemplary data were generated for a sample size of 40 pairs per treatment arm. The 40 'observations' in each group had the values specified by the anticipated pup counts (30 and 35).

Second, a generalised linear model and Poisson distribution were fitted to the exemplary data in *SAS proc glimmix*. The model output step provides expected F-values and degrees of freedom *df* for calculating the non-centrality parameter, approximated as (numerator *df*) $\bigcirc$ (F-value). The critical F-value was estimated from the inverse function for the χ^2 distribution, given the pre-specified confidence $(1 - \alpha)$, the degrees of freedom for the numerator (here $k - 1 = 2 - 1 = 1$), and $\lambda = 0$ for the central distribution.

Finally, power was calculated by estimating the probability that the computed F-statistic exceeded the critical F-value under the non-central F.

The power for a difference of four extra pups per pair was 0.88 and 0.97 for five extra pups per pair for a total sample size of 80 pairs (or 40 pairs per group). However, the power for an operational difference of one extra pup per pair with this sample size was only 0.114. To achieve power >0.8 with such a small effect size, the study would have required approximately 250 pairs. This was clearly unfeasible in terms of housing space and time. Furthermore, the number of pups produced would have greatly exceeded demand, which would have necessitated the euthanasia of excess animals that could not be used.

14.5 Sample Size Balance and Allocation Ratio

Balanced designs have equal numbers of experimental units in each group. A balanced design has the most power for a given sample size N. The *allocation ratio r* describes the balance of subjects in each arm of the trial. If sample size is balanced, then the allocation ratio $r = n_1/n_2 = 1$.

In general, unequal allocation and unbalanced designs are not recommended. Occasionally more power might be obtained by an unequal allocation ratio if the response in one group is known to be more variable than that in the other. Then the appropriate r can be computed as the ratio of standard deviations: $r = n_1/n_2 = s_1/s_2$, and the group with the larger expected standard deviations is allocated more subjects. However, compared to balanced designs, unbalanced designs require substantially more subjects to detect the same effect size with the same power (e.g. 12% for a 2:1 allocation ratio), making them more costly, exposing more subjects to potential risk (Hey and Kimmelman 2014), and requiring the use of far many more animals.

The effect of design imbalance on sample size can be assessed by computing the relative efficiency of the design. Relative efficiency (RE) is the increase in sample size needed for a candidate design to match the precision of a perfectly balanced completely randomised design. The RE of a design is assessed by comparing an unbalanced to a balanced design as the ratio of respective sample sizes and standard deviations. As group sample sizes become more unbalanced, RE (ratio of variances) decreases and total N increases to achieve the same precision as a design with equal sample sizes per treatment group. Relative efficiencies can be estimated for a variety of designs, methods of analysis, and comparisons (difference, matched, stratified, covariance-adjusted), and for both experimental and observational studies.

Example: Relative Efficiency of a Balanced Versus an Unbalanced Allocation Design

---

Suppose a study design is proposed with two treatments A and B, a fixed total sample size N of 20, and variance σ^2. The sample sizes of each group are n_A and n_B, respectively, and $n_A + n_B = 20$. The standard error of the difference between treatment means is

$$SE_{diff} = \sigma \sqrt{1/n_A + 1/n_B}$$

The RE of the design is the ratio of the variance of the unbalanced design $n_A \neq n_B$ to that for the balanced design $n_A = n_B$:

$$RE = \frac{SE_{new}^2}{SE_{balanced}^2} = \frac{\left(\sigma_{new} \sqrt{1/n_A + 1/n_B}\right)^2}{\left(\sigma_{bal} \sqrt{1/n_A + 1/n_B}\right)^2}$$

If variances are assumed to be equal, then the variances cancel, so RE is seen to depend directly on the ratio of sample sizes for each design.

Table 14.3 shows the relative efficiencies for each allocation ratio, from the most balanced ($r = 1$) to least balanced ($r = 1/19$) condition. In this example, the RE for an unbalanced design with $n_A = 3$ and $n_B = 17$ requires twice as many animals to achieve the same level of precision as a balanced design ($n_A = n_B = 10$). For the unbalanced design to obtain the same precision as the balanced design, the sample size required is $1/RE = 1/0.51 = 1.96$ times that of the balanced design. The adjusted sample sizes are $n_A = (3 \cdot 1.96) \cong 6$, $n_B = (17 \cdot 1.96) = 33.33 \cong 34$, for total $N = 40$.

Table 14.3: Relative efficiency of balanced versus unbalanced allocation designs. The total sample size $N = n_A + n_B$ is 20.

n_A	n_B	$\sqrt{1/n_A + 1/n_B}$	Relative efficiency	Correction	Adjusted sample sizes		
					n_A	n_B	N^a
10	10	0.447	1.00	1.00	10	10	20
9	11	0.449	0.99	1.01	9.09	11.11	22
8	12	0.456	0.96	1.04	8.33	12.50	23
7	13	0.469	0.91	1.10	7.69	14.29	23
6	14	0.488	0.84	1.19	7.14	16.67	25
5	15	0.516	0.75	1.33	6.67	20.00	27
4	16	0.559	0.64	1.56	6.25	25.00	32
3	17	0.626	0.51	1.96	5.88	33.33	40
2	18	0.745	0.36	2.78	5.56	50.00	56
1	19	1.026	0.19	5.26	5.26	100.00	106

^{a}Values are rounded up.

14.A Sample SAS Code for Estimating Power for a Given Sample Size Based on Exemplary Non-Gaussian Data. Modified from Stroup (2011, 2012) and Littell et al. (2006)

1. Create Exemplary Data. The Primary Outcome was Number of Pups Per Pair with an Expected Number of 30 for the Control and 35 for the Test Group. The Do Loop Assigns the Number of Experimental Units for Each Treatment Arm

```
data mice;
/* trt = treatment identifiers, pups =
expected total pup count for each arm;*/
input trt $ pups;

/*best guess sample size per treatment arm.
Compare different values until the desired
power is obtained */

n=40;
do obs=1 to N;
output;
end;
```

```
/* Input expected pup counts for each arm;*/

datalines;
control    30
test    35
;
run;
proc print data=mice;
run;
```

2. Perform the Analysis on the Exemplary Model Using the Anticipated Statistical Model. This Extracts the Non-Central F Value and Degrees of Freedom to Enable Calculation of the Non-Centrality Parameter

```
proc glimmix data= mice;
class trt;
model mean = trt / chisq link=log
dist=poisson;
contrast 'control vs experimental' trt 1 -1 /
chisq ;
ods output tests3=F_overall
contrasts=F_contrasts;
run;
```

3. Calculate Power. The Non-Centrality Parameter (ncp) is Calculated from the F Statistics Extracted from the Output. F_crit Defines the Critical Value for F Under the Central Distribution

```
data power;
set F_overall F_contrasts;

/*calculate non-centrality parameter;*/
 ncp=numdf*Fvalue;

/* calculate power for a pre-specified
confidence (alpha) */
 alpha=0.05;

/* for count data the inverse χ² is used
(Stroup 2012)*/
 f_crit=cinv(1-alpha,numdf,0);
 Power=1-probchi(F_crit,numdf,ncp);

/* the inverse F is: */
f_crit=finv(1-alpha,numdf,dendf,0);
power=1-probF(f_crit,numdf,dendf,ncp);

/* output power and compared to desired value */

proc print data=power;
run;
```

References

Altman, D.G. (1991). *Practical Statistics for Medical Research*. Chapman & Hall/CRC.

Altman, D.G. and Bland, J.M. (1995). Absence of evidence is not evidence of absence. *BMJ* 311 (7003): 485. https://doi.org/10.1136/bmj.311.70014.485.

Amrhein, V., Greenland, S., and McShane, B. (2019). Scientists rise up against statistical significance. *Nature* 567 (7748): 305–307. https://doi.org/10.1038/d41586-019-00857-9.

Button, K.S., Ioannidis, J.P., Mokrysz, C. et al. (2013). Power failure: why small sample size undermines the reliability of neuroscience. *Nature Reviews Neuroscience* 14 (5): 365–376. https://doi.org/10.1038/nrn3475.

Carlberg, C. (2014). *Statistical Analysis: Microsoft Excel 2013*. Pearson Education.

Castelloe, J.M. and O'Brien, R.G. (2001). Power and sample size determination for linear models. *SUGI* 26: 240–226. https://support.sas.com/resources/papers/proceedings/proceedings/sugi26/p240-26.pdf.

Catalogue of Bias Collaboration, Spencer, E.A., Brassey, J., Mahtani, K., Heneghan, C. (2017). Wrong sample size bias. *Catalogue of Bias*. https://catalogof-bias.org/biases/wrong-sample-size-bias/ (accessed 2022).

Cumming, G. (2012). *Understanding the New Statistics: Effect sizes, Confidence Intervals, and Meta-Analysis*. New York, NY: Routledge.

Dollins, A.B. (1995). A Computational Guide to Power Analysis of Fixed Effects in Balanced Analysis of Variance Designs. Report DODPI95-R-0003. Department of Defense Polygraph Institute, Fort McClellan, AL.

Dudek, B. (2022). Test statistics, null and alternative distributions: type II errors, power, effect size, and non-central distributions. https://bcdudek.net/power1/non_central.html#power-facilities-in-r (accessed 2023).

Fenschel, M.C., Amin, R.S., and van Dyke, R.D. (2011). A macro to estimate sample-size using the non-centrality parameter of a non-central F-distribution SA18-2011. https://www.mwsug.org/proceedings/2011/stats/MWSUG-2011-SA18.pdf (accessed 2019).

Gardner, M.J. and Altman, D.G. (2012). Confidence intervals rather than P values: estimation rather than hypothesis testing. *British Medical Journal* 292 (6522): 746–750. https://doi.org/10.1136/bmj.292.6522.746.

Greenland, S., Senn, S.J., Rothman, K.J. et al. (2016). Statistical tests, *P* values, confidence intervals, and power: a guide to misinterpretations. *European Journal of Epidemiology* 31 (4): 337–350. https://doi.org/10.1007/s10654-016-0149-14.

Harrison, D.A. and Brady, A.R. (2004). Sample size and power calculations using the noncentral t-distribution. *The Stata Journal* 4 (2): 142–153.

Head, M.L., Holman, L., Lanfear, R. et al. (2015). The extent and consequences of P-hacking in science. *PLoS Biology* 13 (3): e1002106.

Hey, S.P. and Kimmelman, J. (2014). The questionable use of unequal allocation in confirmatory trials. *Neurology* 82 (1): 77–79. https://doi.org/10.1212/01.wnl.0000438226.10353.1c.

Hull, M.A., Reynolds, P.S., and Nunamaker, E.A. (2022). Effects of non-aversive versus tail-lift handling on breeding productivity in a C57BL/6J mouse colony. *PLoS ONE* 17 (1): e0263192. https://doi.org/10.1371/journal.pone.0263192.

Marín-Franch, I. (2018). Publication bias and the chase for statistical significance. *Journal of Optometry* 11 (2): 67–68. https://doi.org/10.1016/j.optom.2018.03.001.

Kirk, R. (2012). *Experimental Design: Procedures for Behavioral Sciences*. SAGE.

Littell, R.C., Milliken, G.A., Stroup, W.W. et al. (2006). *SAS for Mixed Models*, 2e. Cary, NC: SAS Institute, Inc.

Liu, X. and Raudenbush, S. (2004). A note on the non-centrality parameter and effect size estimates for the

F test in ANOVA. *Journal of Educational and Behavioral Statistics* 29 (2): 251–255. https://doi.org/10.3102/10769986029002251.

O'Brien, R.G. and Muller, K.E. (1993). Unified power analysis for *t*-tests through multivariate hypotheses, chap 8. In: *Applied Analysis of Variance in Behavioral Science* (ed. L.K. Edwards), 297–344. New York: Marcel Dekker.

Pearson, E. and Hartley, H.O. (1958). *Biometrika Tables for Statisticians*, vol. 1. Cambridge: Cambridge University Press.

Prodinger, P.M., Bürklein, D., Foehr, P. et al. (2018). Improving results in rat fracture models: enhancing the efficacy of biomechanical testing by a modification of the experimental setup. *BMC Musculoskeletal Disorders* 19: 243. https://doi.org/10.1186/s12891-018-2155-y.

Sena, E.S., van der Worp, H.B., Bath, P.M. et al. (2010). Publication bias in reports of animal stroke studies leads to major overstatement of efficacy. *PLoS Biology* 8 (3): e1000344. https://doi.org/10.1371/journal.pbio.1000344.

Stroup, W.W. (2011). Living with generalized linear mixed models. *SAS Global Forum 2011: Statistics and Data Analysis*. https://support.sas.com/resources/papers/proceedings11/349-2011.pdf (accessed 2019).

Stroup, W.W. (2012). *Generalized Linear Mixed Models: Modern Concepts, Methods and Applications*. Boca Raton: Chapman & Hall/CRC.

Szucs, D. (2016). A tutorial on hunting statistical significance by chasing N. *Frontiers in Psychology* 7: 1444. https://doi.org/10.3389/fpsyg.2016.01444.

Ware, J.J. and Munafò, M.R. (2016). Significance chasing in research practice: causes, consequences, and possible solutions. *Addiction* 110 (1): 4–8. https://doi.org/10.1111/add.126714.

Wasserstein, R.L. and Lazar, N.A. (2016). The ASA's statement on P-values: context, process, and purpose. *The American Statistician* 70: 129–314. https://doi.org/10.1080/00031305.2016.1154108.

Winer, B.J., Brown, D.R., and Michels, K.M. (1991). *Statistical Principles in Experimental Design*, 3e. New York: McGraw-Hill.

Zar, J.H. (2010). *Biostatistical Analysis*, 5e. Upper Saddle River: Prentice-Hall.

15 A Bestiary of Effect Sizes

CHAPTER OUTLINE HEAD

15.1   Introduction

An effect size is an estimate that describes the magnitude and direction of the difference between two groups or the relationship between two or more variables. For interpretation of results, effect sizes are preferred to 'statistical significance' or *p*-values. This is because an effect size uses all information about size and direction of the difference, and is the best summary of the 'effect' of the test intervention, along with a measure of precision (Hojat and Xu 2004; Sun et al. 2010; Cumming 2012).

Understanding how effect sizes are constructed for different types of study is essential for both preliminary sample size estimation and interpretation of results (Box 15.1). First, there is no one definition of 'effect size' (Cumming, 2012). For practical purposes, there are four major categories or families of effect sizes: *group difference* effect sizes (**d** *family*), *strength of association* effect sizes (**r** *family)*, and *risk* and *time-to-event* effect sizes (Table 15.1). However, in practice, over 70 different effect size indices have been identified (Schober et al. 2018). Second, different software programmes may calculate the 'same'

A Guide to Sample Size for Animal-based Studies, First Edition. Penny S. Reynolds.
© 2024 John Wiley & Sons Ltd. Published 2024 by John Wiley & Sons Ltd.

BOX 15.1
Effect Sizes

Choose an effect size (or effect size family) most appropriate for testing study hypotheses.

The effect size must be relevant to study objectives (*context*).

Effect size does not infer causality.

Effect size does not infer statistical significance.

Effect size does not infer practical significance.

Table 15.1: Categories of effect size indices.

Effect size index	Definition
Group difference indices	
Mean	Unadjusted average
Difference between two means	Unadjusted difference
Cohen's *d* family	Standardised difference in two independent sample means, standardised by the average or pooled standard deviation. If samples are paired, use the standard deviation for the paired differences.
Hedge's *g* family	Standardised difference in two independent sample means when $N < 20$, standardised using weighted standard deviation.
Glass *Δ*	Standardised difference in two independent sample means, standardised using the control standard deviation.
Strength of association indices	
Correlation *r*	The strength of linear association between two or more independent variables
Regression slope *β*	The change in the dependent variable with a unit change in the predictor
Proportion of variation R^2	Also called coefficient of variation. The strength of association between one dependent variable and one or multiple predictors
η^2	Correlation ratio. Estimates the proportion of variance in the response variables accounted for by the explanatory variables. Ratio of variance terms $SS_{between}/SS_{total}$
η_p^2	Partial correlation ratio
ω^2	Less-biased alternative to η^2 for small *N*.
Cohen's f^2	Proportion of variance effect size appropriate for multiple regression, multilevel and repeated-measures data.
Cramer's V	Strength of association between two nominal variables, obtained from χ^2 analyses
Risk estimate indices	
Relative risk, RR	Ratio of proportion of one group to proportion of the second group p_1/p_2
Odds ratio, OR	Ratio of the odds of occurrence of an event (or 'success') *p* relative to the odds of failure $(1-p)$ in a test group relative to those for a comparator group: $p_1/(1-p_1)\ /p_2/(1-p_2)$
Cohen's *h*	Difference between two arcsin-transformed proportions
Time to event indices	
Hazard ratio, *HR*	Ratio of events *p* in each group $HR = \dfrac{\ln(p_1)}{\ln(p_0)}$ Ratio of median survival times $HR = \dfrac{M_0}{M_1}$

effect size differently (for example, G∗Power and SPSS; Lakens 2013), and terminology can be idiosyncratic. Therefore, to avoid misunderstanding and misinterpretation, the particular effect size to be used and how it is to be calculated must be clearly described in the protocol and methods. Effect sizes used and presented without context can result in sample size approximations and subsequent interpretations that are extremely misleading.

15.2 Effect Size Basics

Choosing the most appropriate effect size depends on the study objectives and the study context: the *study design* or statistical model, and the *type of effect* most relevant to the research question and study objectives. How the effect size is calculated depends on whether or not the effect size is to be standardised, and if standardised, how the standardiser is to be calculated (Cumming 2012).

Regardless of which family of effect sizes is selected, all effect size calculations require the following information:

The *primary outcome*. The type of outcome variable dictates which class or family of effect sizes can be used.

The pre-specified difference between groups most likely to be of practical or biological significance.

Number of groups to be compared.

Desired *probability* for type I error (α) and type II error (β).

Whether comparisons are *one-sided or two-sided*.

Allocation ratio, or sample size balance among groups.

During planning and initial sample size calculations, careful thought should be given to choosing preliminary values for first-pass approximations. Reasonable initial values can be obtained from systematic reviews and meta-analyses, targeted literature reviews, pilot data, and data from previous experiments (Rosenthal 1994; Ferguson 2009; Cummings 2012; Lakens 2013). Lakens (2013) provides an excellent tutorial on calculating and reporting effect sizes.

> **BOX 15.2**
> *The 'Best' Effect Size*
>
> - results in maximum power for a given sample size
> - is based on continuous variables and equal sample sizes per group,
> - is determined by study objectives, study context, and relevance to the research question.

The 'best' effect size is that which provides the most power for a given sample size (Box 15.2). This is usually an effect size based on a continuous variable (rather than binary or time-to-event variables) and a balanced design (equal number of experimental units per treatment arm). Studies using binary or time-to-event outcomes or designs with unequal allocation require many more subjects than do balanced studies on continuous outcomes.

15.3 *d* Family Effect Sizes

The *d* family of effect sizes describe *differences between independent groups*. For continuous variables, these are expressed as the differences between group means. Effect sizes based on group differences may be *unstandardised* or *standardised* (Box 15.3).

An unstandardised effect size expresses the difference between groups without adjusting for variation in the sample. Raw effect sizes retain the original units of measurement and thus have the enormous advantage of being easy to interpret when units of measurement are themselves meaningful. For these reasons, raw effect sizes are often used in meta-analyses of health research (Vedula and Altman, 2010). Examples of unstandardised effect sizes

> **BOX 15.3**
> *d-Family Effect Sizes*
>
> Effect sizes describing differences between independent groups
>
> *Unstandardised* effect size is a simple difference.
> *Standardised* effect size is the difference scaled by the sample variation.

include a simple difference between means or proportions and sample standard deviation.

A *standardised* effect size is the difference in means scaled or adjusted by some measure of variation in the sample. In the simplest form, the difference is scaled by the within-group standard deviation. The best-known measure is Cohen's *d*. This gives the effect size as the difference in means expressed as a multiple of standard deviation units in the sample. As such, standardised effect size is a dimensionless ratio and independent of units of measurement. This allows a readier comparison of effect sizes across diverse studies with different measurement scales. Standardised effect sizes are thus used extensively in meta-analyses (Vedula and Altman 2010). Standardised effect sizes will also be useful when units of measurement do not have an intrinsic meaning, e.g. ordinal scores, such as client satisfaction (Schober et al. 2018). However, standardised effect sizes have several crucial disadvantages. Practical interpretation is extremely difficult because there is no information about the measurement units. When used in meta-analyses, standardised effect sizes may distort diagnostics, such as funnel plots, and result in increased false positives (Zwetsloot et al. 2017).

15.3.1 The Basic Equation for Continuous Outcome Data

The basic equation is also referred to as the *fundamental equation* (Machin et al. 2006) or the *basic formula* (van Belle 2008).

When outcome data are continuous and normally distributed, the power of a two-sided test to detect a difference between groups is:

$$z_{1-\beta} = \frac{d}{\sigma}\sqrt{\frac{n}{2}} - z_{1-\alpha/2}$$

where *d* is the difference between group means ($\mu_1 - \mu_2$). Assuming equal group sizes ($N = 2n$) then the *sample size per group* is

$$n \geq 2\frac{\left(z_{1-\alpha/2} + z_{1-\beta}\right)^2}{(d/\sigma)^2}$$

The unstandardised effect size is *d*. The standardised effect size is d/σ.

15.3.2 Two-Group Comparisons, Continuous Outcomes, Independent Samples

Cohen's d_s (Cohen 1988) is the best-known metric of standardised effect size. This is the difference between two means, divided by the pooled standard deviation:

$$d_s = \frac{(\bar{x}_1 - \bar{x}_2)}{s_p}$$

That is, effect size measures the difference in means in units of the pooled sample standard deviation.

Hedges's g should be used when n is small ($n < 20$). Cohen's d_s is a biased estimate of the population effect size when sample sizes are small. The results for d_s and *g* are similar if $n > 20$. Hedge's *g* uses a sample size based on the weighted pooled standard deviation:

$$g = \frac{(\bar{x}_1 - \bar{x}_2)}{s_p}$$
$$\cdot \left(1 - \frac{3}{4 \cdot N - 9}\right) = d_s\left(1 - \frac{3}{4 \cdot N - 9}\right)$$

where $N = n_1 + n_2$. (Hedges and Olkin 1985; Cumming 2012; Lakens 2013).

Glass's Δ is used if the intent is to compare one group against a baseline, or if standard deviations differ substantially between groups. This statistic uses the standard deviation of the control sample only, s_0 rather than the pooled standard deviation s_p. Therefore effect size based on Glass's Δ measures the difference in means in units of the control sample standard deviation (Glass 1976).

$$\Delta = \frac{\bar{x}_2 - \bar{x}_1}{s_0},$$

Two-group comparisons, paired samples. Paired designs compare observations taken on the same experimental unit. The observations are therefore correlated within each experimental unit. Common applications include before-after assessments and paired measurements taken on both the right and left sides of a study animal. The effect size for a paired sample $d_{s,\,pair}$ is estimated as the difference of two means divided by the standard deviation of the paired difference s_D:

$$d_{s,pair} = (\bar{x}_2 - \bar{x}_1)/s_D$$

where $s_D = \sqrt{(s_1^2 + s_2^2)/2}$. Accounting for the pairing between differences reduces the variation. Therefore s_D will be smaller, estimates more precise, and effect sizes larger, than if the standard deviation was calculated assuming independence of observations.

15.4 r Family (Strength of Association) Effect Sizes

The r family effect sizes describe the relationship between variables as ratios of shared variance. They are used for correlation, regression, and analysis of variance (ANOVA) models (Box 15.4). The total variance of the data is partitioned into several different pieces or *sources of variation*.

15.4.1 Correlation

The most familiar metric from the r family is Pearson's correlation coefficient r. Correlation is a measure of the strength of linear association between variables. It is estimated as the ratio of shared variance between two variables:

$$r_{x,y} = \frac{cov(x,y)}{s_x s_y}$$

where s_x and s_y are the sample standard deviations for each variable, and the covariance between two sample variables x and y is

$$cov(x,y) = \frac{\sum (X_i - \bar{X})(Y_j - \bar{Y})}{n-1}$$

BOX 15.4
r *Family Effect Sizes*

Effect sizes that describe the relationship between variables as ratios of shared variance. (strength of association)

The *effect size* is specified by the difference between the correlation expected for the population for the null r_0 and the correlation r hypothesised under the alternative hypothesis (the postulated effect of the test intervention), such that

$$f = r - r_0$$

15.4.2 Regression

The *linear regression model* describing the relationship of Y in relation to X is

$$Y = \beta_0 + \beta_1 X + \varepsilon$$

where β_0 is the intercept, β_1 is the change in Y with a unit change in X (the slope) and ε is the random error, which is normally distributed with mean zero and variance σ^2.

Effect size. For regression models, the effect size can be quantified by the regression coefficient, or the coefficient of determination R^2.

Regression coefficients β provide the most immediately obvious information about the magnitude of the effect. In the simplest case with two treatment groups and no other predictors, the groups can be defined by a single indicator variable such that $X = 0$ for the comparator and $X = 1$ for the test group. Then the effect d is the estimate of the true difference in means between the two groups: $d = \beta_1$. The standardised effect size is the regression coefficient divided by the pooled within-group standard deviation of the outcome measure (Feingold 2015). The residual standard deviation (root mean square error) of the regression is used when there are two or more predictors. However, the estimated size of the effect for models with multiple predictors will depend on the scale of each independent variable. Therefore effect sizes based on different variables in the same study, or variables across multiple studies, cannot be compared directly (Lorah 2018).

Regression models have at least three sources of variation: the *total sum of squares*, the *regression sum of squares*, and the *error sum of squares*. The structure of the model and how variance is partitioned are shown by constructing an ANOVA table (Table 15.2).

Table 15.2: Analysis of variance table for the simple linear regression with one predictor variable X.

Source of variation	Degrees of freedom (df)	Sum of squares (SS)	Mean square (MS)	F
Due to explanatory variables (regression)	1	SSR	MSR = SSR/1	MSR/MSE
Residual	$N-2$	SSE = SST_c – SSR	MSE = SSE/$(n-2) = \sigma^2$	
Total variation corrected for mean	$N-1$	SST_c		
Total variation	N	SST		

The *total sum of squares (SST)* is the sum of squared deviations of observations y_i from the mean of those observations $\bar{y}$.

The *regression sum of squares (RSS)* describes the amount of variation among the y values that can be attributed to the linear model. The least squares estimate for the coefficient β_1 is calculated as

$$\hat{\beta}_1 = \frac{\sum (x_i - \bar{x})\left(y_j - \bar{y}\right)}{\sum (x_i - \bar{x})^2}$$

and the SS_{reg} as $\hat{\beta}_1 \frac{\sum (x_i - \bar{x})(y_j - \bar{y})}{\sum (x_i - \bar{x})^2}$ with one degree of freedom. The *regression mean square* MS_{reg} is $SS_{reg}/1 = SS_{reg}$.

The *error sum of squares (SSE)* is the difference between the SST and RSS. The residual mean square, or *mean square error (MSE)*, is the estimate of the variance s^2, with $n-2$ degrees of freedom (the loss of 2 df is because two parameters β_0 and β_1 are estimated). It is a measure of the amount of variation attributable to the regression model.

The *F-statistic* tests the null hypothesis that $\hat{\beta}_1 = 0$ and is calculated as the ratio of MSR/MSE.

The *coefficient of determination* R^2 is the proportion of variance explained by the linear relation of Y and X:

$$R^2 = 1 - \frac{SSE}{SST}$$

where SST is the total sum of squares, and SSE is the error sum of squares, or the residual amount remaining after subtracting the contribution made by the linear fit.

Correlation r is related to the slope b_1 as

$$r = b_1 \cdot \frac{s_x}{s_y}$$

The *adjusted R^2* $\left(R^2_{adj}\right)$ is a corrected goodness-of-fit metric that adjusts for the number of terms in the model:

$$R^2_{adj} = 1 - \left(1 - R^2\right)\frac{N-1}{N-k}$$

The unadjusted R^2 increases as more predictor terms are added to the model, regardless of their predictive or explanatory effect. In contrast, R^2_{adj} will increase only as useful predictors are added and will decrease if variables with no predictive value are added. R^2_{adj} will always be less than or equal to the unadjusted R^2. Therefore, R^2_{adj} is useful for assessing if the appropriate variables are included in the model (Draper and Smith 1998).

The correlation r (the shared variation between Y and X) is not interchangeable with the unadjusted R^2 (the variation in Y attributable to the linear variation in X) as a measure of effect. The square root of the unadjusted R^2 is a biased measure of effect, especially when the sample size is small (Nakagawa and Cuthill 2007). The square root of the adjusted R^2, $r_{adj} = \sqrt{R^2_{adj}}$ is the appropriate effect size measure if correlation is required (Nakagawa and Cuthill 2007).

The *effect size f^2* for the fixed effects model is obtained from the adjusted R^2 as follows:

$$f^2 = \frac{R^2_{adj}}{1 - R^2_{adj}}$$

15.4.3 *Analysis of Variance (ANOVA) Methods*

ANOVA models belong to the *r* family and are mathematically equivalent to regression. The *F-statistic* is the ratio of the estimate of the between-groups, or treatment, mean square to the error mean square. The single-factor ANOVA partitions variation into between-group variation with $k-1$ degrees of freedom (where k is the number of groups) and residual (within group, or error) variation with $k(n-1)$ degrees of freedom. Between-group variation assesses treatment effects.

Effect size. The effect size f for ANOVA designs is

$$f = \sqrt{\eta^2/(1-\eta^2)}.$$

where η^2 is the *correlation ratio*. The correlation ratio η^2 is analogous to R^2 and for a single-factor ANOVA, $\eta_p^2 = R^2$. It describes the proportion of the total variation (total sum of squares SS_{Total}) in the response variable accounted for by differences between groups (between-group sum of squares, SS_{betw}):

$$\eta^2 = \frac{SS_{betw}}{SS_{Total}}$$

For example, η^2 of 0.2 means that 20% of the total variance can be attributed to group membership. It is the uncorrected effect size estimate determined by the variance explained by group membership in the sample for a single study (Lakens 2013). However, studies based on η^2 will be considerably underpowered; several authors recommend strongly against using η^2 as an ANOVA effect size estimator (Troncoso Skidmore and Thompson 2013; Albers and Lakens 2018). If there are only two independent groups, f is related to Cohen's d_s as $f = d_s/2$.

The *partial correlation ratio* η_p^2 is the 'standardised' effect size. It is appropriate only for fixed factor designs without covariates. Software programs such as G*Power require η_p^2 for sample size calculations. It is calculated as

$$\eta_p^2 = \frac{SS_{betw}}{SS_{betw} + SS_{error}} = \frac{SS_{betw}}{SS_{total}}$$

The partial correlation ratio η_p^2 can also be estimated from existing or exemplary *F*-values and degrees of freedom (*df*) as

$$\eta_p^2 = \frac{F \cdot df_{num}}{F \cdot df_{num} + df_{denom}}$$

where df_{num} is the numerator, or between-groups, degrees of freedom, and df_{denom} is the denominator or error degrees of freedom (Cohen 1988).

A major drawback to both η^2 and η_p^2 is that they are biased. Their application in power calculations can lead to underpowered studies because the sample size estimates will be too small. Estimates of η_p^2 become more complicated to calculate as study design becomes more complex because η_p^2 is determined by the different sources of variation contributed by additional factors, factor interactions, nesting, and blocking (Lakens 2013). η_p^2 can be used for comparisons of effect sizes across studies, but only if study designs are similar and there are no covariates or blocking variables (Lakens 2013). When reporting effect sizes for ANOVA, it is recommended to report both the generalized η^2 and the standardised η_p^2.

When sample size is small, the partial omega-squared ω_p^2 is a less-biased alternative to η^2 and η_p^2 (Olejnik and Algina 2003; Albers and Lakens 2018). It estimates the proportion of variance in the response variables accounted for by the explanatory variables. For the completely randomised single-actor ANOVA, ω_p^2 can be calculated from the *F*-statistic and associated degrees of freedom for the treatment and error:

$$\omega_p^2 = \frac{F-1}{\left[F + \frac{(df_{denom}+1)}{df_{num}}\right]}$$

Warning: This formula cannot be used for repeated measure designs. Neither η^2 or ω_p^2 effect sizes are applicable to ANOVA with observational categorical explanatory variables (e.g. sex) or if covariates are included.

15.5 Risk Family Effect Sizes

When outcome measures are binary (0/1; yes/no; present/absent), the effect size is assessed as a comparison between proportions. It is common in epidemiological investigations for effect sizes for proportions to be expressed as *risk difference, relative risk* (RR), or the *odds ratio* (OR) (Sánchez-Meca et al. 2003, Machin et al. 2018). Cramer's *V* is a measure of the strength of association between two nominal variables. It is typically obtained from χ^2 contingency tables (Box 15.5).

15.5.1 Risk Difference, Relative Risk, and Odds Ratio

Table 15.3 shows how proportions are structured for comparisons between the test group and a comparator, or standard. Sample SAS code for calculating RR and OR is provided in Appendix 15.A.

Effect size metrics are calculated as follows:

The *risk difference* or *absolute risk reduction ARR*, is the difference between proportions:

$$ARR = b/n_1 - a/n_0 = p_1 - p_0$$

If the absolute risk difference is zero, this indicates there is no effect.

BOX 15.5
Risk Family Effect Sizes

Effects sizes for binary outcomes: *risk difference, relative risk, odds ratio.*

Effect size for nominal variables: *Cramer's V.*

The number needed to treat (NNTT) is the reciprocal of the risk difference:

$$NNTT = 1/ARR$$

This is an estimate of the number of subjects that need to be treated for one extra subject to benefit.

The *RR* or *risk ratio,* is the ratio of two proportions. It compares the probability, or risk, of an event occurring in one group in one group relative to that for a comparator group:

$$RR = (b/n_1)/(a/n_0) = p_1/p_0$$

When *RR* is 1, risk for the two groups is the same; that is, an *RR* of 1 indicates no effect. If $RR > 1$, there is increased risk in group 1 relative to group 2, and $RR < 1$ risk is decreased in group 1 relative to group 2. Therefore an *RR* of 0.5 means that the risk has been reduced by one-half or 50%. Confidence intervals should be reported to obtain an idea of the precision of the estimate.

There are two possible ways to compute RR depending on whether the presence or absence of an event is of interest. Scientific or clinical judgment is required to determine which RR is appropriate, as they are not simple reciprocals of one another.

The OR is the ratio of the odds of occurrence of an event (or 'success') p relative to the odds of failure $(1 - p)$ in a test group relative to those for a comparator group:

$$OR = \frac{p_1/(1-p_1)}{p_0/(1-p_0)}$$

In case–control studies, the OR is the measure of association between an exposure or risk factor and occurrence of disease.

Table 15.3: Calculation of proportions for comparisons of a test group against a control.

Group	Event 'yes' ('success')	Event 'no' ('failure')	Total	Proportion of events ('successes')
Comparator group ('standard')	a	c	$a + c = n_0$	$p_0 = a/n_0$
Test group	b	d	$b + d = n_1$	$p_1 = b/n_1$
Total	$a + b$	$c + d$	$n_0 + n_1 = N$	

When OR is 1, odds associated with success are similar in the two groups; that is, an OR of 1 indicates no effect. If the OR > 1 odds are increased in group 1 relative to group 2, and OR < 1 odds are decreased in group 1 relative to group 2. Rule of thumb values for OR effect sizes has been reported as 'small' for OR = 1.68, 'medium' for OR = 3.47, and 'large' for OR = 6.71 (Chen et al. 2010). If the ORs are obtained from logistic regression coefficients, they must be standardised using the root mean square before use. ORs obtained from logistic regression are usually unstandardised, as they depend on the scale of the predictor.

15.5.2 Interpretation

Both RR and OR are relative measures of effect, and therefore unaffected by changes in baseline risk. However, proper interpretation requires that both raw numbers and baseline risk should be reported. Absolute changes in risk are more important than relative changes in risk. Sensible interpretation requires consideration of both RR and OR as ratio problems; that is, assessment of magnitudes is based on reciprocals, not simple arithmetic differences.

When the outcome event is rare, RR and OR will be approximately equal. This is because both $(1 - p_1)$ and $(1 - p_2)$ will be close to 1, so OR will approach RR. As a result, OR can be regarded as an estimate of RR when incidence of an event is low in both groups (usually <10%). As incidence or baseline risk increases, RR and OR become more dissimilar. If there are zero events in the comparator group $(p_0 = 0)$, neither the RR or OR can be calculated (because the denominator is zero). If the occurrence of the event in the intervention group is 100% ($p_1 = 1$), OR cannot be calculated.

To more readily interpret OR as a change in number of events, Higgins et al. (2022) recommend converting the OR to RR and then interpreting RR relative to a range of assumed comparator or baseline group risks (ACR):

$$RR = \frac{OR}{1 - ACR(1 - OR)}$$

RR is more easily understood than OR. Disadvantages of OR include overestimation of risk, and

difficulties with interpretation, especially for assessing benefits and harms. For example, an OR = 0.20 cannot be interpreted as an 80% reduction in risk. It means that the odds of the event occurring in the first group (e.g. the test group) are 0.2 times the odds in the second group (e.g. the control). However, OR has several advantages over RR. ORs are *symmetrical* with respect to the outcome definition; that is, interpretation does not depend on which group is used as the numerator or denominator, or whether the emphasis is on the occurrence of an event or its failure to occur. In contrast, interpretation of the RR does depend on outcome definition. Because choice of the reference comparison can greatly affect interpretation, it is important to clearly identify which group is the 'control' or reference comparator group and whether that group is the numerator or denominator of the ratio. For example, Cochrane Reviews use *OR* < 1 as favouring the treatment group, not the control (Higgins et al. 2022).

For observational studies, the study design (e.g. case-control or cohort studies), sampling strategy, case type (incident or prevalent) and source population (fixed or dynamic) will determine which measure of association can be used (Knol et al. 2008, Labrecque et al. 2021).

Example: Survival or Mortality: Which?

A study assessed the number of subjects that survived or died following a course of experimental treatment A relative to control treatment B. Results were as follows:

Intervention	Alive	Dead	Total
Group A	75	25	100
Group B	50	50	100

The *RR* of *death* for group B relative to group A is

$$RR = \frac{50/100}{25/100} = \frac{0.50}{0.25} = 2.0$$

However, the *RR* of *survival* is

$$RR = \frac{50/100}{75/100} = \frac{0.50}{0.75} = 0.67$$

When expressed as RR, survival and mortality are not reciprocals of one another and not substitutable.

In contrast, the *OR* of *death* for group B relative to group A is

$$OR = \frac{(50/50)}{25/75} = \frac{1.0}{0.333} = 3.0$$

The *OR* of *survival* for group B relative to group A is

$$OR = \frac{(50/50)}{75/25} = \frac{1.0}{3} = 0.333$$

The OR of survival is the reciprocal of OR for mortality. OR does not depend on whether the emphasis is on occurrence of an event or its failure to occur.

15.5.3 Nominal Variables: Cramer's *V*

Cramer's *V* measures the strength of association between two nominal variables (categorical variables with no natural ordering or ranking). It is calculated as:

$$V = \sqrt{\frac{\chi^2/N}{\min(c-1, r-1)}}$$

where N is total sample size, c is the number of columns, r is the number of rows, and χ^2 is the chi-squared statistic. V ranges between 0 and 1, with 0 indicating no association and 1 indicating a strong association.

15.6 Time to Event Effect Size

Effect size for time-to-event data is most appropriately expressed as a *hazard ratio* (HR). The HR accounts for both the number and the timing of

---

BOX 15.6
Time to Event Effect Size

Hazard ratio
Odds ratio

events and adjusts for *censoring* of all subjects which did not have the event (Tierney et al. 2007). When 'survival' is measured as counts or proportions, the effect size can be estimated as an OR. However, the OR provides information only on the number of events per group and does not account for timing or duration of events (Box 15.6).

The HR is the ratio of the 'risks' (or proportion of events) in the control and the test groups and may be expressed either as the ratio of events p in each group or the ratio of median survival times M:

$$\text{HR} = \frac{h_1(t)}{h_0(t)} = \frac{\ln(p_1)}{\ln(p_0)} = \frac{M_0}{M_1}$$

HRs are interpreted similarly to RR: that is, *HR* of 1 indicates no effect, $HR < 1$ indicates the chance of the event is reduced, and $HR > 1$ that the chance of the event occurrence is increased. For example, $HR = 0.75$ indicates the chance of having an event is reduced by 25%. HR will always be a more reliable effect size than *RR* because of the incorporation of time effects. If *HR* and *RR* in a study are very similar, this may indicate that time effects are not important (Machin et al. 2006; Machin et al. 2018; Abberbock et al. 2019). Tierney et al. (2007) describe methods for obtaining HRs from published summary data.

15.7 Interpreting Effect Sizes

... the reader is counseled to avoid the use of these conventions, if he can, in favour *of exact values provided by theory or experience in the specific area in which he is working.* (Cohen 1988; italics mine)

Various rule-of-thumb benchmark values have been proposed to indicate how 'meaningful' effect size is, usually in subjective terms such as 'small',

BOX 15.7
Interpreting Effect Size

Unless you know what the effect size index is and how it was calculated, you cannot interpret it.

Benchmark effect sizes should be avoided because they are

- Irrelevant to animal-based research
- Too strict
- Subject to technical problems
- Easily gamed

BOX 15.8
Determining and Interpreting Effect Sizes

- Do not use subjective effect size benchmarks ('small', 'medium', 'large') for planning studies or interpreting results.
- The magnitude of the effect size may be determined more by study design and variation than the anticipated difference in intervention groups.
- Consider effect sizes like any other ratio (both numerator AND denominator are important).

'medium', and 'large'. Benchmarks may be useful as first-pass sample size approximations for study planning purposes. They can also be useful for assessing results in comparison results already well-understood or validated. *However, in most circumstances, effects sizes should be directly calculated from the individual components* (Box 15.7). Predetermined ('canned') benchmarks should be avoided for the following reasons (Box 15.8):

Predetermined effect sizes are irrelevant to animal-based studies. By far, the leading objection to benchmark criteria is that they were originally developed for social and psychology studies and are meaningless without a practical frame of reference (Thompson 2002). Therefore, the criteria defining the 'size' of an effect may be neither appropriate nor relevant for pre-clinical or veterinary clinical studies.

Predetermined effect sizes are too strict. This was recognised by Cohen himself (Cohen 1988) and others (Thompson 2002, Funder and Ozer 2019). Effect size must always be determined and interpreted by study context and practical significance. For example, an effect size $R^2 = 0.1$ in large clinical trial with death and major morbidity as a primary outcome is of far greater practical significance than it would be for a small laboratory study measuring weakly validated biomarker concentrations or behaviour characterised by large between- and within-subject variability.

Predetermined effect sizes have numerous technical problems. The magnitude of an effect (therefore any resulting interpretation of its biological importance) will be affected by presence of covariates, data with non-normal error structure and/or variances, non-independence of observations, and small sample sizes (Nakagawa and Cuthill 2007). Frequently, conversion of one type of effect size metric to another can result in discordant and nonsensical benchmark values, for example conversion of *d* effect sizes to *r* (Lakens 2013). Sampling distributions will generally be non-normal and highly skewed, so the smaller the sample size, the less reliable the effect size estimate will be (Albers and Lakens 2018).

Predetermined effect sizes are subject to 'gaming'. An unfortunately common practice among investigators is manipulating effect size calculations to obtain a preferred sample size. This will often result in estimates of implausibly large effect sizes so that future studies will be extremely underpowered for more biologically realistic and plausible effect sizes.

15.7.1 Interpreting Effect Sizes as Ratios

Cumming (2012) points out that sensible construction and interpretation of effect sizes should consider effect size as a ratio problem. That is, both the numerator and denominator are important and should be considered equally in the interpretation of effect size. Two studies with the same effect sizes may differ in the magnitude of the difference between groups (numerator), the amount of variation in the two samples (denominator), or both. If variation between experimental units is small, then the denominator is small, resulting in large effect sizes. This may conceal the fact that differences between groups are trivial and of no practical importance. Conversely, large standardised effect sizes may be a reflection of better study design (control of variation), not larger effects *per se* (differences between groups). Neglect of basic design features such as ensuring quality of data sampling, collection, and measurement, and especially incorporation of allocation concealment (or blinding) can also account for non-trivial effect sizes (Rosenthal 1994, Funder and Ozer 2019). For example, a meta-investigation of mouse models of amyotrophic lateral sclerosis provided strong evidence that unattributed sources of variation (such as litter effects) were major contributors to apparent therapeutic efficacy of candidate drugs, rather than true treatment effects (Scott et al. 2008; Perrin 2014).

15.7.2 What Is a Meaningful Effect Size?

A 'meaningful' effect size (Box 15.8) is described by the effect size estimate together with a measure of precision (such as 95% confidence intervals). Meaning is context-dependent and is determined by model validity, goals of the intervention, predefined criteria for what constitutes biological or clinical 'importance' (Kazdin 1999; Thompson 2002; Nakagawa and Cuthill 2007), and evaluation in comparison with estimated effect size in comparison to values reported for other similar studies (Schuele and Justice 2006). Unfortunately, discussions of research results are often framed in terms of 'statistical significance' and *P*-values alone, which do not convey much useful scientific information (Cohen 1990; Thompson 2002; Nakagawa and Cuthill 2007; Cumming 2012; Kelley and Preacher 2012; Schober et al. 2018). 'Statistical significance' does not mean that the observed difference is large enough to be of any practical or biological importance. Common misunderstanding of *P*-values and

significance has been discussed in detail elsewhere (Greenland et al. 2016).

Effect sizes and 95% confidence intervals in original measurement units have the advantage of being more readily interpretable compared to standardised effect sizes. Reporting of effect sizes and measures of precision, whether or not they are 'statistically significant', have the further advantage of allowing inclusion in systematic reviews and meta-analyses (Nakagawa and Cuthill 2007). Systematic reviews and meta-analyses increase data 'shelf life', and minimise waste of animals and resources in additional non-informative experiments (Reynolds and Garvan 2020).

15.A Using SAS *Proc Genmod* to Calculate OR and RR (Adapted from Spiegelman and Hertzmark 2005). As an alternative to fitting a logistic regression model, OR and RR can be calculated using the log-binomial maximum likelihood estimators.

```
*to calculate OR, use the logit link;
proc genmod descending;
 class x;
 model y = x / dist = binomial link = logit;
estimate 'Beta' x 1 -1/ exp;
run;

*to calculate RR use the log link;
proc genmod descending;
 class x;
 model y = x/ dist = binomial link = log;
 estimate 'Beta' x 1 -1/ exp;
run;
```

References

Abberbock, J., Anderson, S., Rastogi, P., and Tang, G. (2019). Assessment of effect size and power for survival analysis through a binary surrogate endpoint in clinical trials. *Statistics in Medicine* 38: 301–314. https://doi.org/10.1002/sim.7981.

Albers, C. and Lakens, D. (2018). When power analyses based on pilot data are biased: inaccurate effect size estimators and follow-up bias. *Journal of Experimental Social Psychology* 74: 187–195. https://doi.org/10.1016/j.jesp.2017.09.004.

Chen, H., Cohen, P., and Chen, S. (2010). How big is a big odds ratio? Interpreting the magnitudes of odds ratios in epidemiological studies. *Communications in Statistics: Simulation and Computation* 39: 860–864.

Cohen, J. (1988). *Statistical Power Analysis for the Behavioral Sciences*. Hillsdale: Lawrence Erlbaum Associates.

Cohen, J. (1990). Things I have learned (so far). *American Psychologist* 45 (12): 1304–1312. https://doi.org/10.1037/0003-06615.45.12.1304.

Cumming, G. (2012). *Understanding the New Statistics: Effect sizes, Confidence Intervals, and Meta-Analysis*. New York: Routledge.

Draper, N., and Smith, H. (1998). *Applied Regression Analysis*, 3rd ed. New York: Wiley.

Feingold, A. (2015). Confidence interval estimation for standardized effect sizes in multilevel and latent growth modeling. *Journal of Consulting and Clinical Psychology* 83 (1): 157–168. https://doi.org/10.1037/a0037721.

Ferguson, C.J. (2009). An effect size primer: a guide for clinicians and researchers. *Professional Psychology: Research and Practice* 40 (5): 532–538.

Funder, D.C. and Ozer, D.J. (2019). Evaluating effect size in psychological research: sense and nonsense. *Advances in Methods and Practices in Psychological Science* 2: 156–168.

Glass, G.V. (1976). Primary, secondary, and meta-analysis of research. *Educational Researcher* 5: 3–8. https://doi.org/10.2307/1174772.

Greenland, S., Senn, S.J., Rothman, K.J. et al. (2016). Statistical tests, *P* values, confidence intervals, and power: a guide to misinterpretations. *European Journal of Epidemiology* 31: 337–350. https://doi.org/10.1007/s10654-016-0149-3.

Hedges, L.V. and Olkin, I. (1985). *Statistical Methods for meta-analysis*. San Diego, CA: Academic Press.

Higgins, J.P.T., Li, T., and Deeks, J.J. (2022). Chapter 6: Choosing effect measures and computing estimates of effect. In: *Cochrane Handbook for Systematic Reviews of Interventions version 6.3 (Updated February 2022)* (ed. H. JPT, J. Thomas, J. Chandler, et al.). Cochrane Available from www.training.cochrane.org/handbook.

Hojat, M. and Xu, G. (2004). A visitor's guide to effect sizes – statistical significance versus practical (clinical) importance of research findings. *Advances in Health Sciences Education* 9 (3): 241–249.

Kazdin, A.E. (1999). The meanings and measurement of clinical significance. *Journal of Consulting and Clinical Psychology* 67: 332–339.

Kelley, K. and Preacher, K.J. (2012). On effect size. *Psychological Methods* 17 (2): 137–152. https://doi.org/10.1037/a0028086.

Knol, M.J., Vandenbroucke, J.P., Scott, P. et al. (2008). What do case-control studies estimate? Survey of methods and assumptions in published case-control research. *American Journal of Epidemiology* 168 (9): 1073–1081.

Labrecque, J.A., Hunink, M.M.G., Ikram, M.A., and Ikram, M.K. (2021). Do case-control studies always estimate odds ratios? *American Journal of Epidemiology* 190 (2): 318–321. https://doi.org/10.1093/aje/kwaa167.

Lakens, D. (2013). Calculating and reporting effect sizes to facilitate cumulative science: a practical primer for t-tests and ANOVAs. *Frontiers in Psychology* 2013: 4. https://doi.org/10.3389/fpsyg.2013.00863.

Lorah, J. (2018). Effect size measures for multilevel models: definition, interpretation, and TIMSS example. *Large-Scale Assessments in Education* 6: 8. https://doi.org/10.1186/s40536-018-0061-2.

Machin, D., Cheung, Y.B., and Parmar, M. (2006) *Survival Analysis: A Practical Approach* 2nd edition, Wiley.

Machin, D., Campbell, M.J., Tan, S.B., and Tan, S.H. (2018) *Sample Sizes for Clinical, Laboratory and Epidemiology Studies*. 4th Ed. Wiley.

Nakagawa, S. and Cuthill, I.C. (2007). Effect size, confidence interval and statistical significance: a practical guide for biologists. *Biological Reviews of the Cambridge Philosophical Society* 82 (4): 591–605. https://doi.org/10.1111/j.1469-18515.2007.00027.15. Erratum in: *Biol Rev Camb Philos Soc.* 2009 84(3):515.

Olejnik, S. and Algina, J. (2003). Generalized eta and omega squared statistics: measures of effect size for some common research designs. *Psychological Methods* 8 (4): 434–447.

Perrin, S. (2014). Make mouse studies work. *Nature* 507: 423–425.

Reynolds, P.S. and Garvan, C.S. (2020). Gap analysis of swine-based hemostasis research: "Houses of brick or mansions of straw?". *Military Medicine* 185 (Suppl 1): 88–95. https://doi.org/10.1093/milmed/usz249.

Rosenthal, R. (1994). Parametric measures of effect size. In: *The Handbook of Research Synthesis* (ed. H. Cooper and L.V. Hedges), 231–244. New York: Sage.

Sánchez-Meca, J., Marín-Martínez, F., and Chacón-Moscoso, S. (2003). Effect-size indices for dichotomized outcomes in meta-analysis. *Psychological Methods* 8 (4): 448–467. https://doi.org/10.1037/1082-98915.8.4.448.

Schober, P., Bossers, S.M., and Schwarte, L.A. (2018). Statistical significance versus clinical importance of observed effect sizes: what do *p* values and confidence intervals really represent? *Anesthesia and Analgesia* 126 (3): 1068–1072. https://doi.org/10.1213/ANE.0000000000002798.

Schuele, C.M. and Justice, L.M. (2006). The importance of effect sizes in the interpretation of research: *primer on research*: part 3. *The ASHA Leader.* 11 (10): https://doi.org/10.1044/leader.FTR4.11102006.14.

Scott, S., Kranz, J.E., Cole, J. et al. (2008). Design, power, and interpretation of studies in the standard murine model of ALS. *Amyotrophic Lateral Sclerosis* 9: 4–15.

Spiegelman, D. and Hertzmark, E. (2005). Easy SAS calculations for risk or prevalence ratios and differences. *American Journal of Epidemiology* 162: 199–205.

Sun, S., Pan, W., and Wang, L.L. (2010). A comprehensive review of effect size reporting and interpreting practices in academic journals in education and psychology. *Journal of Educational Psychology* 102 (4): 989–1004.

Thompson, B. (2002). Statistical, practical, and clinical: how many kinds of significance do counselors need to consider? *Journal of Counseling and Development* 80 (1): 64–71.

Tierney, J.F., Stewart, L.A., Ghersi, D. et al. (2007). Practical methods for incorporating summary time-to-event data into meta-analysis. *Trials* 8: 16. https://doi.org/10.1186/1745-6215-8-16.

Troncoso Skidmore, S. and Thompson, B. (2013). Bias and precision of some classical ANOVA effect sizes when assumptions are violated. *Behavior Research Methods* 45: 536–546. https://doi.org/10.3758/s13428-012-0257-2.

van Belle, G.G. (2008). *Statistical Rules of Thumb*, 2nd edition. New York: Wiley.

Vedula, S.S. and Altman, D.G. (2010). Effect size estimation as an essential component of statistical analysis. *Archives of Surgery* 145 (4): 401–402. https://doi.org/10.1001/archsurg.2010.33.

Zwetsloot, P.P., Van Der Naald, M., Sena, E.S. et al. (2017). Standardized mean differences cause funnel plot distortion in publication bias assessments. *eLife* 6: e24260. https://doi.org/10.7554/eLife.24260.

16 Comparing Two Groups: Continuous Outcomes

CHAPTER OUTLINE HEAD

16.1 Introduction

The simplest experimental design is for comparison of two groups (Box 16.1). In general, these designs involve a test intervention and a control, randomly allocated to experimental units in each group. Due to their simple structure, two-arm designs are useful for definitive veterinary clinical trials, which require large sample sizes and sufficient power for testing efficacy. However,

---

BOX 16.1
Two-Group Designs

The simplest between-group design usually comparing a test intervention against a control.

Two-arm designs are powerful tools for assessing efficacy, especially for veterinary clinical trials.

Two-arm are not recommended for exploratory studies with multiple variables; *t*-tests are commonly used to compare continuous outcomes for two groups.

A Guide to Sample Size for Animal-based Studies, First Edition. Penny S. Reynolds.
© 2024 John Wiley & Sons Ltd. Published 2024 by John Wiley & Sons Ltd.

two-arm designs are not the best choice for most laboratory rodent studies, which are mostly exploratory with multiple explanatory variables. Alternative designs are discussed in Chapter 19.

When outcome variables are continuous and normally distributed, sample size can be approximated by formulae based on the t or the z distribution. The t-distribution describes the standardised distances of sample means to the population mean. The t-statistic is based on the sample variance. The shape of a t-distribution depends on the degrees of freedom (df), where $df = n - 1$, so the critical t-value t_{crit} is given by $t_{\alpha/2,\,n-1}$ for a two-sided test and $t_{\alpha,\,n-1}$ for a one-sided test. In contrast, the approximation based on the z-distribution assumes the population variance σ is known, and the population is normally distributed. The z-distribution can be used when sample sizes n are large ($n > 30$) because as n increases, the t-distribution approaches the z-distribution. However, the large-scale asymptotic approximation based on the z-distribution should be used with caution. Although sample size estimates may useful for initial planning, sample sizes will be seriously underestimated compared to exact determinations from the t-distribution. A drawback of the exact method is that it must be determined by numerical iteration over a candidate range of n, although this is readily performed with simple computer code. Sample SAS codes are given in Appendix 16.A.

16.2 Sample Size Calculation Methods

Sample size determinations for two-group comparisons require the minimum information for sample size planning (Chapter 14): pre-specified confidence and power, and the minimum biologically relevant difference to be detected. In addition, information is required as to directionality (if the comparison is one-sided or two-sided, as this determines α or $\alpha/2$, respectively), and the type of comparison to be made. These include a single sample compared to a reference value, comparison of two independent samples, or comparison of paired observations on the same subject. The type of comparison will determine how the sample standard deviation is calculated and therefore affect estimates of the effect size.

Commonly, sample sizes for two-group comparisons are computed to detect a pre-specified biologically relevant difference d with a given power

and confidence. However, sample sizes can also be determined to obtain a pre-specified precision or tolerance for the difference, or for a range of sample variation that can detect the pre-specified difference with a given power.

16.2.1 Asymptotic Large-Scale Approximation

Sample size based on the z-distribution for a two-group comparison is:

$$N = \frac{\left(z_{1-\alpha/2} + z_{1-\beta}\right)^2}{(d/s)^2}$$

where the difference d is the difference between sample means ($d = \bar{x}_1 - \bar{x}_2$), s is the sample standard deviation, d/s is the effect size, and $z_{1-\alpha/2}$ and $z_{1-\beta}$ are the z-scores for confidence and power, respectively.

To detect an *increase in the average by one standard deviation*, the relation becomes

$$N = \frac{\left(z_{1-\alpha/2} + z_{1-\beta}\right)^2}{(1/1)^2} = \left(z_{1-\alpha/2} + z_{1-\beta}\right)^2$$

or the square of the summed z-scores for confidence and power.

16.2.2 Sample Size Based on the t-Distribution

When the study is small, estimating sample size directly from the t-distribution will provide more exact approximations. The sample size calculation for a two-sided comparison of two samples is

$$N = t_{\alpha/2,n-1}^2/(d/s)^2$$

where t is the critical t-value for confidence $\alpha/2$ and $n - 1$ degrees of freedom (df).

The degrees of freedom that define the t-distribution is calculated from the total sample size itself. Therefore, sample size has to be estimated by iterating over a range of candidate sample sizes, and the final sample size is obtained by solving for β:

$$1 - \beta = tinv\left(t_{\alpha/2,n-1} \cdot \frac{d\sqrt{n}}{s}\right)$$

The inverse function *tinv* computes the pth quantile from the t-distribution, with degrees of freedom df

and non-centrality parameter $\lambda = d/s$. Because $t^2 = F$, the critical value for the test statistic can be computed from either the non-central t or F-distribution (Chapter 14).

16.2.3 Sample Size Derived From Percentage Change in Means

Investigators sometimes define a 'minimum clinically or biologically important difference' between two groups as a percentage change in mean values. A 20% difference between the control group and the intervention group as an indication that the test intervention 'works' is a popular rule of thumb, although there may be no particular clinical or biological rationale for this number. However, use of percentage differences can be used for crude sample size estimates when few data are available.

The difference between two group means $(\bar{x}_0 - \bar{x}_1)$ expressed as a percentage change, is

$$d = (\bar{x}_0 - \bar{x}_1)/\bar{x}_0$$

This is rearranged to express the ratio of the means:

$$d = 1 - (\bar{x}_1/\bar{x}_0) = (1 - p).$$

where p is the per cent (or proportion) change.

Estimates of the sample variation are obtained from the coefficient of variation (CV). The CV is the ratio of the standard deviation (s) to the mean $(\bar{x})$, so that $CV = s/\bar{x}$ (and so can be thought of as the inverse of a simple standardised effect size).

Using a modification of Lehr's formula (Lehr 1992; van Belle, 2008 and assuming equal CVs for the two groups, $s_0/\bar{x}_0 = s_1/\bar{x}_1$, then sample size is

$$n = \frac{16 \cdot (CV)^2}{(\ln \bar{x}_0 - \ln \bar{x}_1)^2} = \frac{16 \cdot (CV)^2}{\ln (1 - p)^2}$$

where n is the sample size per group. The ln-transformation is used to stabilise the variance because the CV assumes the standard deviation is proportional to the mean.

The advantage of this method for preliminary planning is that it is simple and does not require much detailed information. However, this method should not substitute for formal sample size calculations and without careful consideration of the specifics for the planned study.

Example: Mouse Two-Arm Drug Study: Percent Difference

An investigator wished to obtain a preliminary sample size estimate for a two-arm study in mice testing a new drug against an inert control. Based on literature data, the investigator thought that a 20% improvement in response in the drug group compared to control would be clinically meaningful and that CV would be between 20% and 50% of the mean.

The desired per cent change difference is $(1 - 0.2) = 0.8$, and the ratio of CV to the mean is 0.3/1–0.5/1. Then

If $CV = 20\%, n = 16(0.20)^2/[\ln(1 - 0.20)]^2 \cong 13$

If $CV = 50\%, n = 16(0.50)^2/[\ln(1 - 0.20)]^2 \cong 80.$

The investigator will require ~13–80 subjects per group to detect a 20% difference in the group means, depending on the amount of variation in the sample. It is apparent that if the response is expected to be highly variable (and hence less precise), more subjects will be needed.

16.2.4 Sample Size Rule of Thumb

A crude estimate for sample size can be obtained from the standardised effect size for the difference between two groups (van Belle 2008) as

$$N = \left(\frac{4 \cdot s}{d}\right)^2 = \frac{16}{(d/s)^2}$$

and for a single sample against a reference value as

$$N = \left(\frac{2 \cdot s}{d}\right)^2 = \frac{4}{(d/s)^2}$$

16.3 Which Standard Deviation?

Solving for sample size requires an estimate of effect size. Effect size for two groups is measured by the standardised difference between the two means,

d/s where d is the difference between means $(\bar{x}_1 - \bar{x}_2)$, and s is the sample standard deviation. However, estimates of s will be determined by whether the comparison is between a single sample and a reference value, a comparison of two independent groups, or a comparison of paired observations.

16.3.1 One-Sample Comparison

If the sample consists of a single group compared to some reference value, then s is the standard deviation of that sample. The standard error of the mean is

$$SE(\bar{x}) = s/\sqrt{n}$$

16.3.2 Two Independent Samples

For a comparison of two independent samples, x_1 and x_2, the standard deviations for the two samples must be pooled. The pooled sample variance is calculated from existing data for the two groups as

$$s^2_{pool} = \frac{(n_1 - 1)s^2_1 + (n_2 - 1)s^2_2}{(n_1 + n_2 - 2)}$$

If sample sizes are equal for the two groups, then the pooled variance is the average of the two variances:

$$s^2_{pool} = \frac{s^2_1 + s^2_2}{2}$$

and the pooled standard deviation is

$$s_{pool} = \sqrt{\frac{s^2_1 + s^2_2}{2}}$$

Example: Anaesthesia Duration in Mice

(Data from Dholakia et al. 2017.) Anaesthesia duration was reported for 16 mice randomly assigned to receive intraperitoneal injections of either ketamine-xylazine (KX) or ketamine-xylazine plus lidocaine (KXL). Mean and standard deviation for each group were KX: 39 (SD 8) min, and KXL: 30 (SD 9) min. What sample size is required to estimate a difference between groups with a 95% confidence interval that is no wider than 15 min, with power of 80%?

In this example, $\alpha = 0.05$, and $\beta = 0.2$. The pooled standard deviation is

---

$$\sqrt{s^2_{pool}} = \sqrt{\frac{(8-1)8^2 + (8-1)9^2}{(8+8-2)}} = \sqrt{\frac{8^2 + 9^2}{2}} = 8.515$$

The specified width of the confidence interval is 15 min, so the precision d is $15/2 = 7.5$ min and the effect size is $7.5/8.515 = 0.88$. Using the large-scale approximation, $z_{1-\alpha/2} = 1.96$, and $z_{1-\beta} = 0.8416$, and sample size per group is approximately

$$N = \frac{(1.96 + 0.8416)^2}{(0.88)^2} = 10.14$$

$\cong 11$ mice per group for a total of 22 mice. Sample size derived from direct approximation based on the non-central t (Chapter 14) results in 23 per group, for a total of 46 mice. This total is more than double that of the simpler large-scale approximation. Therefore, studies based on the z-approximation may be considerably underpowered to detect a true effect.

16.3.3 Paired Samples or A/B Crossover Designs

The paired t-test is equivalent to a one-sample t-test on the differences. The mean and standard deviation for the paired differences $\bar{d}$ is computed from the differences for each paired observation.

The standard deviation for the differences is

$$s = \sqrt{\frac{\sum (d_i - \bar{d})^2}{n-1}}$$

Within-subject correlation is a feature of many association-type effect sizes, such as ANOVA–type models, nested designs, and regression. Because the variance of the paired differences of the n subjects is

$$s^2 = \text{var}(x_1) + \text{var}(x_2) - 2 \cdot \text{cov}(x_1, x_2)$$

and the within-subject correlation r is

$$r = \frac{\text{cov}(x_1, x_2)}{\text{var}(x_1) \cdot \text{var}(x_2)}$$

(deVore 1987), the variance s^2 is also computed as $(s_1 + s_2 - 2 \cdot r \cdot s_1 \cdot s_2)$. Standard statistical packages such as SAS require an estimate of r to estimate power for paired t-tests. This is not required in G*Power if the mean and standard deviation of the differences are used to determine effect size.

Calculating r on raw data using standard statistical packages (e.g. SAS *proc corr*, R *stats* cor.test) ignores non-independence of the observations and will result in inaccurate estimates. Instead, summarise the correlation at the subject level by calculating the averages for each pair of observations $(x_1 + x_2)/2$ for each subject, then calculating the correlation between these averages and the difference for each pair. This is the foundation of the Bland–Altman method used for method comparison.

and 11.7 (SD 3.64) ppm at t_1. The mean and standard deviation for the paired differences $(t_1 - t_0)$ were -12 ppm and 5.8 ppm, respectively. The correlation r between baseline and t_1 measurements was -0.33.

What sample size is required to detect a minimum difference of 10 ppm 90% of the time with 95% confidence?

This is a paired design. Here, $\alpha = 0.05$ and the desired power is $1 - \beta = 0.9$. The 95% confidence interval for the difference is

$$\bar{d} \pm t_{\alpha/2, n-1} \cdot s_{\bar{d}} = -12 \pm (2.262 \cdot 5.8)$$
$$\cong (-25, +1) \text{ ppm}$$

The effect size is $10/5.8 = 1.724$. The sample size to detect a minimum paired difference of 10 ppm is 12 paired measurements, or 6 cows with realised power of 0.97 (Appendix 16.A).

Example: Urinary Fluoride Concentrations in Cattle: Paired Comparisons

(Data from Debackere and Delbeke 1978.) Urinary fluoride concentrations were measured at two different time points for the same 10 cows randomly selected from a herd located in a contaminated area (Table 16.1). Fluoride concentrations at baseline t_0 averaged 23.7 (SD 5.1) ppm

16.4 Sample Size for Two-Arm Veterinary Clinical Trials

Sample size for a definitive clinical trial can be obtained from pilot data (Machin et al. 2018). External pilot data are not included in later data analyses

Table 16.1: Urinary fluoride concentrations (ppm) of 10 cows measured at two different times.

Cattle ID	Urinary fluoride concentrations ppm		Difference ($t_1 - t_0$)	Average ($t_0 + t_1$)/2
	Time 0	Time 1		
1	24.7	12.4	−12.3	18.6
2	18.5	7.6	−10.9	13.1
3	29.5	9.5	−20.0	19.5
4	26.3	19.7	−6.6	23.0
5	33.9	10.6	−23.3	22.3
6	23.1	9.1	−14.0	16.1
7	20.7	11.5	−9.2	16.1
8	18.0	13.3	−4.7	15.7
9	19.3	8.3	−11.0	13.8
10	23.0	15.0	−8.0	19.0

Source: Data from Debackere and Delbeke (1978).

and are not used to test hypotheses about the intervention. Instead, pilot data are used to estimate effect size components, specifically the mean clinically relevant difference between groups and sample variation. However, pilot trials are small, so the standard deviation s estimated from pilot data will be noisy and imprecise. If the pilot s is used in subsequent sample size calculations, the main trial is likely to be considerably underpowered.

Conventional method: Uncorrected s. The conventional method of estimating sample size for a two-arm trial is

$$N = \frac{(r+1)^2}{r} \cdot \frac{(z_{1-\alpha/2} + z_{1-\beta})^2}{(d/s)^2} + \left(\frac{z_{1-\alpha/2}^2}{2}\right)$$

where r is the allocation ratio (Chapter 14). If sample sizes are equal, r is 1, and $(r+1)^2/r = 4$.

More realistic estimates of s can be obtained by applying a correction factor to adjust for small sample size of the pilot. Two methods are adjustment by the upper confidence limit of the standard deviation (Browne 1995; Machin et al. 2018) and adjustment by the non-central t-distribution (Julious and Owen 2006; Whitehead et al. 2016). SAS code for each method is given in Appendix 16.B.

Upper confidence limit. The correction factor k is computed from the upper confidence limit (UCL) of the standard deviation obtained from the pilot (s_{pilot}). The confidence limit is computed from the χ^2 distribution and the pilot study degrees of freedom, $df_{pilot} = N_{pilot} - 2$ (Browne 1995; Machin et al. 2018).

The correction factor k for the pilot standard deviation is

$$k = \sqrt{\frac{df_{pilot}}{\chi^2\left(1-\alpha, df_{pilot}\right)}}$$

The adjusted standard deviation for the definitive trial is

$$s_{new} = k \cdot s_{pilot} = UCL = \sqrt{\frac{df_{pilot}}{\chi^2\left(1-\alpha, df_{pilot}\right)}} \cdot s_{pilot}$$

The standardised effect size for the new trial is then calculated from the desired mean difference to be detected (d) divided by the new standard deviation: d/s_{new}. The total sample size for the new trial is computed as before:

$$N = \frac{(r+1)^2}{r} \cdot \frac{(z_{1-\alpha/2} + z_{1-\beta})^2}{(d/s_{new})^2} + \left(\frac{z_{1-\alpha/2}^2}{2}\right)$$

Non-central t-distribution. The correction factor θ is computed from the inverse function of the t-distribution: $\theta = tinv\,(1-\beta, df_{pilot};\, t_{1-\alpha/2,df,\lambda})$. The inverse function *tinv* computes the pth quantile from the t-distribution, with degrees of freedom df and non-centrality parameter λ. Then the sample size for the definitive trial is

$$N = \frac{(r+1)^2}{r} \cdot \frac{\theta^2}{(d/s_{pilot})^2}$$

Solving requires multiple iterations until the estimate for N converges. However, a good approximation is obtained by substituting $z_{1-\alpha/2}$ for $t_{1-\frac{\alpha}{2},df,\lambda}$. Then

$\theta = tinv\,(1-\beta, df_{pilot}; z_{1-\alpha/2})$, and 1 is added to the final estimate (Julious and Owen 2006; Whitehead et al. 2016).

Example: External Pilot: Dogs With Degenerative Mitral Valve Disease

(Data from Karlin et al. 2019.) A pilot study was performed to assess the potential of circulating trimethylamine N-oxide (TMAO, μmol/L) as a marker for dogs with degenerative mitral valve disease (DMVD). In this example, sample size for a new main trial was estimated to compare dogs with asymptomatic DMVD to those with both DMVD and congestive heart failure. The difference in TMAO between symptomatic and asymptomatic dogs was approximately 4 μmol/L, with a standard deviation of 5.65 and 10 dogs in each group.[1] The effect size was approximately 0.71, with $(10 + 10) - 2 = 18$ degrees of freedom. What sample size would be required for the main trial to detect a mean difference of 4 μmol/L, with $\alpha = 0.05$ and power 0.9?

With the *conventional* method of estimating sample size for a two-arm study, sample size is

$$N = \frac{(1+1)^2}{1} \cdot \frac{(1.96 + 1.282)^2}{(4/5.65)^2} + \left(\frac{1.96^2}{2}\right)$$

$\cong 43$ per arm, for a total of 86 dogs.

With the *upper confidence limit method,* the correction factor k is

$$k = \sqrt{\frac{18}{9.3905}} = 1.3845$$

Therefore, the corrected effect size is 0.51. The sample size for the new trial is

$$N = \frac{(1+1)^2}{1} \cdot \frac{(1.96+1.282)^2}{(4/7.82242)^2} + \left(\frac{1.96^2}{2}\right)$$

$\cong 82$ per arm, for a total of 164 dogs.

With the *simplified non-central t-method,* $\theta = 3.485$. The sample size for the new trial is

$$N = \frac{(1+1)^2}{1} \cdot \frac{3.485^2}{(0.71)^2}$$

$\cong 49$–50 per arm, for a total of 98–100 dogs.

16.A Sample SAS Code for Calculating Sample Size for Two-Group Comparisons

16.A.1 Sample Size Based on z-Distribution

```
data zsample;
      meandiff= [value];
      sd= [value];
      ES= meandiff/sd;
    alpha = 0.05;
     beta= 0.1; [0.1 for 90% power, 0.2 for
80% power]
        z = quantile("Normal", 1-alpha/2);
        zbeta=quantile("Normal", 1-beta);
        N = ((z + zbeta)**2)/(ES**2);
run;
proc print data=zsample; run;
```

16.A.2 Total Sample Size Based on the Non-Centrality Parameter for t

```
*set maximum sample size for iterations;
%let numSamples = 100;
data tsample;

*iterate through range of sample sizes;
do n = 4 to &numSamples by 1;
```

```
*calculate difference to be detected;
      mean1= [VALUE];
      mean2= [VALUE];

  meandiff = mean2-mean1;
      s2= [VALUE]; *calculate pooled variance;

alpha=0.05;

* Calculate non-centrality parameter ncp;
  ncp = n*(meandiff**2)/(4*s2);
   F_crit = finv(1-alpha,1,n-2,0);
  *Solve for power for each n;
     power = 1-probF(F_crit,1,n-2, NCP);
   output;
 end;
run;
     *output realised power and corresponding N.
     Pick N with power that ≥ target power;
proc print data=tsample; run;
```

16.A.3 Sample Size for a Fixed Power: Crossover Design (Cattle Example)

```
proc power;
 pairedmeans test=diff
 alpha=0.05
 meandiff = 10
 std = 5.8
 corr = -0.33
 npairs = .
 power = .9;
run;
```

16.B Sample SAS Code for Calculating Sample Size for a Veterinary Clinical Trial. The standard deviation obtained from pilot data is corrected by computation of its upper confidence interval (UCL) or by the inverse function of power and pilot degrees of freedom

16.B.1 Conventional (Uncorrected Standard Deviation)

```
data power;
 *pilot data;
    meandiff=4;
     spilot=5.65;
    alpha = 0.05;
```

```
 beta=0.1;
 z = quantile("Normal", 1-alpha/2);
zbeta=quantile("Normal", 1-beta);

Nmain= (z*z/2)+4*((zbeta+z)**2)/(meandiff/
Spilot)**2;
 run;
proc print; run;
```

16.B.2 Upper Confidence Interval Correction

```
data power;
 *pilot data;
   meandiff = 4;
   df=18;
   stddev = 5.65;

alpha = 0.05;
beta = 0.1;

z = quantile("Normal", 1-alpha/2);
zbeta = quantile("Normal", 1-beta);
C_crit = quantile('CHISQ',alpha,df );

k=sqrt(df/C_Crit);
UCL=k*stddev;

 Nmain= (z*z/2)+4*((zbeta+z)**2)/(meandiff/
UCL)**2;
    run;
  proc print; run;
```

16.B.3 Simplified Non-Central *t* Correction

```
data power;
 *pilot data;
    df=18; *pilot degrees of freedom;
    meandiff=4;
    spilot=5.65;

ES = meandiff/spilot;

 alpha = 0.05;
 beta=0.1;

 z = quantile("Normal", 1-alpha/2);
 theta = tinv(1-beta,df,z);

Nmain = 4*(theta**2)/ES**2;
    run;
 proc print;
 run;
```

References

Browne, R.H. (1995). On the use of a pilot sample for sample size determination. *Statistics in Medicine* 14 (17): 1933–1940. https://doi.org/10.1002/sim.4780141709.

Debackere, M. and Delbeke, F.T. (1978). Fluoride pollution caused by a brickworks in the Flemish countryside of Belgium. *International Journal of Environmental Studies* 11: 245–252.

DeVore, J.L. (1987). *Probability and Statistics for Engineering and the Sciences*, 2e. Brooks/Cole Publishing.

Dholakia, U., Clark-Price, S.C., Keating, S.C.J., and Stern, A.W. (2017). Anesthetic effects and body weight changes associated with ketamine-xylazine-lidocaine administered to CD-1 mice. *PLoS ONE* 12 (9): e0184911. https://doi.org/10.1371/journal.pone.0184911.

Julious, S.A. and Owen, R.J. (2006). Sample size calculations for clinical studies allowing for uncertainty about the variance. *Pharmaceutical Statistics* 5: 29–37. https://doi.org/10.1002/pst.197.

Karlin, E.T., Rush, J.E., and Freeman, L.M. (2019). A pilot study investigating circulating trimethylamine N-oxide and its precursors in dogs with degenerative mitral valve disease with or without congestive heart failure. *Journal of Veterinary Internal Medicine* 33 (1): 46–53. https://doi.org/10.1111/jvim.15347.

Lehr, R.S. (1992). Sixteen s-squared over d-squared: A relation for crude sample size estimates. *Statistics in Medicine* 11: 1099–1102.

Machin, D., Campbell, M.J., Tan, S.B., and Tan, S.H. (2018). *Sample Sizes for Clinical, Laboratory and Epidemiology Studies*, 4e. Wiley-Blackwell.

van Belle, G. (2008). *Statistical Rules of Thumb*, 2nd edition. New York: Wiley.

Wan, X., Wang, W., Liu, J., and Tong, T. (2014). Estimating the sample mean and standard deviation from the sample size, median, range and/or interquartile range. *BMC Medical Research Methodology* 14: 135. https://doi.org/10.1186/1471-2288-14-135.

Whitehead, A.L., Julious, S.A., Cooper, C.L., and Campbell, M.J. (2016). Estimating the sample size for a pilot randomised trial to minimise the overall trial sample size for the external pilot and main trial for a continuous outcome variable. *Statistical Methods in Medical Research* 25 (3): 1057–1073. https://doi.org/10.1177/0962280215588241.

17 Comparing Two Groups: Proportions

CHAPTER OUTLINE HEAD

17.1 Introduction

Proportions p summarise count data. The proportion is the ratio of the number of events or subjects with the condition of interest divided by the total: $p = x/N$. Count data do not contain as much information as continuous data, so sample size based on proportion outcomes will be much larger. Because proportions are not continuous and normally distributed variables, comparisons cannot involve the use of t-tests or other statistical analysis methods that are standard for continuous data (Box 17.1).

Sample size determinations require prespecification of

Type I and type II errors, α and β

One-sided or two-sided comparison (α or $\alpha/2$ respectively)

The minimum biologically relevant difference to be detected. This is the absolute difference between proportions $d = p_1 - p_0$. This is also known as the *risk difference*.

BOX 17.1
Comparing Proportions

Proportions p are used to summarise binary and count data

Sample size requires specifying both

  absolute difference $p_1 - p_0$
  relative difference with respect to control or baseline proportion

The effect size is

$$\frac{p_1 - p_0}{\sqrt{\bar{p}(1-\bar{p})}}$$

Analysis methods used for continuous data cannot be used for comparisons of proportions.

The value of the control or reference group p_0. Sample size for proportions is also determined by the difference relative to the baseline.

A Guide to Sample Size for Animal-based Studies, First Edition. Penny S. Reynolds.
© 2024 John Wiley & Sons Ltd. Published 2024 by John Wiley & Sons Ltd.

The standardised effect size for comparing two proportions with equal sample sizes is

$$ES = \frac{p_1 - p_0}{\sqrt{\bar{p}(1 - \bar{p})}}$$

where $(p_1 - p_0)$ is the difference d, and $\bar{p} = (p_0 + p_1)/2$ is the mean of the two proportions.

The margin of error, or precision, is one-half of the confidence interval range $(u - l)/2$. It can also be calculated as $z_{1-\alpha/2} \cdot SE(p_{diff})$. Sample size can be obtained for a target precision by iterating over a range of candidate sample sizes, calculating the confidence interval for each, and choosing the sample size that approximates the desired precision.

17.2 Difference Between Two Proportions

17.2.1 One Proportion Known

When one proportion is already known (reference proportion p_0), then sample size for detecting a difference d of the new or anticipated p from the reference level, with a significance level of α and power $1 - \beta$ is:

$$N = \frac{\left[z_{1-\alpha} \sqrt{p_0(1 - p_0)} + z_{1-\beta} \sqrt{p(1 - p)} \right]^2}{d^2}$$

This is usually a one-sided test of the hypothesis $p_{new} < p_0$ or $p_{new} > p_0$, so the significance is α rather than $\alpha/2$.

Example: Before-After Comparison: Mouse Surgery Success

A series of surgeries performed on mice had a loss rate of 15%. The investigator initiated a rigorous series of retraining and procedural changes, with the goal of reducing losses to 5% or less before beginning the new protocol. How many surgeries need to be observed to be 90% confident that the change in procedures has reduced attrition to 5%?

This tests the hypothesis that $p_{new} < p_0$, with proportions $p_0 = 0.15$ and $p_{new} = 0.05$, $\alpha = 0.05$,

and power $1 - \beta = 0.9$. The number of surgeries that will need to be observed is

$$N = \frac{\left[1.645 \sqrt{0.15(1 - 0.15)} + 1.282 \sqrt{0.05(1 - 0.05)} \right]^2}{(0.15 - 0.05)^2}$$

$$= 300.5$$

or approximately 301 surgeries.

17.2.2 Sex Ratios

Sex ratio, and factors inducing sex ratio shifts, are important information for a variety of disciplines such as evolutionary biology, ecology, reproductive biology, genetics, animal husbandry, and management of laboratory populations. In most species with separate sexes, males and females are produced in approximately equal numbers, so the expected sex ratio is 1:1; that is, $p_0 = p_1 = 0.5$. However, sex ratio can vary in response to environmental factors such as temperature, food supply, population density, and parental effects.

The sex ratio R is calculated as a proportion of the total number of animals:

$$R = n_1/N$$

where the total sample size $N = n_1 + n_2$. Calculating direct ratios of one sex to the other ($R = n_1/n_2$) is not recommended. If there are no representatives of one sex in the sample, the resulting 'ratio' will be either undefined or zero depending on which is the numerator or denominator (Wilson and Hardy 2002).

Comparison of observed sex ratio when the expected sex ratio is 1:1. Under the null hypothesis, the sex ratio is 1:1, and the probability p of obtaining a member of the target sex will be exactly one-half, or 0.5 ($p_0 = p_1 = 0.5$). Usually, the hypothesis of interest is a test of a shift in sex ratio from $p_0 = 0.5$ (the null value) to an alternative value p_1 with power $1 - \beta$.

The equation for sample size is a modification of the former equation for a one-sample test:

$$N = \frac{\left(z_{1-\alpha} \cdot 0.5 + z_{1-\beta} \sqrt{0.25 - d^2} \right)^2}{d^2}$$

where the reference proportion *is* $p_0 = 0.5$, p_1 is the new proportion, and $d = (p_0 - p_1)$ is the difference from 0.5 under the alternate hypothesis.

Example: Butterfly Sex Ratios

For many years, the sex ratio of butterflies collected in the field was 1:1; that is, the proportion of female butterflies was approximately 50%. However, recent studies have reported an apparent decline in females (possibly the result of climate change and habitat loss). How many butterflies would need to be collected to detect a shift in the proportion of females from 0.50 to 0.45 with 90% power and 95% confidence?

The null hypothesis of the expected proportion of $p_0 = 0.50$ will be tested against the one-tailed alternative $p_1 = 0.45$, with $\alpha = 0.05$ ($z_{1-\alpha} = 1.645$) and power $1 - \beta = 0.9$ ($z_{1-\beta} = 1.282$). The number of butterflies N to be collected to detect this shift will be approximately:

$$N \geq \frac{\left(0.5(1.645) + 1.282\sqrt{0.25 - (0.50 - 0.45)^2}\right)^2}{(0.50 - 0.45)^2}$$

$$= 350$$

17.2.3 Two Independent Samples

When sample sizes in each group are equal, then the large sample approximation (Wald equation) for total sample size is

$$N \geq (z_{1-\alpha/2} + z_{1-\beta})^2 \cdot \frac{p_0(1 - p_0) + p_1(1 - p_1)}{d^2}$$

where subscripts 0 and 1 indicate the control (or reference) and intervention groups, respectively, and the difference of interest is $d = (p_1 - p_0)$. A conservative estimate for sample size is obtained by assuming that $p = 0.5$, because $p(1 - p)$ is a maximum at $p = 0.5$.

Better estimates for sample size are based on the exact method: solving for power and iterating over a

range of candidate sample sizes. Sample sizes n_0 and n_1 should be equal for simplicity and to minimise the total sample size required for a target power. The equation is

$$Power = 1 - \beta = Pr\left[z \leq \frac{-z_{\alpha/2}\sqrt{\bar{p}\bar{q}/n_0 + \bar{p}\bar{q}/n_1} - d}{\sqrt{p_0 q_0/n_0 + p_1 q_1/n_1}}\right]$$
$$+ Pr\left[z \geq \frac{z_{\alpha/2}\sqrt{\bar{p}\bar{q}/n_0 + \bar{p}\bar{q}/n_1} - d}{\sqrt{p_0 q_0/n_0 + p_1 q_1/n_1}}\right]$$

where $\bar{p} = (p_0 + p_1)/2$; $q_0 = 1 - p_0$; $q_1 = 1 - p_1$; $\bar{q} = 1 - \bar{p}$, and $d = (p_1 - p_0)$ (Zar 2010). SAS code for a two-sided test is provided in Appendix 17.A.

Example: Mouse Infections: Sample Size for Risk Difference

(Data from Zar 2010). Two species of mice were examined for the presence of a certain parasite. For species 1, 18 of 24 mice were infected ($p_0 = 0.75$), and for species 2, 10 of 25 were infected ($p_1 = 0.40$). What sample size in a future study would be required to obtain the same risk difference $p_1 - p_0 = 0.35$, for equal sample sizes $n_1 = n_2$, confidence of 95% and power of 80%?

Sample size based on the Wald large-scale approximation is

$$N = (1.96 + 0.8416)^2 \cdot$$
$$\frac{0.75(1 - 0.75) + 0.4(0.4)}{(0.75 - 0.4)^2} = 27.4$$

$\cong 28$ for each group, for a total of 56.

With the exact method, the sample size is 32 per group, for a total of 64 and power of 80%. For power of 90%, the sample size is 42 per group, for a total of 84 (Figure 17.1).

17.2.4 Confidence Intervals

The conventional Wald confidence interval for one sample is

$$p \pm z_{1-\alpha/2} \cdot \sqrt{p(1 - p)/n}$$

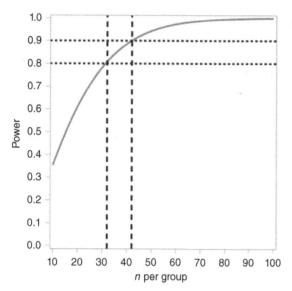

Figure 17.1: Sample size for evaluating risk difference in a mouse infection model estimated by the exact method. For power of 80%, the sample size is 32 per group, and for power of 90%, the sample size is 42 per group.

For the difference between two independent proportions, the confidence interval is

$$(p_1 - p_0) \pm z_{1-\alpha/2} \cdot SE(p_1 - p_0)$$

where the standard error for the difference SE is

$$SE(p_1 - p_0) = \sqrt{\frac{p_1(1-p_1)}{n_1} + \frac{p_0(1-p_0)}{n_0}}$$

However, the Wald equation is not recommended for small samples ($n < 30$ per arm) because it is unstable and usually very inaccurate even for very large samples. It cannot be used when proportions are extreme ($p < 0.1$, $p > 0.9$).

Two preferred alternatives discussed here are the Newcombe (Wilson score interval) and Agresti-Caffo methods. Newcombe (1998) describes 11 methods for constructing confidence intervals for two-group comparisons of proportions. SAS code for these and six additional methods are described by Garner (2016).

Newcombe (Wilson score intervals). This method incorporates the lower and upper confidence limits (l, u) estimated for each proportion (Newcombe 1998; Newcombe and Altman 2000). This method

performs well for all sample sizes. The width of the confidence interval is

$$w = \sqrt{(p_0 - l_0)^2 + (u_1 - p_1)^2} + \sqrt{(p_1 - l_1)^2 + (u_0 - p_0)^2}$$

The lower (l) and upper (u) confidence limits are $(A - B)/C$ and $(A + B)/C$, respectively, where $A = 2 \cdot n_i \cdot p_i + z_{1-\alpha/2}^2$; $B = z_{1-\alpha/2} \cdot \sqrt{z_{1-\alpha/2}^2 + 4 n_i \cdot p_i (1 - p_i)}$; and $C = 2 \cdot \left(n_i + z_{1-\alpha/2}^2 \right)$.

Agresti-Caffo method. The unadjusted proportions are $p_i = x_i / n_i$. The adjusted confidence interval 'adds two successes and two failures' where $\tilde{n}_i = (n + 4)$ trials and $\tilde{p}_i = (x + 2)/(n + 4)$. The adjusted x_i are $\tilde{x}_i = x_i + z_{1-\alpha/2}^2/4$, adjusted n_i are $\tilde{n}_i = n_i + z_{1-\alpha/2}^2/2$, and adjusted p_i are $\tilde{p}_i = \tilde{x}/\tilde{n}_i$. These adjusted values are substituted into the confidence interval and sample size equations (Agresti and Caffo 2000).

Example: Mouse Infections: Confidence Intervals for Difference Between Proportions

(Data from Zar 2010). Two species of mice were examined for the presence of a certain parasite. For species 1, 18 of 24 mice were infected ($p_0 = 0.75$), and for species 2, 10 of 25 were infected ($p_1 = 0.40$).

The 95% confidence intervals for the risk difference ($p_1 - p_0$) of 0.35 were constructed by the conventional Wald, Newcombe, and Agresti-Caffo methods (SAS code in Garner 2016) and are summarised in Table 17.1. The Wald interval is wider than either the Newcombe or Agresti-Caffo intervals, indicating less precision. The

Table 17.1: 95% Confidence intervals constructed for mouse infection data $p_0 = 0.75$ and $p_1 = 0.40$ with a risk difference ($p_1 - p_0$) of 0.35.

Method	95% confidence interval	
Wald	0.091	0.609
Agresti-Caffo	0.072	0.576
Newcombe (Wilson score)	0.073	0.561

Source: SAS code adapted from Garner (2016).

Wald interval limits are symmetrical around the mean; the Newcombe and Agresti-Caffo intervals are not.

17.3 Relative Risk and Odds Ratio

Relative risk (RR), risk differences, and odds ratio (OR) are commonly used in epidemiological studies to describe the probability of an event occurring in one group relative to another. Study designs can be randomised controlled trials or observational cohort and case-control studies.

17.3.1 Relative Risk

The *RR* or *risk ratio* is the ratio of two proportions, usually expressed as the ratio of two 'success' proportions. The RR compares the probability or risk of an event occurring in one group relative to that of a comparator group:

$$RR = p_1/p_0$$

Subscripts 0 and 1 indicate the control (reference or unexposed) and intervention (exposed) groups, respectively.

Confidence intervals. The confidence interval is obtained by first calculating the natural log (ln) of the RR:

$$\ln(RR) \pm z_\alpha \sqrt{\frac{1-p_0}{n_0\,p_0} + \frac{1-p_1}{n_1\,p_1}}$$

then back-transforming to the linear scale. For the 95% confidence interval $z_\alpha = 1.96$, and the confidence limits are

$$\text{Lower limit}\,(l) = \exp\left[\ln RR - 1.96\sqrt{\frac{1-p_0}{n_0 p_0} + \frac{1-p_1}{n_1 p_1}}\right]$$

$$\text{Upper limit}\,(u) = \exp\left[\ln RR + 1.96\sqrt{\frac{1-p_0}{n_0 p_0} + \frac{1-p_1}{n_1 p_1}}\right]$$

Sample size. When prevalence p_0 in the control group is relatively low (<20%) the sample size to achieve a target RR can be approximated by a Poisson-based formula:

$$n_{group} = \frac{4}{p_0\left(\sqrt{RR}-1\right)^2}$$

Alternatively, it can be estimated by the ln-based formulation

$$n_{group} = \frac{8(RR+1)/RR}{p_0\left[ln(RR)\right]^2}$$

This formula is more conservative than the simple Poisson-based formulation, so estimated sample sizes will be larger. As baseline prevalence increases, the required sample size decreases.

Example: Detecting a Given Relative Risk

To detect an RR of at least 5 with prevalence of 1% in the unexposed group, the required sample size is

$$n = \frac{4}{0.01\left(\sqrt{5}-1\right)^2}$$
$$= 261.8 \approx 262, \text{for a total of } 524.$$

Alternatively,

$$n = \frac{8(5+1)/5}{0.01\left[\ln(5)\right]^2} = 370.6 \approx 371, \text{for a total of } 742$$

Rule of thumb. RR reduction is the difference between two RRs ($RR_1 - RR_0$). Studies evaluating a rare binary outcome require approximately 50 events in the control group and an equal number in the exposed group to detect a 50% RR reduction (or halving of risk) with 80% power (Glasziou and Doll 2006).

Example: Pre-specified Risk Reduction

Suppose the risk of incurring a given disease is 10%, and a researcher wanted to determine if a new intervention will reduce risk to 5%. How many subjects are needed?

$$n_{group} = 50/0.1 = 500 \text{ subjects per}$$
$$\text{group or 1000 subjects total}$$

17.3.2 Odds Ratios

The OR is the ratio of the odds of occurrence of an event ('success', p) relative to the odds of it not occurring ('failure', $1-p$) for a test group versus the comparator group. Subscripts 0 and 1 indicate the control (reference or unexposed) and intervention (exposed) groups, respectively.

$$OR = \frac{p_1/(1-p_1)}{p_0/(1-p_0)}$$

Sample size. The sample size to achieve a target OR (OR_{new}) is approximately

$$n_{group} = \frac{8 \cdot \sigma^2_{lnOR}}{(ln \ OR_{new})^2}$$

where $\sigma^2_{lnOR} = \dfrac{1}{p_0} + \dfrac{1}{1-p_0} + \dfrac{1}{p_1} + \dfrac{1}{1-p_1}$

More precise estimates are obtained by

$$N = \frac{(r+1)^2}{r} \cdot \frac{(z_{1-\alpha/2} + z_{1-\beta})^2}{\bar{p}(1-\bar{p}) \cdot (\ln OR_{new})^2}$$

where $\bar{p}$ is the mean proportion $(p_0 + p_1)/2$. The allocation ratio $r = n_1/n_2$. If sample sizes are equal, r is 1, and $(r+1)^2/r = 4$ (Machin et al. 2018).

Confidence intervals. The $100(1-\alpha)\%$ confidence interval is calculated from $\ln(OR)$:

$$\ln(OR) \pm z_\alpha \sqrt{\frac{1}{p_0} + \frac{1}{1-p_0} + \frac{1}{p_1} + \frac{1}{1-p_1}}$$

then back-transformed to the linear scale. The 95% confidence interval is

Lower limit = exp

$$\left[\ln(OR) - 1.96 \sqrt{\frac{1}{p_0} + \frac{1}{1-p_0} + \frac{1}{p_1} + \frac{1}{1-p_1}} \right]$$

Upper limit = exp

$$\left[\ln(OR) + 1.96 \sqrt{\frac{1}{p_0} + \frac{1}{1-p_0} + \frac{1}{p_1} + \frac{1}{1-p_1}} \right]$$

Confidence intervals for more complex regression model-based estimates for ORs can be derived from simulation or percentiles of bootstrap distributions (Greenland 2004).

Example: Mouse Infections: Sample Size for a Change in Odds Ratio

Mice in two groups were examined for parasites. In group 1, 18 of 24 mice were infected ($p_1 = 0.75$), and in group 2, 10 of 25 were infected ($p_0 = 0.40$). In this example, group 1 is the test group p_{test} and group 2 is the comparator group p_0. The OR is

$$OR = \frac{0.75/(1-0.75)}{0.40/(1-0.40)} = 4.5$$

and the mean proportion $\bar{p} = (0.75 + 0.40)/2 = 0.575$. What sample size would be required to detect an OR of 3 with $\alpha = 0.05$ and power of 0.9 and equal sample sizes in each group?

For planning a new study, it can be assumed the baseline proportion in the new study will be the same as the old. For a new OR = 3 and the same baseline proportion, the anticipated proportion for the test group must be estimated first:

$$p_{test,new} = \frac{OR \cdot p_0}{1 - p_0 + OR \cdot p_0} = \frac{3 \cdot 0.40}{1 - 0.40 + 3 \cdot 0.40} = 0.67$$

The new mean proportion is $\bar{p} = (0.67 + 0.40)/2 = 0.533$. Then the new sample size is

$$N = \frac{(1+1)^2}{1} \cdot \frac{(1.96 + 1.2816)^2}{0.533(1-0.533) \cdot (\ln 3)^2} = 203.8$$

≈204 mice, with 102 per group.

When expected proportions are unknown. When there is no information as to the proportions expected for the two groups, effect size can be approximated from Cohen's h for ORs (Cohen 1988). The effect size h is

$$h = \varphi_1 - \varphi_2$$

where $\varphi_i = 2 \cdot \arcsin(\sqrt{p_i})$. The arcsine transformation is used to stabilise the variance for the binomial distribution (Rücker et al. 2008, 2009). The total sample size N is then approximated by

$$N = 4\left(\frac{z_{1-\alpha/2} + z_{1-\beta}}{h}\right)^2$$

Because it is easy to calculate (Rücker et al. 2009), h may be appropriate for first-pass approximations. However, this approximation should be used with considerable caution. Effect sizes based on standardised 'small', 'medium', and 'large' benchmark values are not appropriate for animal-based studies. Apart from that, calculated values of h are difficult to interpret. The transformation is biased because the transformed values do not match the difference between proportions $(p_1 - p_2)$ (Cohen 1988). Bias increases with small sample size and large imbalances between groups (Rücker et al. 2009).

17.4 Skewed Count Data

When count data are concentrated in relatively few categories, the data distribution will be skewed or imbalanced. These data are often characterised by large number of zeros or counts in a few categories and many outlying values. Sample size calculations for two-group comparisons of skewed data are based on a number of underlying distributions, the two most common distributions being the Poisson and negative binomial (Cundill and Alexander 2015).

17.4.1 Poisson Distribution

The Poisson distribution is appropriate for count data describing event rate or the number of events per unit time. The mean event rate is λ.

Total sample size is

$$N \geq \frac{(z_{1-\alpha/2} + z_{1-\beta})^2 \cdot (a + b)}{[ln(\lambda_0) - ln(\mu\lambda_1)]^2}$$

where $a = \dfrac{1}{p_1}\left(\dfrac{1}{\lambda_1}\right)$ and $b = \dfrac{1}{p_0}\left(\dfrac{1}{\lambda_0}\right)$, and subscripts 0 and 1 indicate reference (or control) and intervention groups, respectively.

Rules of thumb. The square root of a Poisson-distributed sample is approximately normal. With Lehr's approximation (van Belle 2008), sample size per group is approximately

$$n_{group} = \frac{4}{\left(\sqrt{\lambda_0} - \sqrt{\lambda_1}\right)^2}$$

To determine sample size for estimating *a given rate above background*, then sample size is $N = 4\sqrt{\lambda_*}$

Example: Avian Influenza

There has been a global increase in the reported number of avian influenza outbreaks in wild birds and poultry. Low pathogenicity strains occur naturally without causing illness. Suppose baseline mortality for a given population was approximately 25,000 deaths per year. How many excess deaths would have to occur to be flagged up as an unusual occurrence?

$$N \geq 4\sqrt{25,000} \approx 633$$

for an increase of approximately 2.5%.

17.4.2 Negative Binomial Distribution

The negative binomial is a generalisation of the Poisson distribution, with mean count μ and an additional parameter k describing over-dispersion. This distribution is used frequently to describe distributions of parasites across hosts.

Total sample size is

$$N \geq \frac{z_{1-\alpha/2}^2 \left[\left(\frac{1}{\mu_0} + \frac{1}{k_0}\right)\left(\frac{1}{p_1} + \frac{1}{p_0}\right)\right] + z_{1-\beta}^2 \left[\frac{1}{p_1}\left(\frac{1}{\mu_1} + \frac{1}{k_1}\right) + \frac{1}{p_0}\left(\frac{1}{\mu_0} + \frac{1}{k_0}\right)\right]}{[\ln(\mu_0) - \ln(\mu_1)]^2}$$

A good approximation can be made if proportions are set to 0.5 ($p_0 = p_1 = 0.5$), sample sizes are balanced ($n_0 = n_1$), and the negative binomial parameter is fixed ($k_0 = k_1 = k$). Then the equation reduces to

$$N \geq \frac{(z_{1-\alpha/2} + z_{1-\beta})^2 \cdot 2 \left[\left(\frac{1}{\mu_1} + \frac{1}{k}\right) + \left(\frac{1}{\mu_0} + \frac{1}{k}\right)\right]}{[\ln(\mu_0) - \ln(\mu_1)]^2}$$

(Brooker et al. 2005). Estimates for μ and k can be based on prior values or calculated from raw data using maximum likelihood (Chapter 10).

Example: Soay Sheep Parasite Burden

The mean lungworm burden of 67 Soay sheep found dead on St Kilda was 47.5 worms per sheep (Gulland 1992). Suppose a new vaccine is proposed with expected efficacy of 50% when the new vaccine tested against a control in a two-arm trial. How many sheep will be required to detect this level of efficacy with 90% power and two-sided confidence of 95%?

Reference worm burden μ_0 is 47.5, and with 50% efficacy the new worm burden μ_1 is 23.8. The negative binomial parameter k reported in the prior study was 0.841. Then with $p_0 = p_1 = 0.5$ and $k_0 = k_1 = 0.841$, sample size for the new study is

$$N \geq \frac{(1.96 + 1.6449)^2 \cdot 2 \left[\left(\frac{1}{23.8} + \frac{1}{0.841}\right) + \left(\frac{1}{47.5} + \frac{1}{0.841}\right)\right]}{[\ln(47.5) - \ln(23.8)]^2}$$

$\cong 132.1$. Therefore, a total of approximately 133 sheep (or 66–67 per treatment arm) will be required.

17.A SAS Code for Computing Sample Size: Difference Between Two Independent Proportions. Proportions are calculated over a range of candidate *n*, then solved for power

```
%let numSamples = 100;
data proppower;
do n1 = 10 to &numSamples by 2;
        alpha = 0.05;
        *input proportions;
            p1=0.75;
            p2=0.40;

        *for maximum power set n1=n2;
        n2=n1;

*calculate difference in proportions and
related metrics;
        pdiff=p1-p2;
            q1=1-p1;
            q2=1-p2;
            pmean=(n1*p1+n2*p2)/(n1+n2);
            qmean=1-pmean;
        f1=pmean*qmean/n1;
        f2=pmean*qmean/n2;

        z=quantile('normal',1-alpha/2);

*calculate power twosample test;
            pow = (z*sqrt(f1+f2)-(pdiff))/
            (sqrt(f1+f2));
            power=1-probnorm(pow);
        output;
    end;
run;
proc print data=proppower;
    run;

*Plot power vs sample size per arm;
```

```
proc sgplot data=proppower aspect=1;
    series x=n1 y=power / lineattrs=
    (thickness=3);
yaxis label = "Power" values = (0 to 1 by 0.1)
valueattrs=(color=black size=12) labelattrs=
(color=black family=arial size=14);
xaxis Label = "n per group" values = (10 to
100 by 5) labelattrs=(color=black family=arial
size=14) valueattrs=(color=black size=12);
refline 0.8 0.9 /axis = y lineattrs=
(thickness=3 color=black pattern=shortdash);
refline 32 42 / axis=x lineattrs=(thickness=3
color=black pattern=dash);
run;
```

References

Agresti, A. and Caffo, B. (2000). Simple and effective confidence intervals for proportions and differences of proportions result from adding two successes and two failures. *The American Statistician* 54: 280–288. https://doi.org/10.2307/2685779.

Brooker, S., Bethony, J., Rodrigues, L. et al. (2005). Epidemiologic, immunologic and practical considerations in developing and evaluating a human hookworm vaccine. *Expert Review of Vaccines* 4: 35–50.

Cohen, J. (1988). *Statistical Power Analysis for the Behavioral Sciences*. Hillsdale: Lawrence Erlbaum Associates.

Cundill, B. and Alexander, N.D. (2015). Sample size calculations for skewed distributions. *BMC Medical Research Methodology* 15: 28. https://doi.org/10.1186/s12874-015-0023-0.

Garner, W. (2016). Constructing confidence intervals for the differences of binomial proportions in SAS®. https://www.lexjansen.com/wuss/2016/127_Final_Paper_PDF.pdf.

Glasziou, P. and Doll, H. (2006). Was the study big enough? Two café rules. *Evidence-Based Medicine* 11 (3): 69–70. https://doi.org/10.1136/ebm.11.3.69.

Greenland, S. (2004). Model-based estimation of relative risks and other epidemiologic measures in studies of common outcomes and in case-control studies. *American Journal of Epidemiology* 160 (4): 301–305. https://doi.org/10.1093/aje/kwh221.

Gulland, F.M.D. (1992). The role of nematode parasites in Soay sheep (*Ovis aries* L.) mortality during a population crash. *Parasitology* 105 (Pt 3): 493–503. https://doi.org/10.1017/s0031182000074679.

Machin, D., Campbell, M.J., Tan, S.B., and Tan, S.H. (2018). *Sample Sizes for Clinical, Laboratory and Epidemiology Studies*, 4e. Wiley-Blackwell.

Newcombe, R.G. (1998). Interval estimation for the difference between independent proportions: comparison of eleven methods. *Statistics in Medicine* 17 (8): 873–890. https://doi.org/10.1002/(sici)1097-0258(19980430), Erratum in: *Statistics in Medicine*, 18(10):1293.

Newcombe, R.G. and Altman, D.G. (2000). Proportions and their differences. In: *Statistics with Confidence*, 2nd ed. (ed. D.G. Altman, D. Machin, T.N. Bryant, and M.J. Gardner), 45–56. Bristol: BMJ Books.

Rücker, G., Schwarzer, G., and Carpenter, J. (2008). Arcsine test for publication bias in meta-analyses with binary outcomes. *Statistics in Medicine* 27 (5): 746–763.

Rücker, G., Schwarzer, G., Carpenter, J., and Olkin, I. (2009). Why add anything to nothing? The arcsine difference as a measure of treatment effect in meta-analysis with zero cells. *Statistics in Medicine* 28 (5): 721–738. https://doi.org/10.1002/sim.3511. PMID: 19072749.

Van Belle, G. (2008). *Statistical Rules of Thumb*, 2nd ed. Wiley.

Wilson, K. and Hardy, I. (2002). Statistical analysis of sex ratios: an introduction. In: *Sex Ratios: Concepts and Research Methods* (ed. I. Hardy), 48–92. Cambridge: Cambridge University Press https://doi.org/10.1017/CBO9780511542053.004.

Zar, J.H. (2010). *Biostatistical Analysis*, 5th ed. Upper Saddle River: Prentice-Hall.

18 Time-to-Event (Survival) Data

CHAPTER OUTLINE HEAD

18.1 Introduction

A time to event outcome combines two separate pieces of information: whether an event did or did not occur in the designated time (Y/N) and the time at which the event occurred. The 'event' of interest may be adverse or classified as a 'failure' (death of a subject, time to humane endpoint, time to tumour appearance), neutral (pollinator visits, duration of shelter animal stay), or positive (time to complete wound healing, time to conception; Box 18.1). *Censoring* is when an individual subject does not experience the event of interest in the study time frame, and is a distinguishing feature of time-to-event data.

Sample size calculations for time-to-event data are more complicated than those for continuous or binary endpoints and usually require large sample sizes. This makes 'survival' as a primary outcome unacceptable for most laboratory studies

---

BOX 18.1
Examples of Time-to-Event Applications

Time from disease diagnosis to death.

Time to nest failure

Number of nestlings surviving to fledging.

Comparison of dogs completing guide dog training to those withdrawn early.

Time from disease induction to palpable tumour appearance

Length of pregnancy in cattle following artificial insemination.

Time from disease diagnosis to disease resolution.

Time from initial disease exposure to first major symptom.

Time from completion of cancer treatment to tumour recurrence.

Time between seizures.

Time between flower visits by pollinators

A Guide to Sample Size for Animal-based Studies, First Edition. Penny S. Reynolds.
© 2024 John Wiley & Sons Ltd. Published 2024 by John Wiley & Sons Ltd.

Differences in fish survival caused by haul net differences.

Time of shelter animal stay, from shelter entry to adoption

Time to complete wound healing.

BOX 18.2
Time to Event Study 'Landmarks'

There must be clear, clinically or biologically relevant, repeatable (or standardised), and measurable definitions for:

Event of interest;
Time 'start';
Time 'end' (for subject monitoring);
End of the study;
Time scale for biological effect
Definition of censoring
Humane endpoints.

using rodents. Specification of time to event variable as a primary outcome is more suited to analysis of large data sets or multi-centre veterinary randomised clinical trials, where there is the potential for observing many more subjects than is feasible for laboratory studies.

18.2 Methodological Considerations

18.2.1 Define the Research Question

The research question will determine both how sample size will be determined and how time-to-event data will be analysed. The three broad categories of research questions are determining the proportion of individuals that will *remain free of the event* after a certain time; the proportion of individuals that will *have the event* after a certain time; and the *risk of the event* at a particular point in time, among those who have survived until that point (Machin et al. 2006).

18.2.2 Define All Survival-Related Items

After formulation of the research question, clear working definitions must be developed for all items used in the calculations. These include the event of interest, the time start, the time scale (hours, days, weeks, months, years), the end of the monitoring or follow-up period, and the end of the study (Box 18.2). Typically, there is a single target event, but more advanced analytical methods are available that can accommodate either multiple or recurrent events.

Death as an experimental endpoint in laboratory animals is usually not ethically justifiable. Humane endpoints for research animals are the earliest scientifically justified point at which 'pain or distress

in an experimental animal can be prevented, terminated, or relieved' (National Research Council 2011). Appropriate humane endpoints must be selected to accurately predict or indicate pain, distress, morbidity, or death, with welfare plans for monitoring, recognition, and appropriate end-stage actions for example, euthanasia when a pre-specified humane endpoint is attained[1].

The time scale for biological effects is critical for assessing the biological or clinical relevance of any differences between groups. For example, a 3-day difference in survival will be clinically relevant for acute infections or diseases with high mortality occurring in 7–10 days. However, for chronic diseases, or conditions that may take weeks to months to evolve, a 3-day difference will have no biological importance (Furuya et al. 2014).

18.2.3 Randomisation Schedule

Subjects may enter a trial sequentially (for example, veterinary clinical trials or laboratory rodents with uncommon genotypes that depend on the timing of litter drop). As a result, recruitment, enrolment,

---

[1] Guidelines for humane endpoint references include https://oacu.oir.nih.gov/system/files/media/file/2022-04/b13_endpoints_guidelines.pdf; National Research Council (US) Committee on Recognition and Alleviation of Pain in Laboratory Animals. Recognition and Alleviation of Pain in Laboratory Animals. Washington (DC): National Academies Press (US); 2009. 4, Effective Pain Management; https://nc3rs.org.uk/3rs-resources/humane-endpoints

and randomisation of eligible subjects may be staggered over long periods of time, varying from weeks, months, or years. Randomisation strategies should be planned for uncertainties in availability to avoid unintentionally uneven allocation ratios. Likewise, follow-up times should be planned with respect to the time allotted to enrolment period. If the study is terminated at a fixed time regardless of study entry time, individual follow-up times will be extremely variable, with the first subject enrolled with the longest follow-up time and the last enrolled subject the shortest. Staggered entry does not affect sample size calculations for continuous or binary endpoints. However, staggered entry will profoundly affect sample size calculations for time-to-event data, as they depend on the number of events observed, and therefore both patterns of study entry and whether or not event rates are low or high (Witte 2002).

18.2.4 Data Structure

Time-to-event data are characterised by differing *time-dependent patterns* of survival (time-dependent *trajectories* or *survivorship curve shape*) and *censoring*.

Survivorship curves illustrate the distribution of survival over time, measured as the number, or proportion, of individuals surviving at each time t. The shape of the curve will be modified by the strength of the test intervention or agent, genotype, and environmental effects. Graphical examination of patterns can provide insight into potential mechanisms affecting survival in laboratory and clinical investigations. The three main patterns (Figure 18.1) were described in the last century (Pearl and Miner 1935; Demetrius 1978).

  Type I survival indicates a constant proportion of individuals dying over the course of the study.

  Type II survival exhibits high survival for most of the study. Mortality increases near the end of the monitoring period.

  Type III survival exhibits high early mortality followed by low mortality for relatively few individuals over the remainder of the monitoring period.

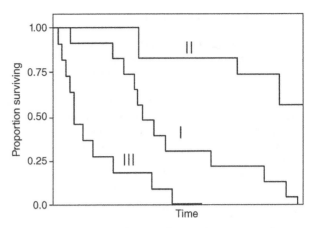

Figure 18.1: Patterns of survival over time. Type I: Constant mortality throughout the study duration. Type II: Low early mortality; high survival over most of the study period. Type III: High early mortality for most subjects, with a few longer-term survivors

Pattern interpretation should be made with caution, especially when sample sizes are small, because very small numbers in the tails of the curves make survival estimates unstable (Altman 1991; Altman and Bland 1998).

Censoring occurs when the event of interest is not observed for a subject, either because the event has not occurred by the time the study ends or because it occurred before the subject entered the study. Censored observations are not 'missing' in the conventional sense. Subjects may be *right-censored* (subjects do not have the event before the study ends), *left-censored* (subjects experience the event before the study begins), or *interval-censored* (the event occurs within some known time interval, but the exact time cannot be precisely identified) (Table 18.1; Figure 18.2).

Because of censoring, time-to-event data cannot be analysed by conventional summary statistics used for continuous data. Attempting to impose a normal distribution on these data will result in large apparent outliers. If conventional analysis methods are used, data from censored subjects will be excluded because exact survival duration cannot be calculated for censored individuals; times are known only for those subjects for which the event occurred. Omitting censored data from analyses contributes to survivorship bias.

Table 18.1: Types of Censoring.

Right censoring: The most common type. Occurs when a subject completes the study without experiencing the event.
  Type I right censoring occurs when *study duration* is fixed beforehand.
  Type II right censoring occurs when the study is completed when a *fixed number of events* has occurred (study duration is variable).
Left censoring occurs when the subject experiences the event before enrolment or monitoring begins.
Examples. In oncology trials, an animal presents with a tumour before the study begins.
Interval censoring: The subject experiences the event within some pre-specified time interval, but the exact time is unknown. Interval censoring occurs when subjects can be monitored only periodically (for example, at weekly or monthly intervals).
Examples. Veterinary studies of client-owned animals followed up at scheduled clinic visits; disease progression studies (Finkelstein and Wolfe 1985; Radke 2003; Rotolo et al. 2017); monitoring of bird nest boxes (Heisey et al. 2007; Fieberg and DelGiudice 2008).

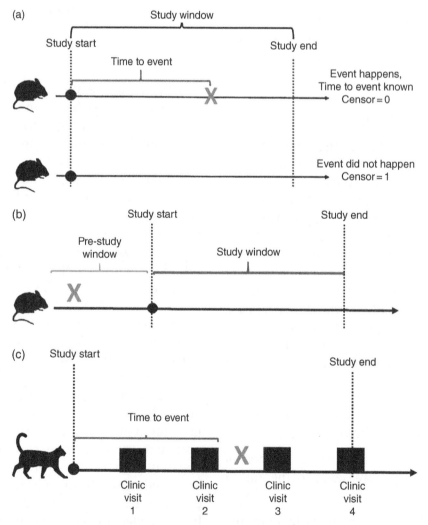

Figure 18.2: Types of censoring. (a) *Right censoring* occurs when a subject completes the study without experiencing the event. (b) *Left censoring* occurs when the event occurs before the study begins. (c) *Interval censoring* occurs when the subject experiences the event within some pre-specified time interval, but the exact time is unknown (for example, death of the animal between clinic visits).

18.2.5 'Responders' Versus 'Non-Responders' and Post Hoc Dichotomisation

After the data are examined, investigators sometimes categorise subjects as 'responders' or 'non-responders', then reanalyse survival differences between these new 'groups'. Senn (2018) lists five reasons for discouraging this practice. First, the operational definitions of 'responder' versus 'non-responder' are arbitrary, because they are usually based on some *post hoc* threshold or measure of 'improvement'. Second, it is implicitly assumed that subgroup responses will be consistent, predictable, and stable for each subject. Third, it is assumed that subgroup responses are somehow representative of population response rather than being an artefact of subgrouping composition and subject variability. Fourth, to be valid, statistical tests for group comparisons require random allocation of treatments to experimental units. Finally, because *post hoc* groups are not randomised, statistical inference tests based on *post hoc* dichotomisation are inappropriate (Weiss et al. 1983; Atkinson et al. 2019 and will be biased and misleading. If such comparisons are unavoidable, descriptive statistics are recommended rather than inference testing.

18.3 Outcome Data

Time-to-event data can be summarised as both counts and survival times. The choice of outcome variable will determine both the effect size and the method of calculating sample size.

18.3.1 Count Data

Dichotomous categorical outcomes (e.g. 'Dead/survived', 'tumour present/absent'; 'yes/no') can be summarised as *counts* (number of subjects) and *proportions p* (where p is the number of subjects with the event/total number of subjects in the group). Effect size can be estimated as *relative risk* (or risk ratio, RR) or *odds ratio* (OR). Parameter estimates of odds ratios obtained from logistic regression can be used as a measure of effect size. Because they are dependent on the scale of the predictor

variables, ln(odds ratios) are unstandardised effect sizes. Problems associated with odds ratios or relative risk for analysing time-to-event outcomes in meta-analyses have been described by Tierney et al. (2007).

Statistical comparisons of simple proportions at several times *t* are not recommended for survival studies. Proportions do not contain time and duration information, and estimates are unstable because of subject attrition over time (Machin et al. 2006).

18.3.2 Survival Times

The *Kaplan Meier survival curves* show the trajectory of events occurring over time. The Kaplan-Meier method is most appropriate for categorical predictor or grouping variables (e.g. experimental drug versus placebo control), or when there are a small number of fixed values (e.g. drug doses 0, 10, 20 mg) that can be considered categorical.

The median survival time *M* is the time at which half (50%) of the subjects have died or showed the occurrence of the event of interest (Box 18.3). The median is reliable as an effect size estimate when events occur fairly regularly over the study time period. If survival times are highly irregular, the median will tend to be biased and unreliable. Median survival cannot be calculated if survival >50% by the end of follow-up (Machin et al. 2006).

The Kaplan-Meier method calculates the *survival function S(t)*. The probability of surviving each time interval is the proportion of patients alive at each time *t*:

$$d_t = \frac{n_t - n_d}{n_t}$$

where *d* is the difference between n_t the number alive at the start of time *t* and n_d the number that died in the interval. The probability of survival to the next time interval t_i is $p_i = 1 - d_i/n_i$. The survival

---

BOX 18.3
Time to Event Effects Size: Median Survival Time

Median survival M is the time at which 50% of the subjects have died and 50% remain alive.

function $S(t)$ is obtained by multiplying the probabilities at each time point.

$S(t) = p_1 \cdot p_2 .. = (1 - d_1/n_1)(1 - d_2/n_2)....$ This is formally expressed as

$$S(t) = \prod_{i-1}^{j}\left(1 - \frac{d_i}{n_i}\right)$$

where Π is the product multiplier and j is the number of times an event is observed. $S(t)$ changes at the time of each event and so is typically presented as a series of steps. Kaplan-Meier curves can also be used to describe the accumulation of events or failures over time. The cumulative distribution function gives the probability that the survival time is less than or equal to a given time t, $S = 1 - S(t)$ (Figure 18.3). Examples include time to complete healing of wounds or ulcers (Moffatt et al. 1992) and probability of malaria recrudescence over 28 days of follow-up time (Dahal et al. 2017).

18.3.3 Hazard Rate and Hazard Ratio

The survival function $S(t)$ is related to the *hazard rate* at time t as

$$S(t) = \exp[-h(t) \cdot t]$$

The hazard rate $h(t)$ is the probability that an individual at time t has an event at that time. It represents the instantaneous event rate for an individual who has already survived to time t. The hazard rate is a measure of instantaneous risk. The

relationship between the proportion surviving (p_s) and the hazard rate $h(t)$ at a given time t, is therefore

$$h(t) = -\ln(p_s/t)$$

The relationship between p_s, $h(t)$ and the median survival time M is

$$h(t) = \frac{-\ln(p_s)}{t} = \frac{-\ln(0.5)}{M} = \frac{\ln(2)}{M}$$

The hazard ratio (HR) is the ratio of the hazard rate at time t in one group compared to hazard rate for the comparator group:

$$HR = \frac{h_1(t)}{h_0(t)} = \frac{\ln(p_1)}{\ln(p_0)}$$

Here p_0 and p_1 are the proportions of subjects with the event for the comparator and test groups, respectively. The hazard ratio can be interpreted in a similar way to a risk ratio (Box 18.4).

When two or more groups are compared, it is often assumed that the hazard ratio is constant over the follow-up period. This is the *proportional hazards assumption*. If true, the hazard ratio provides an average relative treatment effect over time. The proportional hazards assumption must be checked before the data are analysed, as the log-rank test cannot be used to test the equality of survival functions in the presence of non-proportionality. Survival

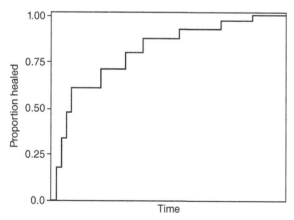

Figure 18.3: Kaplan-Meier curve of cumulative failure probability.

BOX 18.4
Time to Event Effect Size: Hazard Rate and Hazard Ratio

Hazard rate h(t): instantaneous event rate for a subject already surviving to time t

Hazard ratio (HR): ratio of the hazard rate at time t in one group compared to hazard rate for the comparator group

- HR = 1: Both groups (treatment and control) are experiencing an equal number of events at any time t.
- HR > 1: More events occurred in the treatment groups at any time t compared to the control group.
- HR < 1: Fewer events occurred in the treatment groups at any point in time compared to the control group.

curves that cross each other are a strong indicator of non-proportionality. Peto et al. (1977) describe common errors in survival analyses.

18.4 Time-to-Event Sample Size Calculations

Sample size estimates for time-to-event data require an estimate of the effect size (hazard ratio), the expected proportion of subjects in each group that might experience the event, and the expected number of subjects showing the event of interest (Box 18.5).

In effect, two sample sizes must be estimated, the number of subjects expected to have the event (n_E), and the total number of subjects (N). The number of events that occur determines the variance used in the sample size calculation. If censoring does not occur for any subject, then $E = N$. The prevalence of the disease (or proportion of subjects expected to have the event) is P_E. Then the total sample size is

$$N = \frac{n_E}{P_E}$$

Other information needed for time-to-event sample size calculations is the allocation ratios or number of subjects allocated to each group, the desired confidence and power. Two-sided tests are recommended because the existence of a survival difference to be detected or the direction of that difference is usually not known beforehand. Balanced group sizes with a 1:1 allocation ratio result in the most power. For a given power, unbalanced designs will require much large sample sizes compared to balanced designs: approximately 12% for 2:1 and 33% for 3:1 (Hey and Kimmelman 2014). If unequal ratios are planned, sample size estimates must be corrected by the pre-specified allocation ratio.

Effect size. The ratio of median survival times is often the simplest method for estimating the hazard ratio for a new study (Machin et al. 2018):

$$\text{HR} = \frac{h_1(t)}{h_0(t)} = \frac{\ln 2/M_1}{\ln 2/M_0} = \frac{M_0}{M_1}$$

where $h_1(t)$ and $h_0(t)$ are the hazard rates for the test group and the control or comparator group, respectively, and M_1 and M_0 are the corresponding medians.

Number of events. The total number of subjects with the event n_E that need to be observed in a two-group study is

$$n_E = \frac{(r+1)^2}{r} \cdot \frac{\left(z_{1-\alpha/2} + z_{1-\beta}\right)^2}{(\ln HR)^2}$$

The allocation ratio describes the balance of subjects in each arm. The allocation ratio adjustment is $(r+1)^2/r$. For an allocation ratio of 1 (balanced group sample size), the adjustment factor is $(1+1)^2/1 = 4$.

The hazard ratio can be approximated from an expected clinically relevant effect. For example, to determine a 50% decrease in events with a new treatment relative to controls, the planning value for HR would be $1/0.5 = 2.0$.

Number of subjects. The total number of subjects N is determined by the expected proportion of subjects in each group, p_1 and p_0:

$$N = \frac{n_E}{P} = \frac{n_E}{[(1-p_0) + r(1-p_1)]/(1+r)}$$

If the allocation ratio is 1:1, then

$$N = \frac{n_E}{[(1-p_0) + (1-p_1)]/2}$$

BOX 18.5

Sample Size Requirements

- Median survival time or hazard ratio (effect size)
- Proportion of subjects with the event in each group or prevalence
- Number of events
- Probabilities for Type I and II error (α and β)
- Allocation ratio.

The expected value of p_1 can be estimated from the planning values for 'baseline' or control p_0 as

$$p_1 = exp[HR_{new} \cdot \ln(p_0)]$$

If a control value is unknown, then an estimate of prevalence P can be substituted.

Example: Rat Tumour Growth

Mantel et al. (1977) monitored 150 female Sprague Dawley rats for tumour growth; 40 rats developed tumours. Suppose a new study was proposed to test a new drug anticipated to reduce the risk of developing tumours by 50% with 95% confidence ($\alpha = 0.05$) and power of 90% ($1 - \beta = 0.9$). How many rats would be needed?

In the former study $N_{old} = 150$, and $n_{E, old} = 40$. For the new study, the hazard ratio $HR = 2$. 95% confidence ($\alpha = 0.05$) $z_{1-\alpha/2} = 1.96$, and power of 90% $z_{1-\beta} = 1.282$. Then the expected number of subjects with tumours in the new study is:

$$n_{E,new} = 4 \frac{(1.96 + 1.282)^2}{(\ln 2)^2} = 87.5 \cong 88$$

The total sample size required to detect this number of tumours is

$$N = \frac{n_{E,new}}{n_{E,old}} \cdot N_{old} = \left(\frac{88}{40}\right) \cdot 150 = 330$$

What risk of tumorigenicity could be detected with power 80% and α of 5% if fewer animals are used?

The number of events $n_{E, old} = 40$, with $\alpha = 0.05$, $z_{1-\alpha/2} = 1.96$; $1 - \beta = 0.8$, $z_{1-\beta} = 0.8416$

The corresponding number of events n_E and total sample size N required for hazard ratio HR are:

$$N = (n_E/40)150$$

and $n_E = 40 = \dfrac{4 (1.96 + 0.8416)^2}{(\ln HR)^2}$. Substituting

in the values for a range of hazard ratios and calculating sample size N for each shows:

Hazard ratio	n_E	N
1.20	944	3542
1.25	631	2364
1.50	191	716
2	65	245
3	26	98
4	16	61
5	12	46

This example shows that only very large effect sizes for time-to-event data can be detected with the small sample sizes typical of many laboratory studies of rodents. A different primary outcome should be selected, when the number of animals required is too large to be operationally or ethically feasible, or the effect size that could be detected with a given power is implausible. Choose a primary outcome variable that is a reasonable surrogate, continuous, and with small variance.

18.5 Veterinary Clinical Trials

In veterinary clinical trials, sample size estimates will be affected by additional factors such as client non-compliance, losses to follow-up, and duration of the trial. These need to be considered in the planning stages because they will determine the number of events, which in turn determine the variance used in the sample size calculations.

Loss to follow-up. In veterinary clinical studies, subjects that drop out or are lost to follow-up can result in the loss of considerable information because events for these subjects cannot be recorded. Sample size must be increased, or adjusted, to account for these losses (N_{adj}):

$$N_{adj} = N/\left(1 - p_f\right)$$

where p_f is the fraction of subjects anticipated to be lost to follow-up. It is assumed that losses to follow-up occur at random and are not related to the health status of the subject or the intervention (Machin et al. 2006).

Study duration. The minimum or lower limit to study duration T_L is

$$T_L = n_E/R$$

where n_E is the number of events, and R is the anticipated entry rate of subjects into the trial. The entry or recruitment rate R is estimated by multiplying the number of randomised subjects by the recruitment time, $R = N \cdot t$. If there are multiple recruiting sites, R is multiplied by the number of sites n_{sites}, so that $R_{multi} = R \cdot n_{sites}$

The total duration of patient entry into the trial is approximated by the median survival time of the control group (or whatever group has the longer time to develop signs and symptoms) plus the lower limit:

$$T_{total} = M_c + T_L$$

A more complex equation requiring iterative solutions is available, but this method is preferred for its simplicity (Machin et al. 2006).

Example: Clinical Trial of Dogs with Cardiac Disease – Two-Arm Trial

---

Boswood et al. (2016) conducted a randomised, blinded, controlled trial comparing pimobendan to placebo for treating dogs showing echocardiographic signs of cardiac enlargement. One goal was to determine if pimobendan delayed the onset of congestive heart failure (CHF). Subjects were randomised 1:1 to treatment arms. There was a total enrolment of 360 subjects, of which 59 in the pimobendan group and 76 in the placebo group developed CHF. Median times to develop symptomatic CHF were 1228 days and 766 days for pimobendan and control groups, respectively. The power $(1 - \beta)$ of the study was 0.8.

Number of subjects. How many subjects would be needed for a new study, where the goal was to increase the success rate to 70% with α of 5% and power 80%, and assuming a loss to follow-up of 10%?

The 'success rate' for the drug in the previous trial is $(180 - 59)/180 = 0.67$, or 67% (versus 58% for the placebo).

The *hazard ratio* $HR = \ln(0.70)/\ln(0.58) = 0.65$

The number of events $E = \dfrac{4(1.96 + 0.8416)^2}{(\ln 0.65)^2} = 169.4$

$$\cong 170$$

Assuming the proportion of event in the placebo group will be $76/180 = 42\%$, the anticipated *proportion of events* in the intervention group is

$$p_1 = \exp[0.65 \cdot \ln(0.42)] = 0.57$$

Then $N = 170/[(1 - 0.57) + (1 - 0.42)]/2 = 335.5 \cong 336$

With 10% loss to follow-up, $p_f = 0.1$, and

$$N_{adj} = 336/(1 - 0.1) \cong 373$$

Study duration. How much time would a new trial require to recruit the requisite number of subjects?

Suppose the time to recruit 360 subjects in the previous trial was 1.5 years or 548 days. Then $R = 360/548 = 0.67$ subjects/d, and $T_L = 170/0.67 = 259$ days. The total estimated study duration is $1228 + 259 = 1486.5$ days or just over 4 years.

A study of this size and duration usually requires multiple participating sites to recruit enough patients in a reasonable period of time. However, organisation, logistics, coordination, and expense may be considerable.

18.6 Other Study Design Considerations

18.6.1 More Than Two Groups

Sample size calculations can be challenging when time-to-event outcomes must be compared for more than two groups. Most commercial software packages require Cohen's f or η_p^2 for effect sizes, which are not applicable to time-to-event studies. Machin et al. (2018) recommend a three-step procedure for a factorial design where interactions may be of interest. Suppose a study is proposed evaluating two factors A and B. With a 2×2 factorial design, there are four 'groups' to be evaluated: A only, B only, both A and B, and neither A or B. Therefore, to assess the effect of each factor, there are four median survival times to be estimated and two hazard ratios to be compared.

1. Estimate hazard ratios for A and B separately (assuming no interaction). Calculate N for each.
2. If the N for each group is roughly similar, choose the larger N. If hazard ratios (and hence Ns) are very different, prioritise factors based on importance to the research question. For example, suppose A is a comparison of the therapeutic effect of the test drug against a control, versus B which is a comparison of intravenous versus oral drug delivery routes. Factor A will be of higher priority if the goal is to determine efficacy. Power off the most important factor.
3. If interactions ($A \times B$) to assess synergism or antagonism are of greater interest than main effects, then estimate the size of the interaction, and power the study based on the interaction. Be aware that, in general, estimating interactions requires roughly 16 times the sample size compared to that for estimating a main effect for equivalent power.[2]

References

Altman, D.G. (1991). *Practical Statistics for Medical Research*. London: Chapman & Hall/CRC.

Altman, D.G. and Bland, J.M. (1998). Time to event (survival) data. *BMJ* 317 (7156): 468–469.

Atkinson, G., Williamson, P., and Batterham, A.M. (2019). Issues in the determination of 'responders' and 'non-responders' in physiological research. *Experimental Physiology* 104 (8): 1215–1225.

Boswood, A., Häggström, J., Gordon, S.G. et al. (2016). Effect of pimobendan in dogs with preclinical myxomatous mitral valve disease and cardiomegaly: the EPIC study-a randomized clinical trial. *Journal of Veterinary Internal Medicine* 30 (6): 1765–1779. https://doi.org/10.1111/jvim.14586.

Dahal, P., Simpson, J.A., Dorsey, G. et al. (2017). Statistical methods to derive efficacy estimates of antimalarials for uncomplicated *Plasmodium falciparum* malaria: pitfalls and challenges. *Malaria Journal* 16 (1): 430. https://doi.org/10.1186/s12936-017-2074-7.

Demetrius, L. (1978). Adaptive value, entropy and survivorship curves. *Nature* 275: 213–214.

Fieberg, J. and DelGiudice, G. (2008). Exploring migration data using interval-censored time-to-event models. *Journal of Wildlife Management* 72 (5): 1211–1219. https://doi.org/10.2193/2007-403.

Finkelstein, D.M. and Wolfe, R.A. (1985). A semiparametric model for regression analysis of interval-censored failure time data. *Biometrics* 41 (4): 933–945.

Furuya, Y., Wijesundara, D.K., Neeman, T., and Metzger, D.W. (2014). Use and misuse of statistical significance in survival analyses. *MBio* 5 (2): e00904-14. https://doi.org/10.1128/mBio.00904-14.

Heisey, D.M., Shaffer, T.L., and White, G.C. (2007). The ABCs of nest survival: theory and application from a biostatistical perspective. *Studies in Avian Biology* 34: 13–33.

Hey, S.P. and Kimmelman, J. (2014). The questionable use of unequal allocation in confirmatory trials. *Neurology* 82 (1): 77–79. https://doi.org/10.1212/01.wnl.0000438226.10353.1c.

Machin, D., Campbell, M.J., Tan, S.B., and Tan, S.H. (2018). *Sample Sizes for Clinical, Laboratory and Epidemiology Studies*, 4e. Wiley-Blackwell 390 pp.

Machin, D., Cheung, Y.B., and Parmar, M. (2006). *Survival Analysis: A Practical Approach*, 2e. 278 pp.

Mantel, N., Bohider, N.R., and Ciminera, J.L. (1977). Mantel-Haenszel analyses of litter-matched time-to-response data, with modifications for recovery of inter-litter information. *Cancer Research* 37: 3863–3868.

Moffatt, C.J., Franks, P.J., Oldroyd, M. et al. (1992). Community clinics for leg ulcers and impact on healing. *BMJ* 305 (6866): 1389–1392. https://doi.org/10.1136/bmj30568661389.

National Research Council (2011). *Guide for the Care and Use of Laboratory Animals*, 8e. Washington: The National Academies Press.

Pearl, R. and Miner, J.R. (1935). Experimental studies on the duration of life. XIV. The comparative mortality of certain lower organisms. *Quarterly Review of Biology* 10 (1): 60–79.

---

[2] Gelman A (2018) https://statmodeling.stat.columbia.edu/2018/03/15/need-16-times-sample-size-estimate-interaction-estimate-main-effect/

Peto, R., Pike, M.C., Armitage, P. et al. (1977). Design and analysis of randomized clinical trials requiring prolonged observation of each patient. II. Analysis and examples. *British Journal of Cancer* 35 (1): 1–39. https://doi.org/10.1038/bjc.1977.1.

Radke, B.R. (2003). A demonstration of interval-censored survival analysis. *Preventive Veterinary Medicine* 59: 241–256.

Rotolo, M.L., Sun, Y., Wang, C. et al. (2017). Sampling guidelines for oral fluid-based surveys of group-housed animals. *Veterinary Microbiology* 209 (20-29): 2017.

Senn, S. (2018). Statistical pitfalls of personalized medicine. *Nature* 563 (7733): 619–621.

Tierney, J.F., Stewart, L.A., Ghersi, D. et al. (2007). Practical methods for incorporating summary time-to-event data into meta-analysis. *Trials* 8: 16.

Weiss, G.B., Bunce, H. 3rd, and Hokanson, J.A. (1983). Comparing survival of responders and nonresponders after treatment: a potential source of confusion in interpreting cancer clinical trials. *Controlled Clinical Trials* 4 (1): 43–52. https://doi.org/10.1016/s0197-2456(83)80011-7.

Witte, J. (2002). Sample size calculations for randomized controlled trials. *Epidemiologic Reviews* 24 (1): 39–53.

19 Comparing Multiple Factors

CHAPTER OUTLINE HEAD

19.1 Introduction

Comparative experiments are designed to investigate cause-and-effect relationships between one or more independent variables (treatments) on the dependent variables, or response. Typically these are analysed by analysis of variance (ANOVA) methods. Correct application of an ANOVA-type analysis is determined by how the experiment is designed before data are collected (Box 19.1). The ANOVA is a linear statistical regression model where the outcome variable is predicted from one or more quantitative or qualitative categorical independent variables. The structuring of multiple independent variables (or factors) according to the number of independent factors to be compared

A Guide to Sample Size for Animal-based Studies, First Edition. Penny S. Reynolds.
© 2024 John Wiley & Sons Ltd. Published 2024 by John Wiley & Sons Ltd.

BOX 19.1
Comparing Two or More Groups

Key concepts
   Factor: One independent variable
   Level: a group within a factor
   Block: a categorical classifier for a specific nuisance factor used to minimise between-subject variation

Common designs:
   Single-factor completely randomised design
   Randomised complete block
   Factorial
   Split plot
   Crossover
   Nested or hierarchical design.

---

and how variation is to be controlled makes up the experimental design. A good design allows discrimination of the experimental signal from noise caused by extraneous variation (Winer et al. 1991; Box et al. 2005; Montgomery 2012).

Design and analysis of variance methods were developed primarily by Sir Ronald Fisher in the early half of the twentieth century and subsequently extended by George Box and collaborators (Box et al. 2005). Detailed coverage of experimental design is beyond the scope of this book. However, an overview of a few basic designs and methods for estimating sample size are provided. The reader is advised to consult Box et al. (2005), Montgomery (2012), Bate and Clark (2014), Lazic (2016), and Karp and Fry (2021) for more detailed information.

19.2 Design Components

Fisher defined the design structure as the logical strategy by which treatments or treatment combinations are assigned to experimental units. The design structure is based on the number and relationship of independent variables, the *factors*. Each factor has a restricted number of fixed values (or levels). In the simplest case, the single-factor (or one-way) design has one factor and two or more levels. A *blocking factor* is a categorical classifier for a specific nuisance factor. Blocking factors are incorporated to minimise variation between experimental units. Other designs include the randomised block design (RCBD), factorial, crossover, and split-plot designs, repeated measures (within-subject) designs, and hierarchical or nested designs

(discussed in Chapter 20). The most appropriate method for calculating sample size will be closely aligned with the study objectives, the experimental design, and the intended statistical analysis plan (O'Brien and Muller 1993; Hurlbert and Lombardi 2004; Simpson 2015; Gamble et al. 2017).

19.2.1 Constructing the *F*-Distribution

Sample size and power calculations are based on the *F*-distributions for the corresponding null and alternative distribution of each factor in the design (Chapter 14). The total variance of the data is partitioned into different components that discriminate between the different sources of variation. In the simplest case, there are three components: variation between experimental units that receive different treatments (between-group), variation between experimental units receiving the same treatment (within-group), and the residual error variance (or mean square error) that estimates the random variation between experimental units. The ratio of the between-group to within-group mean square is the *F*-statistic for testing the hypothesis of equality of group means. The *F*-value equals one if there is no difference between group means.

> *A note on 'groups'.* The idea of the 'group' as the entity subjected to treatments has occasionally resulted in serious misunderstandings of both the role and method of treatment allocation and randomisation in ANOVA-type designs (Festing 2014; Reynolds 2022). In a properly designed experiment, treatments (experimental or control interventions) are allocated randomly to the experimental units. However, a common practice in the preclinical rodent literature is to refer to experimental 'groups' as 'cohorts'. The term cohort actually has a formal technical meaning. As used in observational studies, a cohort is a group of subjects with group membership defined by some common characteristic or trait. When experimental groups are incorrectly defined as cohorts, this suggests that interventions have been nonrandomly allocated, an invalid and confounded assignment strategy (Festing 2020). The emphasis on 'groups' has also resulted in neglect of more powerful experimental designs more suited to exploratory studies with multiple factors.

A note on strain or species comparisons. If the study design involves comparison of multiple strains, sample size calculations will depend on the inference goals and the types of biological units in the study. Power to detect differences between strains depends on the number of strains. At a minimum, three strains are recommended to test hypotheses of strain (or species) differences (Garland and Adolph 1994). Precision of the estimated effect depends on the number of animals sampled within each strain. Environmental effects due to housing, husbandry, handling, cage and cage stocking density, and even source or vendor can result in large between-animal variation in physiological and behavioural measurements. Variation resulting from the effects of such confounders should be controlled by appropriate blocking strategies. Large formal analyses with multiple strains or species should consider phylogenetic comparative methods (Cooper et al. 2016).

19.3 Sample Size Determination Methods

The investigator should specify beforehand the major components required for sample size determination (Chapter 14). Choice of study design will also depend on the number and type of factors to be studied, the number of levels of each factor, and how they will be spaced; methods for controlling variation (blocking, stratification, covariates), and methods for incorporating time or within-subject dependencies (repeated measures, longitudinal data).

Power is maximised when the number of experimental units n are balanced across the design. Unbalanced sample size allocation results in increased variance (and therefore reduced precision). Unnecessarily large variations can obscure true treatment effects, especially when effects are small. The effects of sample size imbalance on variation and effect size are described in Chapter 14.

19.3.1 Effect Sizes

Effect size for ANOVA-type designs are computed as f, η^2, η^2_p, the coefficient of determination R^2, or adjusted R^2_{adj}. Many commercially available sample size calculation programmes such as G∗Power use estimates of effect size. Effect sizes should be based on best-available evidence, practical significance, and the proposed statistically based design for future experiment. Information for preliminary calculations can be obtained from previous studies, pilot data, or exemplary data. Exemplary data are essentially mock data, constructed to have the characteristics expected for the future sample, such as means and variances or distributional properties. Exemplary data can be determined from previous similar studies, informed and scientifically justifiable guesswork, and/or computer simulations (Goldman and Hillman 1992; O'Brien and Muller 1993; Divine et al. 2010). Researchers should not define 'small', 'medium', and 'large' effect sizes based on general benchmark values. These were originally developed for social and psychology studies and are usually neither appropriate nor relevant for animal-based studies (see Chapter 15).

19.3.2 Non-Centrality Parameter

Sample size formulae may not be available for the more complex study designs. However, sample sizes can be customised using simulations based on the non-centrality parameter λ (Chapter 14). Typically, calculations of λ use values for the minimum biologically important mean difference to be detected and an estimate of the pooled variance of the difference. Other statistics that can be used are summary statistics for each group (means and a measure of variation, SD or SEM), the group means with an estimate of the pooled variance, or summary ANOVA statistics (mean square error or F-values) derived from either prior or exemplary data (O'Brien and Muller 1993; Castelloe and O'Brien 2001). Sample size calculations will require specifying the analysis objectives; that is, if the goal is to determine (i) a pre-specified biologically important *difference* between groups, (ii) the *allowable range of variation* that enables detection of the target difference, or (iii) the *maximum number of groups* that can be tested for a fixed total sample size. Sample SAS codes for simulation-based approximations of sample size are provided in Appendix 19.A.

19.3.3 Mead's Resource Equation

Although originally developed for agricultural research, Mead's resource equation (Mead 1988) has been suggested for laboratory animal studies (Festing 2018). The resource equation is based on the idea that the total information of an experiment with N experimental units can be determined from the total variation based on $N - 1$ degrees of freedom. In a simple experiment, the resource equation partitions the variation among three sources: the treatment component T, the blocking component B, and the error component E:

$$T + B + E = N - 1$$

The treatment degrees of freedom are k, where k is the number of groups. The blocking variable B with $b - 1$ degrees of freedom, is a known categorical variable used to minimise variation from sources not of direct interest to the experimenter. The error component is used for testing hypotheses. Mead (1988) recommends a sample size that allows 10–20 degrees of freedom for the error term.

Mead's resource equation is most appropriate for single-factor designs. A further disadvantage is that the inclusion of a block term seemingly penalises the experiment by reducing degrees of freedom for the error term. However, this is misleading because blocking removes the variation associated with the nuisance variable from the error term and therefore can contribute substantially to the power and precision of the experiment.

Example: Mead's Resource Equation: Simple Experiment

An experiment is planned with four treatments with five mice per group, for a total of 20 mice. Here, $N = 20$, $B = 0$, $T = 4$, so $E = 19 - (4 - 1) = 16$, suggesting the sample size is adequate. If the experiment is to be run in 5 blocks, then $E = 19 - (5 - 1) - (4 - 1) = 12$.

19.3.4 Skeleton ANOVA

The skeleton ANOVA can be thought of as an extension of the resource equation. The skeleton ANOVA table lists the sources of variation and corresponding degrees of freedom for a specific experimental design (Table 19.1). Candidate values for the number of factors, levels of each factor, and number of experimental units can be selected, and corresponding degrees of freedom calculated from those values. The skeleton analysis of variance is therefore a relatively simple method of evaluating the likely

variance for estimates for each factor and factor interactions, the effect of blocking, the magnitude of each relevant residual error term, potential confounding, and information about the power of hypothesis tests. A skeleton ANOVA is especially useful for estimating sample sizes for complex designs. Sample size formulae may not even exist, and determine the optimal allocation of experimental units among multiple sources of variation can be challenging. Examples of complex designs include split-plot and factorial designs, designs with complex blocking and clustering, and repeated measures on different factors (Brien 1983; Draper and Smith 1991; Brien and Bailey 2006, 2009, 2010). The planning rule of thumb is that there should be at least one degree of freedom for all main factors and interactions of interest and 10–20 degrees of freedom for the residual error term for testing hypotheses (Mead 1988).

19.4 Completely Randomised Design, Single Factor

The single factor or one-way ANOVA consists of one factor with k levels or 'groups'. Experimental units are independent of each other, and treatments are randomly allocated to each experimental unit. The main disadvantage of the completely randomised design is that different sources of variation not accounted for by the model will go into the error term. This inflates the mean square error, increasing the error variance and reducing power.

There are three variation components for a single-factor completely randomised design: variation between units receiving different treatments (between-treatment or between-group), the variation between experimental units receiving the same treatment (within-group), and the residual error term (mean square error MSE) that estimates the random variation between experimental units. The ratio of the between-group to within-group mean square is the F-statistic (Table 19.1).

For a balanced design with equal sample sizes for k levels (commonly referred to as 'groups'), the minimal sample size per group n is

$$n \geq \frac{2 \left(F_{1-\alpha, k-1, k(n-1)} \right) \cdot MSE}{\bar{d}^2}$$

Table 19.1: Skeleton ANOVA for common study designs. N is the total number of experimental units (EUs), n is sample size per group (number of replications of EUs), and σ_e^2 is the error or residual variance.

Source of variation	Degrees of freedom
A. Single-factor (one-way) ANOVA, completely randomised design	
Due to 'treatment' (between k groups)	$k - 1$
Residual error	$k(n - 1) = N - k = \sigma_e^2$
Total (corrected)	$kn - 1 = N - 1$
Total	N
B. Single-factor randomised complete block	
Due to 'treatment' (k groups)	$k - 1$
Due to blocks (b blocks)	$b - 1$
Group × block interaction = Experimental error	$(k - 1)(b - 1)$
Within-group variation = Sampling error (MSE)	$kb(n - 1) = \sigma_e^2$
Total (corrected)	$kbn - 1 = N - 1$
C. Two-factor (two-way) ANOVA; 2 × 2 Factorial	
Due to factor A	$a - 1$
Due to factor B	$b - 1$
A × B interaction	$(a - 1)(b - 1)$
Error	$ab(n - 1) = \sigma_e^2$
Total (corrected)	$abn - 1 = N - 1$
D. Crossover design	
Due to 'treatment' (A versus B)	$k - 1$
Period	$p - 1$
Sequence (AB versus BA)	$s - 1$
Subject (sequence)	$n(t - 1)$
Within-group variation	$n - 1$
Error	$(k - 1)(nk - 2)$
Total (corrected)	$nsk - 1 = N - 1$
E. Hierarchical (nested) design B within A	
Due to 'treatment' A (between treatments)	$a - 1$
B nested within A, B(A)	$a(b - 1)$
Error	$ab(n - 1)$
Total (corrected)	$abn - 1 = N - 1$
F. Single factor A with repeated measures on factor time	
A	$a - 1$
Time	$t - 1$

(continued)

Table 19.1: (Continued)

Source of variation	Degrees of freedom
Subjects (A)	$a(n-1)$
Within-subjects	$an(t-1)$
Total (corrected)	$atn - 1 = N - 1$

G. Two-factor design on factors A and B, with EUs randomised to A and repeated measures on B, such that subjects are nested within A: S(A)

Between treatments	
A	$a - 1$
B	$b - 1$
A × B	$(a - 1)(b - 1)$
Error	$ab - 1$
Within treatments	
S(A)	$n - 1$
B × S(A)	$(n - 1)(b - 1)$
Error	$ab(n - 1)$
Total (corrected)	$abn - 1 = N - 1$

H. Split-plot design

Whole 'plot' (effects)	
Replicate on A	$r - 1$
Whole plot A	$a - 1$
Whole plot error	$(r - 1)(a - 1)$
Whole plot error	
Subplot B	$b - 1$
Subplot B × Main A	$(a - 1)(b - 1)$
Subplot error	$rabn - a(r + b - 1)$

where $\overline{d}$ is the mean difference to be detected, MSE is the mean square error (the estimate of the residual variance), and $F_{1-\alpha,v_1,v_2}$ is the critical F-value with confidence $(1 - \alpha)$, and $v_1 = (k - 1)$ for between-treatment degrees of freedom, and $v_2 = k(n - 1)$ for within-group degrees of freedom. The effect size is $\overline{d}/\sqrt{MSE}$. The non-centrality parameter is

$$\lambda = \frac{n \cdot \overline{d}^2}{2 \cdot (MSE)^2}$$

For two groups with group size n, the total sample size is $2 \cdot n$. For multiple treatment levels k, the total sample size is $k \cdot n$.

The following examples show how to calculate sample size from the non-centrality parameter and iterating over a range of candidate n (Appendix 19.A). Pre-specified values for mean difference, variance, and confidence are entered, and the corresponding non-centrality parameter and power are calculated for each n. The appropriate sample size is selected by matching n to that which most closely corresponds to the pre-specified power (for example, 80%).

Example: Guinea-Pig Weight Gain

(Data from Zar 2010.) A small trial was conducted to evaluate the effect of four different diets on weight gain (g) of guinea pigs. The factor was diet with four levels (A, B, C, D). Diets were randomly assigned to each of 20 guinea-pigs. There were five guinea-pigs per diet group ($n = 5$/group). The raw data are:

A	7.0	9.9	8.5	5.1	10.3
B	5.3	5.7	4.7	3.5	7.7
C	4.9	7.6	5.5	2.8	8.4
D	8.8	8.9	8.1	3.3	9.1

The summary statistics (mean, standard deviation) are:

A : 8.2 (2.1); B : 5.4 (1.5); C : 5.8 (2.2); D : 7.6 (2.5)

The ANOVA statistics are:

Source of variation	df	SS	MS	F	P-value
Between groups	3	27.426	9.142	2.034	0.15
Within groups	16	71.924	4.495		
Total	19	99.350			

The mean square error MSE is 4.5 and $R^2 = 0.28$. The largest difference in weight gain was observed between diets A and B. The difference was 2.8 g with 95% confidence limits −0.06–5.62 g. The precision of the difference estimate is $[5.62 − (−0.06)]/2 = 2.84$ g.

19.4.1 Sample Size Approximations Based on Mean Differences

A new study is planned where it is desired to detect a pre-specified mean difference of 2.5 g and 5.0 g between at least two of the four diet groups. What is the minimum sample size that will ensure a power of at least 80% for $\alpha = 0.05$ to detect a

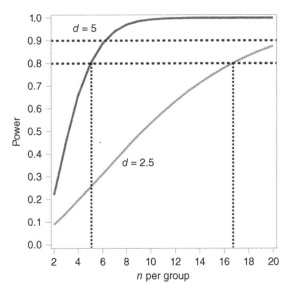

Figure 19.1: Guinea-pig growth: Power of single-factor four-level completely randomised design over a range of sample sizes for two effect sizes (maximum mean difference between two diets). Vertical dotted lines show sample size for a target power of 80%. Horizontal dotted lines indicate the power of 80% and 90%.

difference of at least 2.5 g? A difference of 5.0 g? (These differences correspond to approximate effect sizes of 1 and 2, respectively).

Figure 19.1 shows the power of a single-factor, four-level completely randomised design over a range of sample sizes and two effect sizes: 2.5 g and 5.0 g. For a mean difference of 2.5 g with four groups, 17 guinea-pigs per group ($N = 68$) would be required for a realised power of 0.81 ($\lambda = 11.81$). For a mean difference of 5.0 g, a sample size of 5–6 per group (total sample size $N = 20$–24) would be required for a realised power of 0.80–0.89 ($\lambda = 13.47$). Sample size per group becomes smaller as the size of the effect to be detected increases.

19.4.2 Sample Size Approximations Based on Number of Levels for a Single Factor

The number of levels of a single factor ('groups') that can be realistically tested for a given power can be determined by iteration over a plausible range of experiment sizes. For the guinea pig example, group sample sizes were estimated for experiments consisting of 2–6 groups and a mean difference of 2.5 g with power of 0.8 and confidence of 95% (Figure 19.2). For a two-group ($k = 2$)

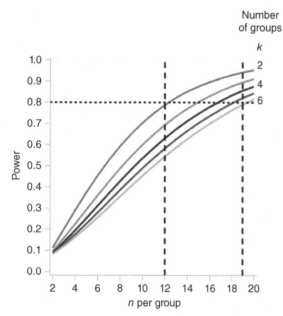

Figure 19.2: Guinea-pig growth: Sample size per group in a single-factor four-level design for two to six levels (or groups). Increasing the number of groups to be compared greatly increases the sample size required to obtain a given power.

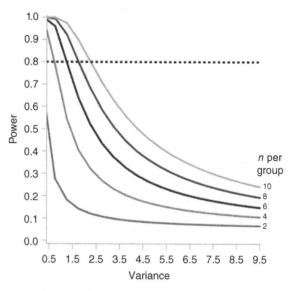

Figure 19.3: Sample size and variance. Increasing error variance greatly increases the sample size required to obtain a given power.

experiment, a sample size n of 12–13 per group is required for a total 24–26 animals. For 6 groups ($k = 6$), the sample size required is 19–20 per group, for a total of 114–120 animals. Sample sizes per group become smaller as the number of groups is reduced.

19.4.3 Sample Size Approximations Based on Expected Range of Variation

Sample size and expected power can be estimated for a range of variances by substituting estimates for the grand mean and the mean square error into the minimal difference term. For the guinea pig data, the grand mean is

$$\overline{Y} = (8.16 + 5.38 + 5.84 + 7.64)/4 = 6.8\,g$$

The non-centrality parameter λ is

$$\lambda = n \cdot \frac{\sum a_i^2}{MSE}$$

where $\sum_{i=1}^{k} a_i^2 = \sum_{i=1}^{k} (y_i - \overline{Y})^2$ the sum of the squared corrected group means, with mean square error MSE. For the guinea-pig data, the non-centrality parameter is:

$$\lambda = n \cdot \frac{(8.16 - 6.8)^2 + (5.38 - 6.8)^2 + (5.84 - 6.8)^2 + (7.64 - 6.8)^2}{MSE}$$

In this example, sample sizes were approximated over a range of MSE from 0.5 to 9.5 (Figure 19.3). These values represent effect sizes of approximately 0.55 (small) to over 10 (implausibly large). In this example, a sample size of 4 per group might be adequate if variance is less than ~20% of the mean difference. However, the required sample size per group increases considerably with increasing variance. Even a sample size of 10 per group will not provide sufficient power to detect the target difference of 2.5 g if variance is much greater than 2.5.

In the presence of large variation, the number of animals required to obtain a desired power of at least 80% power will be unacceptably large or will require implausibly large effect sizes (Figure 19.4). Study refinement to minimise the effects of between-animal variation and alternative designs should be considered. Candidate designs include randomised complete block and factorial designs.

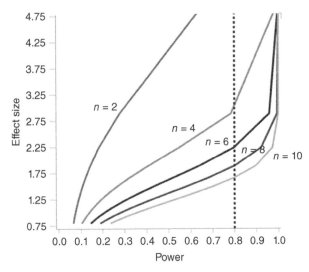

Figure 19.4: Sample size, effect size and power.

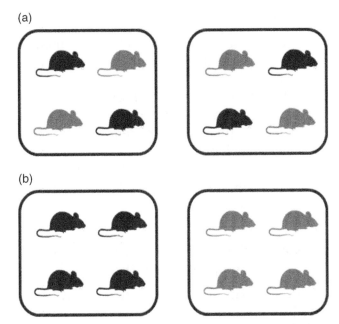

Figure 19.5: Blocking and pseudo-replication. (a) *Blocking*. There are eight mice housed four to a cage. Two mice in each cage receive drug A, and two receive drug B. The cage is the block, and the individual mouse is the experimental and the biological unit. Both levels (A, B) of the treatment (Drug) are represented in each block. The total sample size is eight mice, with four mice per drug. (b) *False block (pseudo-replication)*. There are eight mice housed four to a cage. All four mice in one cage receive drug A, and all mice in the second cage receive drug B. Although individual mice are biological units, the experimental unit is now the cage. The total sample size is now two (not eight). There will be zero degrees of freedom to test treatment effects. Analysing the data from the eight mice as if they are independent experimental units artificially inflates the true sample size.

19.5 Randomised Complete Block Design

The randomised complete block design allocates experimental units according to a predefined characteristic (the blocking variable) into homogeneous groups (blocks), after which treatments are randomly allocated to units within each block. The randomised complete block design is more powerful than a completely randomised design. Blocking improves power and precision by controlling and minimising biological variation, thereby improving the detection of the treatment signal. An added advantage of this class of designs is that the systematic incorporation of blocking ('planned heterogeneity') substantially improves the reliability, validity, and reproducibility of research results (Voelkl et al. 2018; Würbel et al. 2020).

The *block* is a group of experimental units expected to be more homogeneous than others based on their classification by a given nuisance factor. The block term is identified by a categorical classifier. It is not a new factor or treatment. Nuisance or blocking variables can be properties of the animals (age, litter, cage or tank), experimental conditions (cage location, time of day, week of study entry, specific technical staff performing the experiment, donor), and experimental location (different laboratories, different sites). Inappropriate specification of the experimental unit and block results in artificial inflation of the true sample size or *pseudo-replication* (Figure 19.5).

In a balanced block design, every level of the intervention or treatment factor occurs the same number of times within each level of the block. Treatments are randomly assigned to the experimental units within each block. If there are k treatments, then there should be at least k experimental units in each block. Computer-generated block designs can be derived in SAS (e.g. *proc plan* and *proc factex*) and in R (e.g. 'blocksdesign').

The skeleton ANOVA for a randomised complete block design with a single factor (Table 19.1) accounts for two sources of error variation. The experimental error term quantifies the random variation between experimental units and is calculated by the interaction between block and treatment. Within-group variation (the mean square error) is due to replication on the blocks and is a measure of sampling error.

Method

1. Define the blocking variable so that homogeneous experimental units are grouped together.
2. Randomise *treatments* allocated to each unit or animal within each block. Ideally, ensure that all treatments are represented equally in each block.
3. Include the *(treatment x block) interaction* in the analysis to assess the effect of the block as a source of variation.

Example: Guinea-Pig Growth: Randomised Complete Block Design

The 20 animals in the study were actually measured in 'batches' over five weeks. A batch consisted of one complete replicate of all four diets. The blocking variable is the week of study entry. Each block contained four animals, with one of four diets randomly allocated to each guinea-pig in each block.

Diet	Guinea-pig weight gain (g)				
	Block 1	Block 2	Block 3	Block 4	Block 5
A	7.0	9.9	8.5	5.1	10.3
B	5.3	5.7	4.7	3.5	7.7
C	4.9	7.6	5.5	2.8	8.4
D	8.8	8.9	8.1	3.3	9.1

The ANOVA table is

Source of variation	df	SS	MS	F	P-value
Diet (Between groups)	3	27.426	9.142	11.825	0.0007
Block	4	62.647	15.662	20.259	<0.0001
Within groups	12	9.277	0.773		
Total	19	99.350			

Blocking was clearly effective in controlling for variation between animals. Compared to the completely randomised design (where the effect of blocking was ignored), diet was now flagged up as statistically significant. Although the block effect required $(5 - 1) = 4$ degrees of freedom (with correspondingly fewer degrees of freedom for the error term), the mean square error is 0.773, a seven-fold reduction in variation. The largest difference between means is for diets A and B is 2.8 g with 95% confidence interval of (1.6, 4.0). The precision for the estimate of the mean difference is therefore $2.4/2 = 1.2$. The R^2 is 0.91, indicating considerable improvement in model fit.

For a sample size calculation based on this revised estimate of variance, only 4 guinea-pigs per group (total sample size $N = 16$) are required to detect a mean difference of 2.5 g for a realised power of 0.827, and 5 per group ($N = 20$) for realised power of 0.93.

Example: Experiment to be Conducted over Several Days

An investigator wished to conduct a single-factor three-treatment experiment on mice, with two experimental drugs and one control. The experimental unit is the individual animal. A power analysis indicated the total required sample size is 75 with $n = 25$ mice per group. However, only 15 animals can be processed in one day. Therefore, the experiment will have to be conducted over five days.

One solution is to set up the study design as a randomised complete block design, with day as the blocking variable. The experiment is operationalised by randomly assigning animals to one of five blocks (block = run day) so that there are 15 animals for each block. Then treatments are randomly allocated to each animal within block, so that all treatments are represented equally in each block. In this example, there will be 5 animals/treatment group on each day.

Multi-batch designs are a type of randomised controlled block experiment where the entire experiment is replicated, and systematic heterogeneity accounted for by replicate × treatment interactions. Heterogeneity factors can be extended from different time periods within the laboratory to different experimenters and different laboratories (Karp et al. 2020).

Example: Mouse Oncology: Blocking on Donor

An experiment was proposed to test the effect of three candidate drugs on a mouse model of cancer, where recipient mice are inoculated with cells from donor mice. One donor mouse could supply enough cells for three recipients. The study was originally powered with total sample size of 30 recipient mice and 10 donor mice, for a total sample size of 40. However, the investigator was concerned about variation between donors affecting results, and proposed doubling the original sample size 'to account for the variation'.

The study can be designed as a randomised block, with donor mouse as the block. Each drug treatment A, B, and C can be represented once in each block. Then the number of blocks required = 30/3 = 10 donor mice. Donor mice are randomly allocated to 10 blocks each consisting of three recipient mice. The drugs A, B, and C are then randomly allocated to the recipient mice in each block. The experiment is effectively replicated ten times without increasing the total number of mice, and variation is accounted for by including (donor x treatment interaction) in the analysis.

19.6 Factorial Designs

Many laboratory animal-based studies are essentially exploratory in nature. The goal is usually to assess several or many factors of potential interest, with the objective of identifying the most promising and screening out those that appear less promising. Factorial designs are ideal for this type of experimental as they are nimble, efficient, and well-suited to detect interactions between multiple factors, which results in most 'discovery by chance' (Box et al. 2005). For most practical purposes, results are easily quantified by second-degree least squares polynomial regression, with effect sizes of all main effects and two-way interactions quantified by the respective regression coefficients:

$$Y = \beta_0 + \beta_1 X_1 + \beta_2 X_2 + \cdots + \beta_{12} X_1 X_2 + \cdots + e_{ij}$$

A major advantage of factorial designs is that because they require relatively few runs per factor, information is maximised for relatively small sample sizes. This class of designs is extremely sparing of animal use (Bate and Clark 2014, Karp and Fry 2021).

Factorial designs can consist of any combination of factors and number of levels (Table 19.1). The number of possible treatment combinations is found by multiplying the number of factors k by the number of levels in each factor. The fundamental workhorse design (and the basis for screening designs) is the 2^k factorial, with k factors and two levels for each factor. For example, a 2^3 full factorial design consists of three factors, each at two levels, for eight possible treatment combinations.

Example: Factorial Designs and Number of Treatment Combinations

A study was proposed to compare immune response of three strains of mice raised on two temperatures and administered one of 4 drugs (3 test compounds, 1 control). This is a $3 \times 2 \times 4$ factorial with 24 treatment combinations.

For simplicity and ease of interpretation, the number of levels for a quantitative factor is deliberately restricted to two or three values: a low value and a high value that bracket the expected range of possible response and an intermediate, or centre, value that allows testing for curvature in the response. These qualities make them far superior to the conventional one-factor-at-a-time (OFAT) two-group comparisons (Czitrom 1999). Screening designs for *factor reduction* ('screening out the trivial many from the significant few') consist of 2^k factorial designs when a relatively large number of candidate inputs or interventions thought to affect the response can be reduced to a few of the more promising or important based on the main effects and interactions with the largest effect size. Subsequent experiments can then be designed to optimise responses based on the specific factors identified in the preceding phases (Chapter 6; Montgomery 2012; Collins 2018). Multiple factors can be studied in fractional factorial designs when the cost and size of full factorials become too prohibitive (Box et al. 2005), or if certain combinations of factor levels are too toxic or otherwise logistically unacceptable (Collins et al. 2009). For genuine replication (Box et al. 2005) and so that statistical inference is valid, run combination is randomly allocated to each experimental units, and run order randomised for all treatment combinations and replicate runs.

Example: Weight Gain in Mice on Antibiotics

(Adapted from Snedecor and Cochran 1980). An experiment was planned to measure weight gain of newly weaned mice given an antibiotic at a high dose ('High' $=1$), an intermediate dose (intermediate $= 0$) or vehicle control ('low' $= -1$) and vitamin B12 or vehicle control. This is a 3×3 factorial design with nine treatment combinations:

Run	Antibiotic	Vitamin B$_{12}$
1	1	1
2	1	0
3	1	−1
4	0	1
5	0	0
6	0	−1
7	−1	1
8	−1	0
9	−1	−1

One complete replicate consists of nine mice with each treatment combination (run) assigned to one mouse. To obtain an estimate of the experimental variance, the design could be replicated. Two replicates require 18 mice. For genuine replication (Box et al. 2005), so that statistical inference is valid, run combination is randomly allocated to each mouse, and order of all 18 runs is randomised.

Power and sample size for factorial designs. In a factorial design, a replicate is an independent repeat run of each combination of treatment factors (Montgomery 2012). Therefore, rather than thinking of treatments as randomised to 'groups' of animals and basing sample size of number per group, it is more constructive to evaluate sample size for a factorial design as the number of replicate runs required to estimate the error variance. In a balanced factorial design, power to detect the main effect of a factor is determined by the sample size per level of each factor, not the sample size of the 'group'. This means that in a two-factor design, each animal is doing double duty by contributing to the main effect estimates for both factors at the same time. That is, when estimating the main effect for each factor, two means are compared based on $m = 2n$ observations from n experimental units. The minimum number of replicates per treatment combination is two.

The effect size ES is obtained from regression coefficients for each main effect, and standardising them by the expected standard deviation of the response variable Y:

$$ES = \beta / SD_Y$$

Sample size is determined by the smallest effect size for a main effect to be detected at a pre-specified level of power (Collins 2018). In a 2×2 factorial, there are four means to be estimated, based on m observations. Machin et al. (2018) recommend estimating total sample size N for each factor separately, assuming no interaction, and selecting the larger N. If sample size estimates are very different, factors should be prioritised based on study objectives and the study powered with respect to the most important factor.

Non-centrality parameter. For 2×2 factorial, three non-centrality parameters λ are calculated for each effect and the interaction term. The values for each λ are calculated from the a number of levels of factor A, b number of levels of factor B, $\tau(A)_i$ the mean difference of factor A at level i, $\tau(B)_i$ the mean difference of factor B at level j, $\tau(AB)_{ij}$ is the interaction effect of factor A at level i and factor B at level j, and the mean square error MSE (σ^2).

Factor A effects:

$$\lambda_A = \frac{bn\sum_{i=1}^{a}\tau(A)_i^2}{MSE}$$

Factor B effects:

$$\lambda_B = \frac{an\sum_{j=1}^{b} \tau(B)_j^2}{MSE}$$

Factor A × B interaction effects:

$$\lambda_{AB} = \frac{n\sum_{i=1}^{a}\sum_{j=1}^{b}\tau(AB)_{ij}^2}{MSE}$$

Definitive studies based on factorial designs will require formal simulation-based power analyses, especially if interaction effects are of primary interest (Lakens and Caldwell 2021). If the interaction is approximately the same size as the main effect, sample sizes will have to be increased by four-fold to achieve sufficient power, and will be up to 16 times larger if the interaction is only half that of the main effect.

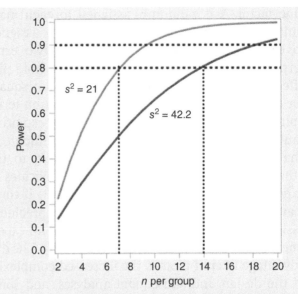

Figure 19.6: Sample size estimates for factorial design. Reducing residual variance minimises sample size.

Example: Effect of Diet on Mouse Tumorigenesis

(Data from Sødring et al. 2015.) Researchers examined the effects of dietary heme and nitrate on intestinal tumorigenesis in 80 A/J min/+ mice. The design was a 2 × 2 factorial on the factor hemin (yes = 1, no = 0) and nitrite (yes = 1, no=0) for four treatment combinations. Mice were randomly assigned to the four diet groups and after eight weeks on each diet, tumorigenesis was quantified as tumour 'load', measured as total lesion area (mm²). Summary statistics were:

Treatment group	Hemin	Nitrite	n	Mean lesion area (mm²)	SD
1	0	1	20	9.7	3.6
2	1	0	21	14.5	9.0
3	1	1	20	13.1	7.3
4	0	0	19	10.9	4.5

When analysed as a single-factor ANOVA with four groups, treatment effect was not statistically significant ($p = 0.096$). Examination of summary data plots and the standard deviations for group means suggest one or more outlying values occur in the hemin-only and hemin + nitrite groups. The residual variance (MSE) was 42.2.

Re-analysis as a 2 × 2 factorial on factors hemin, nitrite, and the (hemin × nitrite) interaction indicates a statistically significant effect of hemin only ($F = 5.67$, $p = 0.02$). The effect of nitrite alone and the (hemin × nitrite) interaction were not statistically significant.

Suppose a new study was planned to further examine the (hemin × nitrite) interaction, and power the study to detect a treatment effect of (hemin × nitrite) interaction of approximately 2.5 mm², with $a = b = 2$ levels. The non-centrality value is:

$$\lambda_{AB} = \frac{n \cdot 2 \cdot 2 \cdot 2.5^2}{42.2} = 0.592 \cdot n$$

Figure 19.6 shows the power curves for estimated sample size per group. The study would require 14 mice per level, for a total of 56. By reducing residual variation by approximately 50% (for example, by identifying and removing the causes for the large outliers observed in the prior study), the revised sample size is 7 mice per level for a total of 28.

19.7 Split-Plot Design

Split-plot designs contain two sizes of experimental unit, the main unit (whole plot) and the subunits. Each main unit contains a number of subunits.

One factor A is randomly assigned to each main unit, and a second factor B is randomly assigned to each subunit (subplots). There are two error terms. The mean square error for the whole plot effect for factor A is larger than the mean square for either factor B or the (A × B) interaction term. Split-plot designs are extremely useful for exploring multiple factors when (1) it is more difficult to change the level of one factor compared to the others, (2) if one factor requires more replicates or experimental units than the others, or (3) if some feature of the experimental conditions precludes assignment of treatment combinations to all experimental units (Kowalski and Potcner 2003). The disadvantage of split plots is the increased complexity of the design and subsequent analyses, and some loss of precision for the main unit treatment comparison (Snedecor and Cochran 1980). Kowalski and Potcner (2003) describe three scenarios where split-plot designs are superior to a completely randomised or factorial design. The skeleton ANOVA (Table 19.1) provides a relatively simple method for preliminary assessments of experimental unit allocation across factors.

Example: Murine Irradiation Study: Skeleton ANOVA for a Split-Plot Design

An experiment was planned to test biomarker levels of mice irradiated at one of three sub-lethal levels of radiation, then administered one of four drugs thought to act as anti-apoptosis mitigators.

Drugs were to be randomly administered to individually marked mice. Batches of 16 mice (4 per drug) were to be irradiated at each radiation level. However, use of the device was very expensive, and changing radiation level was difficult, so changing settings could only be done once per day. The researcher intended to run all drug treatment combinations at each radiation level over three consecutive days. Are experimental units allocated appropriately?

This is a split-plot design, with radiation level as the whole plot factor and drug as the subplot factor. There are $a = 3$ levels of radiation intensity and $r = 1$ replicate. At the subplot level, there are $b = 4$ drugs and $n = 4$ mice per drug.

The skeleton ANOVA constructed from the proposed plan (Table 19.2) shows there are zero error degrees of freedom for testing the effect of radiation level, and therefore no statistical test is possible for this factor. If the effects of radiation are of primary interest, this setup would be a waste of resources and money. If resources and budget allow, each radiation level ($a = 3$) could be replicated so there are two runs at each radiation level ($r = 2$). This setup provides $(3 - 1)(2 - 1) = 2$ degrees of freedom for the error term. If the effects of drug are of higher priority to the study objectives than radiation level, the investigator could consider choosing a single fixed radiation level to test several replicates of drug effects.

Table 19.2: Skeleton ANOVA for a proposed split-plot design with $a = 3$ levels of radiation intensity, $r = 1$ replication of radiation at the whole plot level, $b = 4$ drugs, and $n = 4$ mice per drug. Zero degrees of freedom for the radiation effect indicate that the design must be revised if radiation effects are of interest.

Source			Degrees of freedom	
Whole 'plot' (radiation effects)				
'Plot'	Replicate on radiation	$r - 1$		$1 - 1 = 0$
Radiation level	Whole plot treatment	$a - 1$		$3 - 1 = 2$
Between radiation levels	Whole plot error	$(r - 1)(a - 1)$		$0(2) = 0$
Sub 'plot' (drug treatment effects)				
Drug	Subplot	$b - 1s$		$4 - 1 = 3$
Drug × radiation level	Subplot × Main	$(a - 1)(b - 1)$		$2(3) = 6$
Between mice	Subplot error	$rabn - a(r + b - 1)$		$2 \cdot 3 \cdot 4 \cdot 4 - 4(2 + 2 - 1) = 84$

19.8 Repeated-Measures (Within-Subject) Designs

Repeated-measures designs consist of observations made on the same subject so that observations are correlated (Box 19.2). Types of repeated measures designs included before-after and crossover designs (with paired observations on the same subject), repeated measures on treatment (different treatments randomly applied to the same subject), repeated measures on time (observations on the same subject are obtained for multiple time points), and spatial autocorrelation designs (observations are correlated over space rather than time). Sample size calculations for repeated measures data require information on the number of repeated observations per experimental unit and an estimate of the correlation among the repeated observations (Diggle et al. 2002).

Observations are correlated when pairs of observations close to each other are more similar than pairs of observations that are more widely separated. Recall for a series of n independent observations with mean $\overline{Y}$ and $\text{var}(\overline{Y}) = s^2/n$, the large-sample normal approximation for sample size is:

$$N \geq z_{1-\alpha/2}^2 \cdot (s/d)^2$$

However, if there is a correlation r between observations made at t time points, then

$$N \geq z_{1-\alpha/2}^2 \cdot (s/d)^2 \cdot \left[1 + r\frac{2(t-1)}{t}\right]$$

When comparing the overall difference in response between two or more groups, the group means are averaged over time. The total sample size N increases as correlation r increases. This is because the average group effect size is a between-subjects comparison of averages, and the variance of the estimate increases with r. However, when comparing group differences across time (time × intervention interaction) as correlation r increases, then total sample size N decreases, as the contribution of individual variance to the estimate of the rate of change in response is reduced (Diggle et al. 2002).

Repeated-measures designs can test one or more of three hypotheses for overall group differences (test on main effects), the overall trend over time (time effect), and differences between groups with time (time × intervention interaction) (Winer et al. 1991). Therefore, three sources of variation must be accounted for:

Between-subject variation (random effects): the variation between animals due to the characteristics of the animals in the sample.

Within-subject variation: the variation within a single subject resulting from the effects of time and measurement error. Time dependencies result when measurements are obtained on the same subject over time. Measurements taken at short time intervals will be more highly correlated than those taken further apart.

Interaction effects: the variation attributable to treatment differences across time (time × intervention interaction).

Sample size calculations for repeated–measures designs should include preliminary estimates of the variance and the expected correlation structure for the repeated measurements. The four most common correlation patterns (Guo et al. 2013) are:

1. *zero correlation*, if observations are independent.
2. *compound symmetry,* if any two sequential observations are equally correlated.
3. *autoregressive,* when time points are equally spaced, and correlation decreases with increasing distance between observations.
4. *unstructured,* if correlation is present but without any specific pattern.

BOX 19.2
Repeated Measures Designs

1. *Paired measurements*: Before-after, crossover designs
2. *Repeated measures on treatment:* Treatments randomly applied to the same subject
3. *Repeated measures on time, longitudinal designs*: Measurements taken on the same subject at two or more time points
4. *Spatial autocorrelation*: Measurements are correlated in space rather than time

Choice of a specific correlation structure must be done with care, as the wrong structure or a structure that is too simple will inflate type I error and increase false positives. The conventional Pearson correlation coefficient does not account for the within-subject correlation structure of responses so is unsuited for estimating correlation for a repeated-measures study. More appropriate preliminary estimates can be obtained by calculating the correlation based on subject averages of the two measurements (Bland and Altman 1994, 1995). Irimata et al. (2018) describe several alternative methods for estimating correlation for repeated-measures designs. The best is to obtain relevant variance components from a mixed-model analysis of variance on prior or exemplary data (Castelloe and O'Brien 2001; Hamlett et al. 2004; Irimata et al. 2018). Mixed-model analyses provide the most reliable (and therefore smallest) estimates of total sample size (Hedeker et al. 1999; Guo et al. 2013; Irimata et al. 2018; Shan et al. 2020).

19.8.1 Before-After and Crossover Designs

In these designs, the 'groups' consist of observations paired on the individual subject or experimental unit. Because each subject serves as its own control, between-subject variation is eliminated, and precision is greatly increased. The advantage of these designs is that the same of level of statistical power can be achieved with fewer subjects than would be the case if observations were obtained from independent, parallel arm groups. Disadvantages include carry-over effects (when treatment effects from the first test period persist to the second) and missing data with subject dropout (Senn 2002).

A *before-after* design consists of two sets of observations made in a fixed sequence. The first set of observations is made on each subject at baseline before application of the experimental intervention, and the second set after the intervention.

A *crossover design* has subjects randomly allocated to one of two sequences of two (or more) treatments given consecutively. The sequence order AB or BA is randomised to subject. Subjects allocated to AB receive treatment A first, followed by treatment B, and subjects allocated to BA receive the treatments in reverse order. The great advantage of the crossover design is that two or more factors can be investigated on the same subjects, resulting in a conservable reduction in animal numbers. However, the disadvantage of crossover trials is the potential for *carry-over effect*: measurements obtained for the second intervention may be affected by the previous intervention, resulting in interaction of the two treatments. The study must be designed so that the washout period between treatment applications is sufficiently long to minimise carry-over effects. Senn (2002) argues against multi-stage testing for carry-over effects, as has been recommended in the past.

Sample size in an AB/BA crossover trial can be estimated by using power calculation methods for a paired *t*-test, two-sample *t* test, or equivalence test, depending on the hypothesis to be tested. In SAS, the power analysis for the paired sample *t*-test can be performed with *proc power*. It requires an estimate of the within-subject correlation r. In R, the power analysis can be performed with the pwr.t.test function

---

Example: Crossover Trial: Heart Rate Elevation in Cats

(Data from Griffin et al. 2021.) Heart rates of 21 healthy adult cats presenting for routine wellness exams were measured in a randomised two-period two-treatment crossover trial to assess anxiety levels as result of exam location. The two locations were an isolated exam room with the owner present (A) or a treatment room without the owner (B). Data from the trial indicated a reduction in heart rate of approximately 30 bpm when cats were examined with owner present. How many cats are required for a new study if it was desired to detect a clinically relevant difference of 25 bpm over the baseline heart rate of 185 (SD 34) bpm for a confidence of 95% and power of 80%?

The correlation r was estimated from the variance components from the original data using mixed-model analysis of variance and equal correlation (compound symmetry). The subject variance was 543.59, and residual variance was 343.98, so the correlation was 543.59/ (543.59 + 343.98) = 0.6. The sample size for the new study could then be approximated using the formula for computing power for a paired means t-test with baseline mean heart rate of 185 bpm, mean difference of 25 bpm, and standard deviation 34. The new study would require approximately 15 cats.

19.8.2 Repeated Measures on Time: Continuous Outcome

If the mean difference between levels $\bar{d}$ can be assumed to be constant over time, then the sample size per group n to estimate the time-averaged response is

$$n = \frac{[z_{1-\alpha} \cdot \sqrt{2 \cdot \bar{p}(1-\bar{p})} + z_{1-\beta}\sqrt{p_1(1-p_1) + p_2(1-p_2)}]^2 [1 + r(t-1)]}{t(p_1 - p_2)^2}$$

where p_1 is the proportion responding in group 1 ($q_1 = 1 - p_1$), p_2 = response proportion in group 2, $\bar{p} = (p_1 + p_2)/2$, r is the common correlation across the n observations, and t is number of time points (Diggle et al., 2002). Sample SAS code is provided in Appendix 19.B.

Example: Turtle Hatchling Sex Ratio

Sex ratios of red-eared slider turtle (*Trachemys scripta*) hatchlings are more female-biased in late-season clutches compared to early season clutches, and certain locations shown more sex bias compared to others (Carter et al. 2018). Suppose researchers wish to sample clutches at two locations four times over the season. Approximately how many nests must be sampled to detect a shift between locations in sex ratio from 0.5 to 0.7 with a confidence 95% and power of 80%?

In this example, $t = 4$, $p_1 = 0.5$, and $p_2 = 0.7$. If there is no shift in sex ratio is anticipated over the season, then correlation of repeated outcomes is zero, and $n = 23.2 \cong 24$ per location. If correlation is 0.6 (indicating a positive relationship of sex bias

$$n = \frac{2(z_\alpha + z_{1-\beta})^2 \cdot [1 + r(t-1)]}{t \cdot \bar{d}^2/s^2}$$

where r is the correlation of the repeated measures, s^2 is the assumed common variance in the two groups (or MSE), and t is the number of time points (Diggle et al. 2002). Sample SAS code is provided in Appendix 19.B.

19.8.3 Repeated Measures on Time: Proportions Outcome

Assuming there is a stable difference in proportions $p_1 - p_2$ between two groups across t time points, the number of subjects (n) in each of two groups is

with time), then $n = 65.1 \cong 66$ per location. A more sophisticated design could account for clustering effects of nest within location and hatchlings within nest.

19.8.4 Repeated Measures in Space: Spatial Autocorrelation

Observations that are correlated over space are common in field studies of animal spatial response. For example, animals may be limited in mobility or dispersal, causing aggregation in some areas and avoidance of others, and patterns of aggregation and dispersal may reflect response to environmental gradients, such as salinity or ambient temperature. Other examples of spatial autocorrelation include studies of brain architecture and function, and mapping of genetic distances (with distances based on allele frequencies rather than physical distance).

If correlation occurs between spatial clusters, the large-sample normal approximation for total sample size N is

$$N = (n_s \cdot n_c) \geq z_{1-\alpha/2}^2 \cdot (s/d)^2[(1 + \bar{r}(n_c - 1)]$$

where N is the total number of experimental units, n_s is the number of experimental units within a cluster, n_c is the number of clusters (the repeated points), and $\bar{r}$ is the average correlation between clusters (Conquest 1993).

Example: Spatial Correlation and Mussel Shell Length

(Adapted from Conquest 1993.) A researcher planned a study where N freshwater mussels were to be collected at each of $S = 10$ sampling stations, and shell length of each mussel was to be measured. The researcher wished to estimate shell length with a precision d of 10%, and a level of confidence of 95%. From previous studies, the estimated coefficient of variation (CV) for shell length $(s/\overline{Y})$ is thought to be approximately 30%. Sampling stations are close enough such that the average spatial correlation $\bar{r}$ is approximately 0.25. How many mussels need to be sampled to obtain an estimate of mean shell length with the desired precision?

The total number of mussels to be sampled is

$$N \geq n_s \cdot S$$

where n_s is the number of mussels at each station and S is the number of sample stations. For $CV = (s/\overline{Y}) = 0.3/1 = 0.3$, $d = 0.1$, $\bar{r} = 0.25$, $\alpha = 0.05$, $z_{1-\alpha/2} = 1.96$, and $S = 10$, the total sample size is

$$N \geq z_{1-\alpha/2}^2 \left(\frac{s}{d}\right)^2 [1 + \bar{r}(S-1)]$$

$$N \geq 1.96^2 \left(\frac{0.3}{0.1}\right)^2 [1 + 0.25(10-1)] = 112.3 \cong 113$$

The investigator needs to sample at least 113 mussels. The number of mussels to be sampled at each station is therefore $n_s = N/S = 113/10 = 13$ mussels per station.

19.A Guinea-Pig Data: Sample SAS Code for Calculating Sample Size for a Single-Factor Four-Level (a) Completely Randomised Design; (b) Randomised Complete Block Design

19.A.1 Completely Randomised Design

```
proc glm;
class TRT;
model wt=TRT;
means TRT;
lsmeans TRT / diff pdiff CL;
run;
```

19.A.2 Randomised Complete Block Design

```
proc glm order=data;
class BLOCK TRT;
model wt=BLOCK TRT;
means TRT / waller regwq;
run;


title 'power for single-factor ANOVA';
*set sample size per group limit;
%let numSamples = 20; *maximum sample size per
                        group;
%let kmax = 10;        *maximum number of
                        levels;
data anpow;
do n = 3 to &numSamples by 1;
do k = 2 to &kmax by 1;
    meandiff = 2.5; *biologically meaningful
    difference to be detected;
    sigma2 = 4.95;
    alpha = 0.05;

*calculate noncentrality parameter ncp;
        NCP = (n*meandiff**2)/ (2*sigma2);
*calculate power;
            NumDF=k-1;
            DenDF=k*(n-1);

        F_crit = finv(1-alpha,
```

```
            NumDF, DenDF, 0);
            power = 1- probf(F_crit,
            NumDF, DenDF, NCP);
            output;
        end;
    end;
run;
proc sort;
 by k;
 run;
proc print data=anpow;
run;
```

19.B Sample SAS Code for Calculating Sample Size per Group for a Simple Repeated-Measures Design

```
* determines sample size per group;
* 5 timepoints (icc=.4);* difference in
proportions of 0.5 and 0.67 (odds ratio = 2);
* power = 0.8 for a 2-tailed 0.05 test;

data test;
    za = probit(.975);
    zb = probit(.8);
    n = 5;
    p1 = 0.5; p2 = 2/3;
    q1 = 1-p1; q2 = 1-p2;
pbar = (p1+p2)/2;
qbar = (q1+q2)/2;
    rho = 0.4;
num = ((za*sqrt(2*pbar*qbar) + zb*sqrt(p1*q1 +
p2*q2))**2)*(1 + (n-1)*rho);
den = n*((p1-p2)**2);
npergrp = num/den;

proc print;
    var npergrp;
run;

ans n = 71.87 = 72
data test;
alpha=0.05;
beta=0.8;
za = probit(.975);
za2=quantile('normal',1-alpha/2);
zb = probit(.8);
zb2=quantile('normal',beta);
t = 4;
p1 = 0.5;   p2 = 0.7;
q1 = 1-p1; q2 = 1-p2;
pbar = (p1+p2)/2;
qbar = (q1+q2)/2;
rho = 0.6;
```

```
num = ((za*sqrt(2*pbar*qbar) + zb*sqrt(p1*q1 +
p2*q2))**2)*(1 + (t-1)*rho);
den = t*((p1-p2)**2);
npergrp = num/den;
proc print;
var npergrp;
    run;
```

References

Bate, S. and Clark, R. (2014). *The Design and Statistical Analysis of Animal Experiments*. Cambridge: Cambridge University Press.

Bland, J.M. and Altman, D.G. (1994). Correlation, regression, and repeated data. *British Medical Journal* 308: 896.

Bland, J.M. and Altman, D.G. (1995). Statistics notes: calculating correlation coefficients with repeated observations: part 1—correlation within subjects. *BMJ* 310: 446. https://doi.org/10.1136/bmj.310.6977.446.

Box, G.E.P., Hunter, W.G., and Hunter, J.S. (2005). *Statistics for Experimenters*, 2nd ed. New York: Wiley.

Brien, C.J. (1983). Analysis of variance tables based on experimental structure. *Biometrics* 39: 53–59.

Brien, C.J. and Bailey, R.A. (2006). Multiple randomizations (with discussion). *Journal of the Royal Statistical Society, Series B* 68: 571–609.

Brien, C.J. and Bailey, R.A. (2009). Decomposition tables for experiments. I. A chain of randomizations. *The Annals of Statistics* 37: 4184–4213.

Brien, C.J. and Bailey, R.A. (2010). Decomposition tables for experiments. II. Two-one randomizations. *The Annals of Statistics* 38: 3164–3190.

Carter, A.W., Sadd, B.M., Tuberville, T.D. et al. (2018). Short heatwaves during fluctuating incubation regimes produce females under temperature-dependent sex determination with implications for sex ratios in nature. *Scientific Reports* 8 (1): 3. https://doi.org/10.1038/s41598-017-17708-0.

Castelloe, J.M. and O'Brien, R.G. (2001). Power and sample size determination for linear models. *SUGI* 26: 240–226. https://support.sas.com/resources/papers/proceedings/proceedings/sugi26/p240-26.pdf.

Collins, L.M. (2018). Chapter 3: Introduction to the factorial optimization trial. In: *Optimization of Behavioral, Biobehavioral, and Biomedical Interventions*, Statistics for Social and Behavioral Sciences (ed. L. G. Collins), 67–113. Springer International Publishing AG https://doi.org/10.1007/978-3-319-72206-1_3.

Collins, L.M., Dziak, J.J., and Li, R. (2009). Design of experiments with multiple independent variables: a resource management perspective on complete and reduced factorial designs. *Psychological Methods* 14 (3): 202–224. https://doi.org/10.1037/a0015826.

Conquest, L.L. (1993). Statistical approaches to environmental monitoring: did we teach the wrong things? *Environmental Monitoring and Assessment* 26 (2-3): 107–124. https://doi.org/10.1007/BF00547490.

Cooper, N., Thomas, G.H., and FitzJohn, R.G. (2016). Shedding light on the 'dark side' of phylogenetic comparative methods. *Methods in Ecology and Evolution* 7 (6): 693–699. https://doi.org/10.1111/2041-21019.12533.

Czitrom, V. (1999). One-factor-at-a-time versus designed experiments. *The American Statistician* 53: 126–131. https://doi.org/10.1080/00031305.1999.10474445.

Diggle, P.J., Heagerty, P., Liang, K.-Y., and Zeger, S.L. (2002). *Analysis of Longitudinal Data*, Oxford Statistical Science Series, 2e. Oxford: Oxford University Press.

Divine, G., Kapke, A., Havstad, S., and Joseph, C.L. (2010). Exemplary data set sample size calculation for Wilcoxon-Mann-Whitney tests. *Statistics in Medicine* 29 (1): 108–115. https://doi.org/10.1002/sim.3770.

Draper, N. and Smith, K. (1991). *Applied Regression Analysis*, 2e. New York: Wiley.

Festing, M.F.W. (2014). Randomized block experimental designs can increase the power and reproducibility of laboratory animal experiments. *ILAR Journal* 55: 472–476.

Festing, M.F.W. (2018). On determining sample size in experiments involving laboratory animals. *Laboratory Animals* 52 (4): 341–350. https://doi.org/10.1177/0023677217177382.

Festing, M.F.W. (2020). The "completely randomized" and the "randomized block" are the only experimental designs suitable for widespread use in preclinical research. *Scientific Reports* 10: 17577.

Gamble, C., Krishan, A., Stocken, D. et al. (2017). Guidelines for the content of statistical analysis plans in clinical trials. *JAMA* 318 (23): 2337–2343.

Garland, T. and Adolph, S.C. (1994). Why not to do two-species comparative studies: limitations on inferring adaptation. *Physiological Zoology* 67: 797–828.

Goldman, A.I. and Hillman, D.W. (1992). Exemplary data: sample size and power in the design of event-time clinical trials. *Controlled Clinical Trials* 13 (4): 256–271. https://doi.org/10.1016/0197-2456(92)90010-w.

Griffin, F.C., Mandese, W.W., Reynolds, P.S. et al. (2021). Evaluation of clinical examination location on stress in cats: a randomized crossover trial. *Journal of Feline Medicine and Surgery* 23 (4): 364–369. https://doi.org/10.1177/1098612X20959046.

Guo, Y., Logan, H.L., Glueck, D.H., and Muller, K.E. (2013). Selecting a sample size for studies with repeated measures. *BMC Medical Research Methodology* 13: 100. https://doi.org/10.1186/1471-2288-13-100.

Hamlett A, Ryan L, Wolfinger R (2004). On the use of *proc mixed* to estimate correlation in the presence of repeated measures. SUGI29 Paper 198-29 https://support.sas.com/resources/papers/proceedings/proceedings/sugi29/198-29.pdf (accessed 2022).

Hedeker, D., Gibbons, R.D., and Waternaux, C. (1999). Sample size estimation for longitudinal designs with attrition. *Journal of Educational and Behavioral Statistics* 24: 70–93. https://doi.org/10.3102/10769986024001070.

Hurlbert, S.H. and Lombardi, C.M. (2004). Research methodology: experimental design sampling design, statistical analysis. In: *Encyclopedia of Animal Behavior*, vol. 2 (ed. M.M. Bekoff), 755–762. London: Greenwood Press.

Irimata K, Wakim P, Li X (2018). Estimation of correlation coefficient in data with repeated measures. Paper 2424-2018, *Proceedings SAS Global Forum*, 2018:8-11

Karp, N.A. and Fry, D. (2021). What is the optimum design for my animal experiment? *BMJ Open Science* 5: e100126.

Karp, N.A., Wilson, Z., Stalker, E. et al. (2020). A multi-batch design to deliver robust estimates of efficacy and reduce animal use – a syngeneic tumour case study. *Scientific Reports* 10 (1): 6178. https://doi.org/10.1038/s41598-020-62509-7.

Kowalski, S.M. and Potcner, K.J. (2003). How to recognize a split-plot experiment. *Quality Progress* 36 (11): 60–66.

Lakens, D. and Caldwell, A.R. (2021). Simulation-based power analysis for factorial analysis of variance designs. *Advances in Methods and Practices in Psychological Science* 4 (1): https://doi.org/10.1177/2515245920951503.

Lazic, S.E. (2016). *Experimental Design for Laboratory Biologists*. Cambridge: Cambridge University Press.

Machin, D., Campbell, M.J., Tan, S.B., and Tan, S.H. (2018). *Sample Sizes for Clinical, Laboratory and Epidemiology Studies*, 4e. Wiley-Blackwell.

Mead, R. (1988). *The Design of Experiments*. Cambridge UK: Cambridge University Press.

Montgomery, D.C. (2012). *Design and Analysis of Experiments*, 8th ed. New York: Wiley.

O'Brien, R.G. and Muller, K.E. (1993). Unified power analysis for t-tests through multivariate hypotheses. In: *Applied Analysis of Variance in Behavioral Science* (ed. L.K. Edwards), 297–344. New York: Marcel Dekker.

Reynolds, P.S. (2022). Between two stools: preclinical research, reproducibility, and statistical design of experiments. *BMC Research Notes* 15: 73. https://doi.org/10.1186/s13104-022-05965-w.

Senn, S. (2002). *Cross-Over Trials in Clinical Research*, 2nd ed. New York: Wiley.

Shan, G., Zhang, H., and Jiang, T. (2020). Correlation coefficients for a study with repeated measures. *Computational and Mathematical Methods in Medicine* 2020: 7398324. https://doi.org/10.1155/2020/7398324.

Simpson, S.H. (2015). Creating a data analysis plan: what to consider when choosing statistics for a study. *The Canadian Journal of Hospital Pharmacy* 68 (4): 311–317. https://doi.org/10.4212/cjhp.v68i4.1471.

Snedecor, G.W. and Cochran, W.G. (1980). *Statistical Methods*, 7e. Ames: Iowa State University Press.

Sødring, M., Oostindjer, M., Egelandsdal, B., and Paulsen, J.E. (2015). Effects of hemin and nitrite on intestinal tumorigenesis in the A/J min/+ mouse model. *PLoS ONE* 10 (4): e0122880. https://doi.org/10.1371/journal.pone.0122880.

Voelkl, B., Vogt, L., Sena, E.S., and Würbel, H. (2018). Reproducibility of preclinical animal research improves with heterogeneity of study samples. *PLoS Biology* 16 (2): e2003693. https://doi.org/10.1371/journal.pbio.2003693.

Winer, B.J., Brown, D.R., and Michels, K.M. (1991). *Statistical Principles in Experimental Design*, 3e. Boston MA: McGraw-Hill.

Würbel, H., Voelkl, B., Altman, N.S. et al. (2020). Reply to 'It is time for an empirically informed paradigm shift in animal research'. *Nature Reviews Neuroscience* 21: 661–662. https://doi.org/10.1038/s41583-020-0370-7.

Zar, J.H. (2010). *Biostatistical Analysis*, 5th ed. Upper Saddle River: Prentice-Hall.

20 Hierarchical or Nested Data

CHAPTER OUTLINE HEAD

20.1 Introduction

The focus of many animal-based experiments is the individual subject or biological unit. However, collections of subjects often exhibit some sort of hierarchical structuring, with observational units nested within larger organisational units, or clusters (Box 20.1). For example, pups are located within litters, and litters are clustered within dam. Individual mice are assigned to a given drug intervention, the individual mouse brain is harvested, then cells are harvested from different regions within each brain.

A Guide to Sample Size for Animal-based Studies, First Edition. Penny S. Reynolds.
© 2024 John Wiley & Sons Ltd. Published 2024 by John Wiley & Sons Ltd.

BOX 20.1
Examples of Hierarchical Data

- Nestlings in different nests
- Pups in each litter from each multiparous breeding pair
- One of two drugs is randomly assigned to each mouse, the brain of each mouse is harvested, and cells from each brain region are isolated and plated
- Brains from two strains of mouse are harvested, cells are isolated and plated, then one of two drugs is randomly assigned to cells in each well in a plate
- Multisite veterinary trials where dogs with a particular cancer within each site are randomly assigned to one of two or more treatments
- Synaptic vesicles counts are obtained for each set of neurons harvested from the brains of several strains of mice
- Cows within herds within locations
- Survey studies: Individuals within clinics within city
- Meta-analyses

Two- and three-level models are common in studies of neurobiology (Aarts et al. 2014, 2015), rodent breeding (e.g. pups within litter within dam; Lazic and Essiou 2013; Hull et al. 2022), ecology (e.g. nestlings within nest within site), agriculture (e.g. livestock within herds within locations), multisite veterinary clinical trials (e.g. subjects within clinic within city), and meta-analyses (individuals within studies).

When units are nested, observations for the units within each cluster will not be independent. The individual units within clusters will be more similar than individual units located in different clusters. Treating observations as independent in subsequent analyses is 'pseudo-replication', and results in artificially inflated sample sizes (Lazic 2010; Lazic and Essiou 2013; Lazic et al. 2018). Failure to account for nesting can result in inflated Type I error rates, increased false positives, and incorrect interpretation of results (Julian 2001; Moerbeek 2004; Aarts et al. 2014, 2015). It can also result in a loss of statistical power to detect any experimental effects that actually exist, especially

when both anticipated sample sizes and effect size to be detected are small (Goldstein 2003; Moerbeek 2004).

One solution to these problems is to average across all observations for each cluster, then analysing the means. However, averaging loses a lot of information and does not permit exploration of the effects of predictor variables acting at the different hierarchical levels of organisation.

Hierarchical or multilevel design models (also called nested models, random effects models, mixed models, and variance components models) specifically account for the nested structure of organismal units and the correlation within clusters. Design features adjust for the different variances that occur between and within clusters (thus ensuring estimation of correct standard errors). Predictor or explanatory variables (*covariates*) can be incorporated at both the level of the individual observation and the cluster. There are two main types of multilevel designs: between-cluster and within-cluster (Figure 20.1). Between-cluster

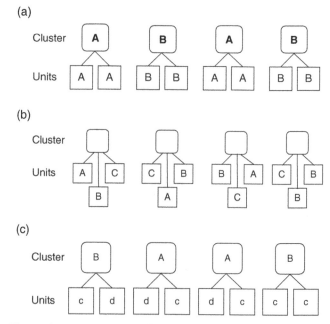

Figure 20.1: Treatment allocation strategies and units of randomisation in multilevel designs. (a) Between-cluster design: The cluster is the experimental unit. Clusters are randomised to intervention, and all observation units within the cluster receive that intervention. (b) Within-cluster design: The unit within the cluster is the experimental unit. Within-cluster units are randomised to intervention, and clusters are a blocking variable. (c) Multi-cluster design. Different treatments are assigned to clusters and units within clusters, respectively. Treatment 1 (A or B) is randomly allocated to level-2, and Treatment 2 (*c* or *d*) is randomly allocated to level-1 units.

designs randomise the factor or treatment to the whole cluster; experimental units within each cluster are essentially subsamples. Within-cluster designs randomise treatment to units within each cluster. Repeated measures designs (where multiple observations are nested within each subject) and randomised block designs (where at least one replicate from each treatment is nested within blocks) are a type of multilevel design.

There are at least two sample size calculations required for multilevel designs: the sample size per cluster (the number of units per cluster) and the total number of clusters. When clustering is present, the correlation between observations within clusters (the *intra-class correlation*, ICC) will affect statistical power. Larger sample sizes will be required if variance components are large and the sample size at the lowest level (e.g. number of individuals per cluster) is very small (Konstantopoulos 2008, 2010; Austin and Leckie 2018).

20.2 Steps in Multilevel Sample Size Determinations

There are four steps for sample size determination for multilevel models:

1. Identify the unit of randomisation
2. Determine design features
3. Decide on the effect size: δ, ICC, f^2, R^2
4. Other considerations: Balance, sparse data, and costs

20.2.1 Identify the Unit of Randomisation

In randomised experimental studies, test or control interventions can be applied to the whole cluster, to individuals within each cluster, or different interventions may be randomised at different levels.

Cluster-level assignment. When interventions are assigned randomly to whole clusters, the cluster is the unit of randomisation. All individuals from the same cluster will therefore belong to the same experimental or treatment group. Machin et al. (2018) refer to this assignment

strategy as *aggregate*. Sample size determinations prioritise the number of clusters k, not the number of individuals n (Aarts et al. 2014, 2015). Cluster sample size is the priority if the focus of the study is testing the effects of an intervention applied at the cluster level, or if the number of subjects contained in each cluster is anticipated to be small, or else must be fixed *a priori* during the design phase. Cluster-level designs have the advantage of relative simplicity (only one treatment is allocated to the entire cluster). However, they are liable to a type of representation bias if there is large between-cluster variation in subject characteristics. The requirement for large number of clusters to be sampled to achieve adequate power may make these studies very expensive, especially for multisite trials or large-scale surveys (Raudenbush and Liu 2000).

Within-cluster assignment. When interventions are assigned randomly to different individuals within the same cluster, the individual is the unit of randomisation, and all treatments are represented in each cluster. Within-cluster designs are sometimes referred to as multilevel blocked trials. Individual subjects are within 'blocks' because all treatments are represented in each cluster (Spybrook et al. 2011; Konstantopoulos 2012). Individual randomisation is often preferable to cluster randomisation because sample sizes required are usually much smaller, and it is easier to control for representation bias. However, individual treatment allocations will not be possible if subjects cannot be individually identified, and 'treatment contamination' (inadvertent application of the wrong intervention) may be a consideration (Torgerson 2001). For testing effects of an intervention applied at the individual level, sample size determinations prioritise the number of individuals n. The number of clusters k will be determined by the anticipated cluster size, or number of individuals within each cluster.

If there are more treatments than individuals per cluster, so that all treatments cannot be represented in each block, incomplete block or fractional factorial designs should be considered. More complex experiments can be designed where different interventions can be separately assigned to cluster and individual units.

Multilevel designs are best understood as a regression model. Features of the regression model to be considered are number of *levels,* or hierarchical units, and specification of *predictor variables* (or *covariates)* at the different levels (if applicable).

20.2.2 No Predictors

The most basic multilevel model has no predictor variables and two hierarchical levels – the individual level (level-1) and the cluster level (level-2). The individual is nested within cluster.

Continuous outcome. The model is described by regression on the two levels as:

$$\text{Level 1: } Y_{ij} = \beta_{0j} + e_{ij}, e_{ij} \sim N\left(0, \sigma_e^2\right)$$

$$\text{Level 2: } \beta_{0j} = \gamma_{00} + u_{0j}, u_{0j} \sim N\left(0, \sigma_u^2\right)$$

Here Y_{ij} is the response measured for the *i*th individual in the *j*th cluster, the intercept β_{0j} is the mean response for the *j*th cluster, and γ_{00} is the grand mean (mean across all clusters). The covariance matrix of the random effects is designated as the G-matrix. It consists of the variance for the intercept, the variance for the slopes, and the covariance between intercepts and slopes. The level-1 random effect, or error, term is e_{ij}, and is assumed to be normally distributed with mean 0 and variance σ_e^2, the variance between individuals. The level-2 random component is normally distributed with mean 0 and variance σ_u^2, the variance between clusters. The total observed variance for the response is thus $\sigma_e^2 + \sigma_u^2$ (Bell et al. 2013). Combining the two equations gives the full model:

$$Y_{ij} = \gamma_{00} + \left(u_i + e_{ij}\right)$$

that is, the response is the sum of the grand mean plus the between- and within-cluster variances. In SAS, the variance components σ_u^2, σ_e^2 are obtained from the variance components option in the random statement of *proc mixed* (SAS code is provided in Appendix 20.A).

Categorical outcome. When outcome variables are categorical (such as binary, proportions, count,

or ordinal data), the assumptions of normally distributed error distributions no longer apply. The outcome must be transformed with the appropriate non-linear link function in a generalised multilevel model. The response η_i is the log odds of the event occurring for the *i*th individual in the *j*th cluster:

$$\eta_{ij} = log\left[\frac{\mu}{1-\mu}\right]$$

where $\mu = 1/[1 + \exp(-\beta_0)]$. The model is

$$\text{Level 1: } \eta_{ij} = \beta_{0j}$$

$$\text{Level 2: } \beta_{0j} = \gamma_{00} + u_{0j}, u_{0j} \sim N(0, \tau_{00})$$

where β_{0j} is the intercept or the average log odds of the event occurring in cluster *j* and the level-2 error variance term is designated by τ_{00}. Because the variance of a non-Gaussian model is determined by the population mean, the 'level-1 error variance term' as such is not computed as it is with Gaussian data but is estimated as $\pi^2 / 3 = 3.29$ (Ene et al. 2015; Hox et al. 2018). The level-2 random component is assumed to belong to the exponential family of distributions: binary, binomial, Poisson, geometric, and negative binomial distributions. Count data are usually modelled with a Poisson distribution. Ordinal data are described by proportional odds, or cumulative logit, models.

In SAS, the variance components are obtained from the variance components option in the random statement of *proc glimmix*. The grand mean (with 95% confidence intervals) is obtained from the fixed effects intercept (Appendix 20.A). Other distributions can be specified by the relevant distribution and link functions.

20.2.3 Multilevel Models with Predictors

The basic model is easily expanded to include one or more predictors, or covariates, at any or all levels. The level to which a treatment, or intervention, predictor is assigned depends on how the unit of randomisation is defined. The associated slope terms (regression coefficients $\beta_1, \beta_2,$ etc.) are of primary interest as they describe the effects of the intervention on the outcome. Interpretation is simplified if

Table 20.1: Dummy variables.

Dummy variables are used to streamline calculations when categorical variables with two or more distinct 'levels' are used as predictors (Draper and Smith 1998).

1. *Two category 'levels' A and B.* One dummy variable Z that takes values of 0 or 1 is required to uniquely code for each group.
   Example. One of two treatments is randomly assigned to individual within a cluster. Treatment is represented by indicators $Z_1 = 0$ (control) and $Z_1 = 1$ (test intervention). The regression equation is

$$Y_{ij} = \beta_{0j} + \beta_1 \cdot Z_1 + eij$$

   Example. One of two treatments is randomly assigned to equal number of individuals of both sexes, male or female. Sex is represented by indicators $Z_2 = 0$ (male) and $Z_2 = 1$ (female). The regression equation is

$$Y = \beta_0 + \beta_1 Z_1 + \beta_2 \cdot Z_2 + e$$

2. *More than two category 'levels'.* More than two levels require $k - 1$ dummy variables $z_1, z_2, ..., z_{k-1}$, with each taking the values of 0 or 1.
   Example. Animals are sampled from each of three locations A, B, C. To code for location, two dummy variables z_1 and z_2 are required:

$$\text{Location A} : z_1 = 1, z_2 = 0$$
$$\text{Location B} : z_1 = 0, z_2 = 1$$
$$\text{Location C} : z_1 = 0, z_2 = 0$$

The regression equation is

$$Y = \beta_0 + \beta_1 X_1 + a_1 z_1 + a_2 z_2 + e$$

The regression coefficients a_1 and a_2 estimate the difference between each group and the reference group (0, 0).

intervention variables are categorical and dummy-coded (Table 20.1). For example, suppose a two-arm experiment is planned comparing a test drug A versus a vehicle control C. Then 'treatment' can be coded with a single dummy variable to indicate the two groups (0 for C and 1 for A). The regression coefficient will be the quantitative difference in responses between groups and is thus a measure of unstandardised effect size.

Additional covariates can be added to increase precision (reduce standard errors) of the estimated intervention effects, and reduce noise, therefore improving signal and increasing power to detect treatment effects if they exist (Hedges and Hedberg 2007; Raudenbush et al. 2007; Konstantopoulos 2008, 2012). Signalment variables (age, sex, reproductive status, weight, etc) are useful covariates.

Level-1 predictors. If the treatment is applied at the level of the individual, the predictor or treatment indicator variable X is incorporated into level-1 regression:

$$Y_{ij} = \beta_{0j} + \beta_j X_j + e_{ij}$$

with the level-2 or cluster-effect:

$$\beta_{0j} = \gamma_{00} + u_{0j}; u_{0j} \sim N(0, \tau_{00})$$

$$\beta_{1j} = \gamma_{10} + u_{1j}; u_{1j} \sim N(0, \tau_{11})$$

Here γ_{00} is the grand mean of the response, γ_{10} is the treatment effect, or the difference in means between the experimental group and the comparator group $(\overline{Y}_t - \overline{Y}_c)$, τ_{00} is the between-cluster variation, and τ_{11} is the variance for the treatment effect between clusters.

Level-2 predictors. Predictors at the cluster level (level-2) are indicated by W, rather than *20*. The cluster-level regressions are

$$\beta_{0j} = \gamma_{00} + \gamma_{01} W_j + u_{0j} \text{ (level} - 2 \text{ intercept)}$$

$$\beta_{1j} = \gamma_{10} + \gamma_{11} W_j + u_{1j} \text{ (level} - 2 \text{ slope)}$$

The combined model is

$$Y_{ij} = (\gamma_{00} + \gamma_{01} W_j) + (\gamma_{10} \cdot X_j) + (\gamma_{11} \cdot X_{1ij} \cdot W_{1j})$$
$$+ (u_{0j} + u_{1j} \cdot X_{ij} + e_{ij})$$

The cross-level term $X_{1ij} \cdot W_{1j}$ indicates the strength of the adjustment of level-2 characteristics on the level-1 (subject) responses within each cluster (Sample SAS code is provided in Appendix 20.B).

20.2.4 Constructing the Model

Bell et al. (2013) and Ene et al. (2015) recommend a 'bottom-up' model-building strategy. The initial model is intercept-only with no predictors, then predictors and random effects are added one at a time. Predictors should be specified *a priori,* and selection based on best available knowledge or scientific justification. Final model specification is based on coefficients that are statistically significant (or 'meaningfully large') and model fit based on the deviance test, which assesses differences in the $-2 \cdot \log$ likelihood values between candidate models.

Model misspecification is indicated by failure to converge, or a non-positive definite g-matrix. The latter indicates that one or more of the random effects variance components are zero. There may be too few observations to properly estimate the random effects (there should be at least $p + 1$ observations to estimate p random effects), too many random effects specified in the model, or there is little to no variation between units at the given level. Dropping the zero variance random effects is an option. An alternative is to use a 'population average' model and dispense entirely with random effects. The disadvantage of this approach is that these models cannot capture between-cluster differences, but this may not matter if cluster effects are not of primary interest or are too small to be of practical importance (McNeish et al. 2016).

20.3 Estimating Effect Size

Effect sizes can be estimated by Cohen's d, unstandardised or standardised regression coefficients, multilevel R^2, Cohen's f^2, and/or the intraclass correlation coefficient ICC (Lorah 2018).

20.3.1 Cohen's d

Estimating a d effect size requires specification of both the average difference between groups δ and the variance of effect sizes within and across clusters, expressed as standard deviation s. With clustered studies, there are multiple sources of variation and therefore several choices of the standard deviation to be used for the standardisation: the total variance, the variance for the subjects nested within clusters, the variance between clusters, and when applicable, the variance for the main effect of treatment.

Initial effect sizes for continuous data can be approximated using Cohen's d and ignoring for level-2 clustering and associated predictors. Recall that effect size is expressed in standard deviation units as $d = (\overline{Y}_1 - \overline{Y}_0)/\sigma_{total}$, where $(\overline{Y}_1 - \overline{Y}_0)$ is the biologically important difference δ between treatment means, and σ_{total} is the standard deviation of the outcome, estimated as the square root of the total variance σ_{total}^2. For the two-level model, the standard deviation of the outcome is $\sqrt{(\sigma_u^2 + \sigma_e^2)}$, and for three levels $\sqrt{(\sigma_u^2 + \sigma_v^2 + \sigma_e^2)}$. Values for δ can be a 'best guess' or hypothesised scientifically meaningful difference $Y_1 - Y_0$ to be detected. Basic descriptive or summary statistics can be used to obtain the standard deviation of the outcome Y.

If there is little or no prior information on anticipated mean differences and variances, Cohen's benchmark values for small, medium, and large effect sizes could be substituted. However, these values (and their interpretation) will be highly unreliable and are not recommended for animal-based studies (see Chapter 15).

When outcome variables are categorical, the effect size is estimated by the difference in log odds ratios:

$$\delta = (\eta_1 - \eta_0) = \ln\left[\frac{p_1(1-p_0)}{p_0(1-p_1)}\right]$$

where η_0 and η_1 are the log odds ratios for the control and experimental groups, respectively, with the average treatment effect defined in terms of the log odds of the proportions of 'successes' (the number of events occurring) in each group.

20.3.2 Fixed Effects Regression Coefficients

For regression-type models, regression coefficients for the fixed effects provide information about the magnitude of the treatment differences (which is of most biological interest). Feingold (2015) recommends that the regression coefficient for slope β is divided by the pooled within-group standard deviation of the response Y. The estimated effect size will be somewhat smaller if the effect size is estimated without accounting for cluster variances. Because the magnitude of the fixed effects depend on the scale of each independent variable, effect sizes based on different variables in the same study, or variables across multiple studies, cannot be compared directly (Lorah 2018).

An alternative measure of effect size for regression models is the proportion of variation explained by the predictors. Cohen's f^2 estimates the proportion of variance in the sample explained by the (categorical) variable for the intervention effect relative to all other covariates in the sample, or R^2. Here f^2 is calculated by running the regression model several times to obtain three different variance components: the variance components for the full model with all fixed effects and covariates included, variance components for the covariates-only model (which estimates the variance in the response without the fixed effects), and variance components for the null model (intercept-only model with no predictors). The respective R^2 for the full and covariate models are calculated from these variance components as

$$R^2_{full} = \frac{\sigma^2_{null} - \sigma^2_{full}}{\sigma^2_{null}}; R^2_{other} = \frac{\sigma^2_{null} - \sigma^2_{other}}{\sigma^2_{null}}$$

For a binary response variable, R^2 is:

$$R^2_{binary} = \frac{\sigma^2_{fixed}}{\sigma^2_{fixed} + \tau^2_{00} + \pi^2/3}$$

where σ^2_{fixed} is the sample variance of the fixed effects linear predictor, and τ_{00} is the cluster (level-2) error variance term (Austin and Merlo 2017).

Cohen's f^2 is then calculated as:

$$f^2 = \frac{R_{full}^2 - R^2_{other}}{1 - R^2_{full}}$$

Sample size N is obtained from f^2 and the non-centrality parameter λ, where $N = \lambda/f^2$ (See Chapter 14).

Selya et al. (2012) describe how to calculating f^2 using SAS *proc mixed*. Bates et al. (2011) and Nakagawa and Schielzeth (2013) have developed R packages for estimating R^2 for multilevel models, further extended by Johnson (2014) and Nakagawa et al. (2017) to include generalised linear models for non-normal outcomes.

Example: Calculating R^2 and f^2: Sparrow Model of Early Life Stress

(Data from Grace et al. 2017.) Nestling house sparrows (*Passer domesticus*) were used as a model of the effects of early life stress on growth and survival. Nestlings were fed mealworms containing either corticosterone or vehicle control to assess the effect of glucocorticoid exposure (a stress indicator) on body mass gain ('treatment', TRT). Measurements were obtained for mass (M) and tarsus length (TL) on day 12, and SEX (male or female). For this example, data were limited to observations for four nestlings per nest, two of which received corticosterone and two of which received control. There were 31 nestlings in the experimental and control groups, respectively, for a total of 72 nestlings in 18 nests.

The model is a two-level design, with individual nestlings within nest as the unit of randomisation to which treatment interventions were applied. Mass M is the dependent variable, TRT is the fixed effect (1 = corticosterone, 0 = control), and TL and sex are covariates. R^2 and Cohen's f^2 were calculated in SAS (adapted from Selya et al. 2012). Variances and corresponding R^2 are shown in Table 20.2. Cohens f^2 is

$$f^2 = \frac{(0.234 - 0.07)}{(1 - 0.234)} = 0.215$$

20.3.3 Intraclass Correlation Coefficient (ICC)

The intraclass correlation ICC is the proportion of total variance in the response Y that can be attributed to the cluster. Because it measures strength of association, the ICC can be interpreted as an effect

Table 20.2: Calculation of R^2 and f^2 for sparrow data.

Source	Variance	R^2
Full model (TRT, SEX, TL)	2.209	$R^2_{full} = (2.886\text{-}2.209)/$ $2.886 = 0.234$
Covariates only (SEX, TL)	2.683	$R^2_{other} = (2.886\text{-}2.683)/$ $2.886 = 0.070$
Intercept-only (no predictors)	2.886	

size in the same way as a conventional correlation coefficient r (Snijders and Bosker 1993, 1999; Lorah 2018). It is also a measure of the 'information content' contained in each cluster. As the correlation between observations increases, the level-1 variance (σ_e^2) becomes smaller, and ICC increases. ICC is zero if responses for all individuals are independent of one another. A non-zero ICC denotes responses are not independent, and ICC of 1 indicates all the responses in all clusters are the same. A large ICC indicates the information contained in individual observations is similar. Therefore, if ICC is large, increasing within-cluster sample size will be redundant and will not contribute much in the way of new information for tests of treatment effects. (When ICC is calculated as a measure of inter-rater agreement, a high value is desirable as a measure of method reliability).

If the design is balanced (equal n within each cluster), the between-cluster ICC in a two-level model with a continuous outcome is

$$ICC_u = \frac{\sigma_u^2}{\sigma_u^2 + \sigma_e^2}$$

That is, the ICC is the ratio of the between-cluster variance to the total variance. For a three-level model, there are two cluster ICCs to be calculated:

$$ICC_u = \frac{\sigma_u^2}{\sigma_u^2 + \sigma_v^2 + \sigma_e^2}$$

$$ICC_v = \frac{\sigma_v^2}{\sigma_u^2 + \sigma_v^2 + \sigma_e^2}$$

where u and v refer to the level-2 and level-3 clusters, respectively.

For non-Gaussian outcomes (e.g, binary, count, proportions), the ICC for a two-level model is

$$ICC = \frac{\tau_{00}}{\tau_{00} + \pi^2/3}$$

where τ_{00} is the intercept variance. For binary and proportion response data, the level-1 error variance term is determined by the population mean. The second term for calculating the total variance is estimated from the logistic link function with scale factor of 1 and is $\pi^2 / 3 = 3.29$ (Ene et al. 2015; Hox et al. 2018).

Initial values of ICC may be obtained from previously published data. If raw data are available, the ICC can be calculated from the sum of the variance components of the intercept-only regression model (Bell et al. 2013; Ene et al. 2015).

Example: Calculating ICC: Magpie Egg Weights

(Data from Reynolds 1996.) Weights were obtained for 118 eggs from 19 Black-billed Magpie (*Pica pica*) nests. The fitted model is an intercept-only no-predictor model:

Egg weight per egg per nest $= a_0 + (u_i + e_{ij})$

where the random intercept term a_0 is an estimate of mean egg weight, u_i is the random effects error term for the between-nest effects, and e_{ij} is the residual variance, or the error term for egg. In SAS, the variance components σ_u^2, σ_e^2 are obtained from the variance components option in the random statement, and mean egg weight (with 95% confidence intervals) from the fixed effects intercept (Appendix 20.A).

The covariance parameter estimates are 0.68 for level-2 nest effects (intercept) and 0.12 for level-2 egg effects (residual). Mean egg weight is 9.3 (95% confidence interval 8.9, 9.7) g. The ICC $= \sigma_u^2/(\sigma_u^2 + \sigma_e^2) = 0.68/(0.68+0.12) = 0.85$. The ICC indicates that 85% of the variation in egg weight can be accounted for by nest of origin.

20.4 Other Considerations: Balance, Sparse Data, Costs

20.4.1 Balanced Versus Unbalanced Designs

Balance (equal sample sizes within each cluster) is rarely possible for multilevel animal-based studies, either by design or (more usually) as a result of natural variation. Some experiments are designed to be unbalanced, for example, studies featuring experimental transfer of offspring to artificially increase or reduce litter or clutch size (e.g. Voillemot et al. 2012). Examples of natural variation leading to imbalance include litter size in rodents, clutch sizes of breeding birds, and unequal losses resulting from differential mortality or other types of attrition.

When cluster sizes are expected to be highly variable, sample size calculation must be adjusted for this variation with the coefficient of variation (*CV*) of the cluster sizes *n*. van Breukelen et al. (2007) suggest that unequal cluster sizes can be adjusted by the minimum relative efficiency of the sample, which works out to an approximate increase in cluster number of 11%.

20.4.2 Sparse Data

Most multilevel animal-based studies are characterised by very small cluster sizes. In general, level-1 sample sizes <10 do not seem to affect Type I error, confidence interval coverage, or statistical bias to any great extent (Maas and Hox 2005; Clarke 2008; Bell et al. 2008, 2010, 2014). At least $n = 5$ observations per cluster may be sufficient to obtain valid and reliable regression coefficient estimates for two-level models (Clarke 2008) with as few as 10 clusters, and sampling variability can be assessed with bootstrapping, or other computer-intensive simulation techniques (Maas and Hox 2005).

However, power to detect differences between treatment and intervention groups may be poor. Simulations indicate that power achieved desired 0.80 power only for relatively large sample sizes ($n > 20$) for level-1 predictors, and level-2 sample sizes of $k > 30$. Statistical power of level-2 predictors never achieved 0.80 even for very large n and k. Small n also increased the number of zero-variance random effects (Bell et al. 2010, 2014). Sample size approximations should be based on number of clusters rather than number of subjects.

20.4.3 Costs

For multisite veterinary and agricultural trials, and large-scale surveys, maximising the number of sites will be more important than the number of subjects per site for maximising power for detecting a treatment effect. However, adding sites will cost considerably more in terms of logistics, time, resources, and money compared to costs of adding more subjects within each site. Alternatively, the number of sites may be fixed. Therefore, choice of the optimal sample size will entail cost-benefit analyses between power and cost trade-offs and whether the primary objectives of the study are to make inferences about the main effect of treatment, assess treatment-by-cluster variation, or effects of site characteristics on treatment response. Raudenbush and Liu (2000) provide an excellent tutorial and associated SAS code. Raudenbush et al. (2007) have developed a user-friendly, free software programme for optimising sample size and power for a variety of designs: `https://sites.google.com/site/optimal designsoftware/home`

20.5 Sample Size Determinations

20.5.1 Rules of Thumb

The classic sample size rule of thumb for multilevel studies is a minimum of 30 groups with 30 subjects in each group (the so-called '30/30 rule'; Hox 2010). However, these sample sizes originated with social science, education, and survey studies, and will be wholly unrealistic for many animal-based studies. Typically, the usual cluster size for animal studies will be fewer than 10 subjects, with large variation in cluster size.

When the cluster is the unit of randomisation, the number of clusters to be sampled is most important for determining power, and cluster size will be relatively unimportant. However, when the treatment intervention is applied to subjects rather than the entire cluster, the individual within cluster is the unit of randomisation. The number of subjects required per treatment arm can be determined using the usual power calculation methods, then approximating the number of clusters required to obtain that number of subjects. However, this approximation will be increasingly unreliable as variation in cluster size increases.

20.5.2 Sample Size Based on Design Effect

If an estimate of ICC is available, sample size can be approximated using the *design effect* formula. The design effect quantifies the increase in sample size expected for a nested design over that of a simple random sample with no clusters (Kish 1987). That is, if a sample size N has been approximated using conventional power calculations, then N must be multiplied by the design effect (D_{eff}) to adjust for the effects of clustering. The design effect increases with both increased ICC and n per cluster, therefore larger N would need to be sampled.

The design effect factor D_{eff} will be determined by whether cluster sample sizes are balanced or unbalanced.

Balanced cluster sizes. If sample sizes can assumed to be approximately equal for each cluster, then the design effect factor is:

$$D_{eff} = 1 + (n - 1) \cdot ICC$$

where n is the within-cluster sample size.

20.5.3 Initial Approximations

Perform power calculations under the assumption of simple random sampling in the usual way. Then multiply that sample size estimate by D_{eff}.

Example: Adjusting Sample Size for Cluster Effects

A new study was proposed where power calculations based on the assumption of simple random sampling indicated that approximately 100 animals would be required. However, it is anticipated that clustering effects will be important. There are 10 subjects per cluster with ICC of 0.25. What is the revised sample size?

The effective sample size would require

$$D_{eff} = 1 + (10 - 1) \cdot 0.25 = 3.25$$

that is, 3.25 times as many subjects as a non-clustered random sample. Then the new study would require 3.25 (100) = 325 subjects. With approximately 10 subjects per cluster, the study requires 325/10 = 32.5 clusters, rounded up to 33 clusters.

20.5.4 Asymptotic Normal Approximation: Balanced Cluster Sizes

Sample size for a balanced two-arm study with level-1 randomisation (non-aggregate design) is

$$N = 4 \cdot D_{eff} \frac{\left(z_{1-\alpha/2} + z_{1-\beta}\right)^2}{(ES)^2} + \frac{z_{1-\alpha/2}^2}{2}$$

where ES is the effect size δ/s_{total}.

For a study with level-2 randomisation (aggregated design),

$$N = 4 \cdot \frac{\left(z_{1-\alpha/2} + z_{1-\beta}\right)^2}{\left[\delta/(s_{\bar{y}})\right]^2} + \frac{z_{1-\alpha/2}^2}{2}$$

where $s_{\bar{y}} = \sqrt{\frac{D_{eff} \cdot s_{total}^2}{n}}$

20.5.5 Asymptotic Normal Approximation: Unbalanced Cluster Sizes

For many types of animal-based studies, there will be a considerable amount of between-cluster imbalance. Common examples are litter sizes in rodents and clutch sizes in birds, reptiles, and amphibians. Relative to balanced cluster designs or unclustered studies, large variation in cluster size n contributes to loss of power. As a result, projected sample sizes for clustered studies need to be adjusted for the anticipated variation in cluster sizes.

Machin et al. (2018) recommend an adjusted D_{eff} based on the coefficient of variation CV for within-cluster sample sizes:

$$CV_n = s_n/\bar{n}$$

where s_n and $\bar{n}$ are the standard deviation and mean cluster sizes, respectively. Then the adjusted design effect will be

$$D_{eff} = 1 + ICC \cdot \left((\bar{n} - 1) \cdot \left[1 + \left(\frac{k-1}{k} \right) \cdot CV^2 \right] \right)$$

where k is the number of clusters.

Example: House Sparrow Clutch Size

(Data From Grace et al. (2017).) In this study, brood size for $k = 28$ house sparrow nests ranged from 2 to 11 nestlings. Mean brood size $\bar{n}$ was 5 (SD 2), so the coefficient of variation is $2/5 = 0.4$, with ICC ≈ 0.49. Then

$$D_{eff} = 1 + 0.49 \cdot \left((5-1) \cdot \left[1 + \left(\frac{18-1}{18} \right) \cdot (0.4)^2 \right] \right)$$

$$= 3.2$$

Suppose a new study was proposed to test a nestling-level intervention (non-aggregate design), to detect an effect size of 0.8 with a two-sided $\alpha = 0.05$ and power of 80%. Then sample size is

$$N \geq 4 \cdot (3.2) \frac{(1.96 + 1.2816)^2}{(0.8)^2} + \frac{(1.96)^2}{2} \approx 212$$

Assuming an average of 5 nestlings per nest, approximately $212/5 \approx 43$ nests would need to be sampled.

20.5.6 Sample Size Based on the Non-centrality Parameter

The F-statistic for a multilevel design follows a non-central F-distribution $F(1, m-1, \lambda)$, where m is the number of clusters, and λ is the non-centrality parameter. Because λ is related to the power of the test, the total sample size N and/or the total number of clusters m can be obtained by iterating over different candidate values for any of N, m, and/or effect size to find the values that will result in the designated power. Alternatively, power can be calculated for prespecified values for any of N, k, effect size, and/

or variation explained by a covariate (R^2, $\hat{f}^2$). Sample SAS code is provided in Appendix 20.A.

The form of λ to be calculated depends on choice of the unit of randomisation (individual or cluster), the number of hierarchical levels, and whether the effect to be tested in the difference between treatments (with or without covariates), the variance of the treatment effect, or the cluster x-treatment effect (Raudenbush and Liu, 2000). Spybrook et al. (2011) give itemised descriptions of calculation methods for λ.

20.5.7 Two-Level Model, Subjects Within Cluster as Unit of Randomisation

When treatments X_j are randomly assigned to n individuals within m clusters, the model is $Y_{ij} = \beta_{0j} + \beta_j X_j + e_{ij}$. The difference between experimental and control groups means is $d = (\bar{Y}_e - \bar{Y}_c)$. Assuming a balanced cluster design, the non-centrality parameter is

$$\lambda = \frac{m \cdot d^2}{\tau_{11} + \frac{\sigma_e^2}{n}}$$

where τ_{11} is the variance for the treatment effect between clusters, and σ_e^2 is the between-subjects variation.

20.5.8 Two-Level Model, Cluster as Unit of Randomisation

When treatments W_j are randomly assigned to m clusters (with n individuals per cluster), the model is $Y_{ij} = \gamma_{00} + \gamma_{01} W_j + u_{0j} + e_{ij}$. The main effect of treatment γ_{01} is the mean difference between experimental and control group means $(\bar{Y}_e - \bar{Y}_c)$. Assuming an equal number of clusters per treatment group, the non-centrality parameter is

$$\lambda = \frac{\gamma_{01}^2}{\frac{4}{m} \left[\tau + \frac{\sigma_e^2}{n} \right]}$$

where m is the number of clusters, n is the number of subjects per cluster, γ_{01} is the regression coefficient for the treatment effect, τ is between-cluster variation, and σ_e^2 is the between-subjects variation. Then the non-centrality parameter can be used to estimate sample size by iteration. If the design is balanced, the F-statistic for testing the main effect of treatment follows the non-central F distribution with 1 and $(m-1)$ degrees of freedom.

Example: Sparrow Model of Early Life Stress

(Data from Grace et al.). Average body masses at 12 days post-hatch of nestling house sparrows (*Passer domesticus*) treated with either corticosterone or vehicle control were 21.5 and 20.2 g, respectively. The between-nestling variation σ_e^2 was 2.228, and between-nest variation τ was 2.209.

A new study was planned to investigate if corticosterone induced a reduction of 1.5 g compared to control, with all nestlings in a nest administered the same treatment. It was assumed that the median clutch size would be 6 nestlings per nest. How many nests need to be sampled to detect this difference with 95% confidence and power 80% and 90%?

The cluster is the unit of randomisation. Figure 20.2 shows the results obtained by calculating the non-centrality parameter over a range

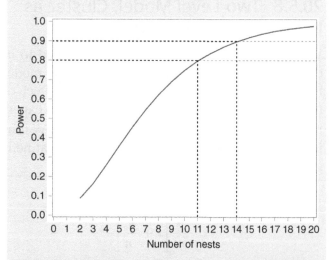

Figure 20.2: Number of nests to be sampled to attain prespecified power in a study of sparrow early life stress.

of sample sizes and solving for m. For power 80%, 11 nests must be sampled, and 14 nests for power of 90%.

20.A Notes on Software

In SAS, regression coefficients and variance components can be estimated with *proc mixed* for continuous normally distributed response variables (Bell et al. 2013), and *proc glimmix* for binary or categorical outcomes (Ene et al. 2015; Kiernan 2018). In R, both marginal and conditional R^2 and ICC for mixed models can be calculated using r2() and icc() in the lme4 package (Bates et al. 2011). Nakagawa and Schielzeth (2013) and Nakagawa et al. (2017) provide R code for obtaining R^2 from generalised linear mixed-effects models. The Centre for Multilevel Modelling, University of Bristol, has developed software MLwiN for complex multilevel modelling (http://www.bristol.ac.uk/cmm/). Spybrook et al. (2011) have developed a user-friendly, free software programme for optimising sample size and power when sampling costs must be factored into the study design, https://sites.google.com/site/optimaldesignsoftware/home.

20.A.1 SAS Codes for Estimating Variance Components

1. *Continuous outcomes.* The variance components σ_u^2, σ_e^2 are obtained from the variance components option in the random statement of *proc mixed*. The grand mean (with 95% confidence intervals) is obtained from the fixed effects intercept. The SAS code is:

```
proc mixed data=twolevel covtest CL
method=ML;
class cluster;
model Y = / s CL;
random intercept / subject=cluster
type=vc;
run;
```

2. *Binary outcome.* The variance components are obtained from the variance components

option in the random statement of *proc glimmix*.

```
proc glimmix data=twolevel
method=laplace CL;
class cluster;
model Y = / s CL dist= binary link=logit
oddsratio;
random intercept /subject=cluster
type=vc CL;
covtest /Wald;
run;
```

3. *Level-2 predictors*. Sample SAS code for the treatment applied to subjects (X_1) and two additional predictors W_1 and W_2 at the cluster level is:

```
proc mixed data=twolevel covtest CL
method=ML;
class cluster;
model Y = X1 W1 W2 / s CL;
random intercept X1 / subject=cluster
type=vc;
run;
```

If the treatment was applied at the level-2 cluster level (W_1), and there were two level-1 covariates X_1 and X_2, the code would be:

```
proc mixed data=twolevel covtest CL
method=ML;
class cluster;
model Y = X1 X2 W1 / s CL;
random intercept X1 X2 / subject=cluster
type=vc;
run;
```

4. Sample SAS code for obtaining total sample size from the non-centrality parameter for a two-level model, and subjects within cluster as unit of randomisation.

```
%let numClust = 100;
data cluster;
*iterate through range of cluster sample
sizes;
do m = 5 to &numClust by 5;
*prespecified difference;
meandiff = 1.5;
s2_subj= 2.228;        *pooled variance;
s2_clus= 2.209;
n = 6;                 *assume median
clutch size of 6;

alpha=0.05;
```

```
* Calculate non-centrality parameter;
  NCP = m*(meandiff**2)/(s2_clus +
s2_subj/n);
   F_crit = finv(1-alpha,1,m-1,0);
     power = 1-probF(F_crit,1,m-1, NCP);
    output;
  end;
 run;

*output realised power;
proc print data = cluster;
run;
```

20.B SAS Code for Estimating Variance Components for Calculations of R^2 and Cohen's f^2 for House Sparrow Data (Code modified from Selya et al. 2012)

```
*1 Calculate variance for full model level-1
treatment effect & covariates;
ods output CovParms = VarFull;
proc mixed data=sparrow covtest CL method=ML;
class nest;
model mass = TRT SEX TARSUS / s CL;
random INT / subject=nest type= vc;
run;
quit;
ods output close;

*2. Calculate variance for covariates (minus
TRT);
ods output CovParms = VarPart;
proc mixed data=sparrow covtest CL method=ML;
class nest;
model mass= SEX TARSUS/ s CL;
random INT / subject=nest type=vc;

run;
quit;
ods output close;

*3. Calculate variance for null model (no
predictors);
ods output CovParms = VarNull;
proc mixed data=sparrow covtest CL method=ML;
class nest;
model mass = / s CL;
random INT / subject=nest type=vc;
run;
quit;
ods output close;
```

```
*Merge datasets;
DATA VarAll;
merge VarFull(rename=(Estimate=VAB))
VarPart(rename=(Estimate=VA))
VarNull(rename=(Estimate=Vnull));
by CovParm;
DROP CovParm;
run;

* Calculate R2 and f2;
DATA results;
set VarAll;
R2full = (Vnull - Vfull)/Vnull;
R2part = (Vnull - Vpart)/Vnull;
f2 = (R2full - R2part)/(1 - R2full);
run;
proc print;
run;
```

References

Aarts, E., Dolan, C.V., Verhage, M., and van der Sluis, S. (2015). Multilevel analysis quantifies variation in the experimental effect while optimizing power and preventing false positives. *BMC Neuroscience* 16: 94. https://doi.org/10.1186/s12868-015-0228-5.

Aarts, E., Verhage, M., Veenvliet, J.V. et al. (2014). A solution to dependency: using multilevel analysis to accommodate nested data. *Nature Neuroscience* 17: 491–496.

Austin, P.C. and Leckie, G. (2018). The effect of number of clusters and cluster size on statistical power and Type I error rates when testing random effects variance components in multilevel linear and logistic regression models. *Journal of Statistical Computation and Simulation* 88 (16): 3151–3163. https://doi.org/10.1080/00949655.2018.1504945.

Austin, P.C. and Merlo, J. (2017). Intermediate and advanced topics in multilevel logistic regression analysis. *Statistics in Medicine* 36 (20): 3257–3277. https://doi.org/10.1002/sim.7336.

Bates, D., Maechler, M., and Bolker, B. (2011). lme4: linear mixed-effects models. R package, version 0.999375-42. http://CRAN.R-project.org/package=lme4 (accessed 2022).

Bell, B.A., Ene, M., Smiley, W., and Shonenberger, J.A. (2013). A multilevel primer using SAS *proc mixed*. *SAS Global Forum 2013 Proceedings*. http://support.sas.com/resources/papers/proceedings13/433-2013.pdf (accessed 2019).

Bell, B.A., Ferron, J.M., and Kromrey, J.D. (2008). Cluster size in multilevel models: the impact of sparse data structures on point and interval estimates in two-level models. In: *Proceedings of the Joint Statistical Meetings (Survey Research Methods section)*, 1122–1129. Alexandria, VA: American Statistical Association.

Bell, B.A., Morgan, G.B., Schoeneberger, J.A. et al. (2014). How low can you go? An investigation of the influence of sample size and model complexity on point and interval estimates in two-level linear models. *Methodology European Journal of Research Methods for the Behavioral and Social Sciences* 10: 1–11.

Bell, B.A., Morgan, G.B., Schoeneberger, J.A., et al. (2010). Dancing the sample size limbo with mixed models: how low can you go? *SAS Global Forum 2010 Proceedings* (11-14 April 2010). Paper 197-2010, Seattle, WA. https://support.sas.com/resources/papers/proceedings10/197-2010.pdf.

Clarke, P. (2008). When can group level clustering be ignored? Multilevel models versus single-level models with sparse data. *Journal of Epidemiology and Community Health* 62 (8): 752–758. https://doi.org/10.1136/jech.2007.060798.

Draper, N.R. and Smith, H. (1998). *Applied Regression Analysis*, 3e. New York: Wiley.

Ene, M., Leighton, E.A., Blue, G.L., and Bell, B.A. (2015). Multilevel models for categorical data using SAS *proc glimmix*: the basics. *SAS Global Forum 2015 Proceedings*. https://support.sas.com/resources/papers/proceedings15/3430-2015.pdf (accessed 2019).

Feingold, A. (2015). Confidence interval estimation for standardized effect sizes in multilevel and latent growth modelling. *Journal of Consulting and Clinical Psychology* 83 (1): 157–168. https://doi.org/10.1037/a0037721.

Goldstein, H. (2003). *Multilevel Statistical Models*, 3e. London: Edward Arnold.

Grace, J.K., Froud, L., Meillère, A., and Angelier, F. (2017). House sparrows mitigate growth effects of post-natal glucocorticoid exposure at the expense of longevity. *General and Comparative Endocrinology* 253: 1–12. https://doi.org/10.1016/j.ygcen.2017.08.011.

Hedges, L. and Hedberg, E.C. (2007). Intraclass correlation values for planning group randomized trials in education. *Educational Evaluation and Policy Analysis* 29 (1): 60–87.

Hox, J.J. (2010). *Multilevel Analysis: Techniques and Applications*, 2nd ed. New York: Routledge.

Hox, J.J., Moerbeek, M., and van de Schoot, R. (2018). *Multilevel Analysis: Techniques and Applications*, 3e. Taylor & Francis Group.

Hull, M.A., Reynolds, P.S., and Nunamaker, E.A. (2022). Effects of non-aversive versus tail-lift handling on breeding productivity in a C57BL/6J mouse colony. *PLoS ONE* 17 (1): e0263192. https://doi.org/10.1371/journal.pone.0263192.

Johnson, P.C.D. (2014). Extension of Nakagawa and Schielzeth's R^2 GLMM to random slopes models. *Methods in Ecology and Evolution* 5 (9): 944–946. https://doi.org/10.1111/2041-21020.12225.

Julian, M. (2001). The consequences of ignoring multilevel data structures in non-hierarchical covariance modeling. *Structural Equation Modeling* 8: 325–352. https://doi.org/10.1207/S15328007SEM0803_1.

Kiernan, K. (2018). Insights into using the GLIMMIX procedure to model categorical outcomes with random effects. Paper SAS2179-2018. https://support.sas.com/resources/papers/proceedings18/2179-2018.pdf.

Kish, L. (1987). *Statistical Design for Research*. New York: Wiley.

Konstantopoulos, S. (2008). The power of the test for treatment effects in three-level cluster randomized designs. *Journal of Research on Educational Effectiveness* 1 (1): 66–88. https://doi.org/10.1080/19345740701692522.

Konstantopoulos, S. (2010). Power analysis in two-level unbalanced designs. *Journal of Experimental Education* 78 (3): 291–317.

Konstantopoulos, S. (2012). The impact of covariates on statistical power in cluster randomized designs: which level matters more? *Multivariate Behavioral Research* 47 (3): 392–420. https://doi.org/10.1080/00273171.2012.673898.

Lazic, S.E. (2010). The problem of pseudoreplication in neuroscientific studies: is it affecting your analysis? *BMC Neuroscience* 11: 5. https://doi.org/10.1186/1471-2202-11-5.

Lazic, S.E., Clarke-Williams, C.J., and Munafò, M.R. (2018). What exactly is 'N' in cell culture and animal experiments? *PLoS Biology* 16 (4): e2005282. https://doi.org/10.1371/journal.pbio.2005282.

Lazic, S.E. and Essiou, L. (2013). Improving basic and translational science by accounting for litter-to-litter variation in animal models. *BMC Neuroscience* 14: 37. https://doi.org/10.1186/1471-2202-14-37.

Lorah, J. (2018). Effect size measures for multilevel models: definition, interpretation, and TIMSS example. *Large-Scale Assessments in Education* 6: 8. https://doi.org/10.1186/s40536-018-0061-2.

Maas, C.J.M. and Hox, J.J. (2005). Sufficient sample sizes for multilevel modelling. *Methodology* 1 (3): 86–92. https://doi.org/10.1027/1614-1881.1.3.86.

Machin, D., Campbell, M.J., Tan, S.B., and Tan, S.H. (2018). *Sample Sizes for Clinical, Laboratory and Epidemiology Studies*, 4e. Wiley-Blackwell.

McNeish, D., Stapleton, L.M., and Silverman, R.D. (2016). On the unnecessary ubiquity of hierarchical linear modeling. *Psychological Methods* 22: 114–140. https://doi.org/10.1037/met0000078.

Moerbeek M (2004). The consequences of ignoring a level of nesting in multilevel analysis. *Multivariate Behavioral Research*, 39, 129–149. https://doi.org/10.1207/s15327906mbr3901_

Nakagawa, S., Johnson, P.C.D., and Schielzeth, H. (2017). The coefficient of determination R^2 and intra-class correlation coefficient from generalized linear mixed-effects models revisited and expanded. *Journal of the Royal Society Interface* 14: 20170214. https://doi.org/10.1098/rsif.2017.0213.

Nakagawa, S. and Schielzeth, H. (2013). A general and simple method for obtaining R^2 from generalized linear mixed-effects models. *Methods in Ecology and Evolution* 4 (2): 133–142. https://doi.org/10.1111/j.2041-21020.2012.00261.x.

Raudenbush, S.W. and Liu, X. (2000). Statistical power and optimal design for multisite randomized trials. *Psychological Methods* 5 (2): 199–213. https://doi.org/10.1037/1082-98920.5.2.199.

Raudenbush, S.W., Martinez, A., and Spybrook, J. (2007). Strategies for improving precision in group-randomized experiments. *Educational Evaluation and Policy Analysis* 29 (1): 5–29. https://doi.org/10.3102/0162373707299460.

Reynolds, P.S. (1996). Brood reduction and siblicide in Black-Billed Magpies (*Pica pica*). *The Auk* 113 (1): 189–199. https://doi.org/10.2307/4088945.

Selya, A.S., Rose, J.S., Dierker, L.C. et al. (2012). A practical guide to calculating Cohen's f^2, a measure of local effect size, from PROC MIXED. *Frontiers in Psychology* 3: 111. https://doi.org/10.3389/fpsyg.2012.00111.

Snijders, T.A. and Bosker, R.J. (1993). Standard errors and sample sizes for two-level research. *Journal of Educational and Behavioural Statistics* 18: 237–259.

Snijders, T.A.B. and Bosker, R.J. (1999). *Multilevel Analysis: An Introduction to Basic and Advanced Multilevel Modeling*. Thousand Oaks, CA: Sage.

Spybrook, J., Bloom, H., Congdon, R., et al. (2011). Optimal design plus empirical evidence: documentation for the "Optimal Design" software. http://hlmsoft.net/od/ (accessed 2022).

Torgerson, D.J. (2001). Contamination in trials: is cluster randomisation the answer? *BMJ* 322: 355. https://doi.org/10.1136/bmj.322.7282.355.

van Breukelen, G.J., Candel, M.J., and Berger, M.P. (2007). Relative efficiency of unequal versus equal cluster sizes in cluster randomized and multicentre trials. *Statistics in Medicine* 26 (13): 2589–2603. https://doi.org/10.1002/sim.2740.

Voillemot, M., Hine, K., Zahn, S. et al. (2012). Effects of brood size manipulation and common origin on phenotype and telomere length in nestling collared flycatchers. *BMC Ecology* 12: 17. https://doi.org/10.1186/1472-6785-12-17.

21 Ordinal Data

21.1 Introduction

Score data are extremely common in animal-based studies, especially for assessment of health- and welfare-related characteristics and grading histopathological images. Scores are also common in survey and questionnaire studies, e.g. the Likert scale (Box 21.1). These variables are ordinal and consist of data that are categorical with a meaningful ordering or ranking.

Ordinal data are relatively simple to collect, as data collection involves counting the number of subjects in each of a number of predefined categories. However, analysis and interpretation can be difficult. Because the difference between any pair of adjacent scores is qualitative, a rank score can be 'better' or 'worse' than another, but the number codes for the ranks and intervals between adjacent ranks do not have the same quantitative values as do continuous data measured on a ratio scale. Distances between intervals cannot be standardised (for example, the difference between 'Very good' and 'Good' is not the same as that between 'Moderate' and 'Poor'). Participant preferences for particular categories may be biased (e.g. respondents may prefer a 'neutral' or extreme categories over intermediate categories).

BOX 21.1
Common Examples of Ordinal Data

Likert scores

    1 = 'Strongly Agree' 2 = 'Agree'; 3 = 'Neutral', 4 = 'Disagree', 5 = 'Strongly Disagree'

Pain scores: 0 = No pain to 10 = Severe pain Quality of life scores

    Exercise tolerance: 1 = Very good; 2 = Good; 3 = Moderate; 4 = Poor
    Demeanor: 1 = Alert, responsive; 2 = Mildly depressed; 3 = Moderately depressed; 4 = Minimally responsive; 5 = Unresponsive

Plumage condition scores

    1 = Fully feathered; 2 = Some small bare patches, 3 = Strongly damaged feathers and large bare patches; 4 = Mostly or completely denuded

Body condition scores

    1 = Emaciated; 2 = Underweight; 3 = Ideal weight; 4 = Overweight; 5 = Obese

Histopathologic scores

    Necrosis: 1 = Rare (<5%); 2 = Multifocal (6–40%); 3 = Coalescing (41–80%); 4 = Diffuse (>80%)

A Guide to Sample Size for Animal-based Studies, First Edition. Penny S. Reynolds.
© 2024 John Wiley & Sons Ltd. Published 2024 by John Wiley & Sons Ltd.

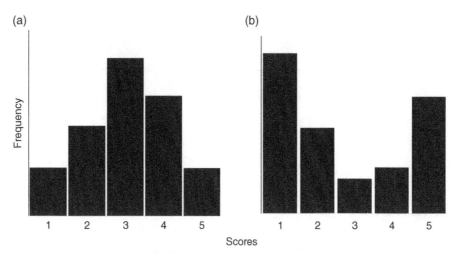

Figure 21.1: Two simulated ordinal data sets on a 5-point scale, both with median score 3. (a) The distribution mirrors the cognitive 'expectation'. (b) The median does not represent the clinical picture, which shows most subjects occurring at either extreme and relatively few occurring at the median.

Basic arithmetic operations (addition, subtraction, multiplication, division) cannot be performed on ordinal data. Descriptive statistics used for continuous data (e.g. mean, median, mode, standard deviation) often do not have a clinically or biologically sensible interpretation when applied to ordinal data. Although it is sometimes suggested that ordinal data can be summarised by the median and mode, such summary statistics can be very misleading. Figure 21.1 shows two data sets with the same median score of 3. However, biological or clinical interpretation will be very different. In the first sample, the distribution mirrors the cognitive expectation that the summary statistic reflects the central tendency of the data. In second sample, most subjects occur at both extremes of the ranking, with relatively few occurring at the median.

Ordinal data can be analysed by conventional rank-based non-parametric tests such as the Mann-Whitney U-test for unpaired two samples, Kruskal-Wallis (more than two unmatched groups) and Friedman tests (more than two matched pair groups or repeated measures). Cumulative ordinal regression or cumulative link models are more powerful alternatives (Agresti 2013).

an odds ratio, the expected distribution of proportions across categories, and the allocation ratio (Campbell et al. 1995; Machin et al. 2018).

Sample size increases substantially as proportion representation becomes more unbalanced toward dominance of one category. If the mean proportions $\bar{p}_i$ are expected to be roughly similar across all categories, then $\bar{p}_i$ depends on the number of categories k and $\bar{p}_i$ is approximately $1/k$ (Table 21.1). For a two-arm study with $i = 4$ categories of response, and responses across all four categories are expected to be similar, then $p_1 = p_2 = p_3 = p_4 = 1/4 = 0.25$. In contrast, if there is one dominant category with sparse representation in the other categories, sample size will be inflated by approximately 43% over the uniform distribution case.

Maximum power will be obtained if sample sizes for each intervention group are equal. To correct for allocation ratio, the sample size formula is adjusted by $(1 + r)^2/r$. If there is an equal number of subjects in each group, then $r = n_1/n_2 = 1$, and the correction is $(1 + 1)^2/1 = 4$. The expected distribution of proportions across categories can be obtained from previous data. If there is no information on the likely response pattern, sample size can be approximated based on the most likely pattern of distribution (Table 21.1).

21.2 Sample Size Considerations

Sample size determinations for ordinal data are based on the proportion of events expected for each category and the anticipated effect size, expressed as

21.3 Sample Size Approximations

Approximating sample size for ordinal categorical data is a four-step process.

Table 21.1: Sample size inflation resulting from proportion imbalance over four categories. Five common scenarios are presented. The sample size for the uniform category distribution is represented by n_0. The ratio n/n_0 is the sample size inflation caused by imbalances. Sample size requirements increase substantially with increasing dominance of a single category.

Category	0	1	2	3	
Proportions	p_1	p_2	p_3	p_4	n/n_0
Equally probable occurrence (uniform)	0.25	0.25	0.25	0.25	1.00
Graduated occurrence	0.10	0.20	0.30	0.40	1.04
Single under-represented category	0.10	0.30	0.30	0.30	1.14
Bimodal representation	0.05	0.05	0.45	0.45	1.15
Single dominant category	0.10	0.10	0.10	0.70	1.43

Source: Adapted from Whitehead (1993).

1. Obtain a best guess for proportions expected for each category in the treatment and control groups.
2. Calculate cumulative proportions $P_{0,i}$ and $P_{1,i}$ for the treatment and control groups.
3. Determine an expected odds ratio OR for the new study, or if pre-existing data are available, calculate the odds ratio for each category based on the cumulative proportions:

$$OR_i = \frac{P_{1,i}/(1 - P_{1,i})}{P_{0,i}/(1 - P_{0,i})}$$

A 'summary' odds ratio for planning purposes can be estimated from the geometric mean of all the odds ratios.

4. Calculate the average proportions $\bar{p}$ of the treatment and control probabilities for each category i.

Then, sample size is

$$N_k = \frac{3}{1 - \sum_{i=1}^{k} \bar{p}^3} \cdot \frac{(1+r)^2}{r} \cdot \frac{(z_{1-\alpha/2} + z_{1-\beta})^2}{(\ln OR)^2}$$

If allocation is equal, then $(r + 1)^2/r$ is 4. If all average proportions $\bar{p}$ are approximately equal, then $3 / \left(1 - \sum_{i=1}^{k} \bar{p}^3\right)$ depends only on the total number of categories k, and reduces to $3/[1 - (1/k^2)]$.

Example: Sample Size from Raw Count Data: Lung Tissue Histopathology Changes in Mice

An experiment tested a new anti-inflammatory compound versus an inert control randomly allocated to 40 mice, with 20 mice per group. Histopathology scores (0 = normal, 1 = mild, 2 = moderate, 3 = severe) describing inflammatory changes in lung tissue were obtained from the mice after a two-week course of treatment. The histopathology scores are as follows:

Test	0	2	1	0	1	2	0	0	0	0
	1	0	0	0	3	0	1	0	0	0

Control	0	0	2	2	2	1	3	1	3	1
	0	0	2	0	0	2	2	0	1	1

Suppose it is desired to conduct a new study, for which it is assumed that the distribution of lung damage scores will be similar to the previous study. There will be equal sample sizes in each of the control and intervention groups. What is the new sample size?

First, reorganise the data in tabular form and calculate counts and proportions for each category. The proportions p for each of $i = 4$ histopathology categories are shown in Table 21.2.

Table 21.2: Mouse lung tissue histopathology: calculation of proportions for each histopathology category.

Histopathology category i	Number of mice		Proportion of mice	
	Control, n_0	Test, n_1	Control, p_0	Test, p_1
0	7	13	0.35	0.65
1	5	4	0.25	0.20
2	6	2	0.30	0.10
3	2	1	0.10	0.05
Total	20	20	1.00	1.00

$$OR_1 = \frac{P_{1,1}/(1-P_{1,1})}{P_{0,1}/(1-P_{0,1})} = \frac{0.85/(1-0.85)}{0.60/(1-0.60)} = 3.78$$

$$OR_2 = \frac{P_{1,2}/(1-P_{1,2})}{P_{0,2}/(1-P_{0,2})} = \frac{0.95/(1-0.95)}{0.90/(1-0.90)} = 2.11$$

The odds ratio for the new study is estimated from the geometric mean of the odds ratios derived from the previous data:

$$OR_{new} = (3.45 \cdot 3.78 \cdot 2.11)^{1/3} = 3.02, \text{ or} \approx 3$$

The average proportion $\bar{p}_i$ is calculated for each category pair: $(p_{oi} + p_{1i})/2$ and summarised in Table 21.3:

Three odds ratios are estimated from these data for comparing baseline risk against each of the remaining three categories:

Then for $\alpha = 0.05$ and $1 - \beta = 0.8$, sample size N is

$$OR_0 = \frac{P_{1,0}/(1-P_{1,0})}{P_{0,0}/(1-P_{0,0})} = \frac{0.65/(1-0.65)}{0.35/(1-0.35)} = 3.45$$

$$N \geq \frac{3}{1-(0.145)} \cdot 4 \cdot \frac{(1.96 + 0.8416)^2}{(\ln 3)^2}$$

$$= 91.3, \text{ or approximately } 92.$$

Table 21.3: Mouse lung tissue histopathology: calculation of expected proportions for a new study based on prior odds ratios.

Histopathology category i	Number of mice		Proportion of mice		Average $\bar{p}$	Cumulative proportion	
	Control, n_0	Test, n_1	Control, p_0	Test, p_1		Control, $P_{0,i}$	Test, $P_{1,i}$
0	7	13	0.35	0.65	0.50	0.35	0.65
1	5	4	0.25	0.20	0.23	0.60	0.85
2	6	2	0.30	0.10	0.20	0.90	0.95
3	2	1	0.10	0.05	0.08	1.00	1.00
Total	20	20	1.00	1.00			

Example: Sample Size for Detecting a Pre-Specified Odds Ratio

Suppose a new study is planned where the distribution of subjects in the control arm is expected to be similar to those in the initial study. Disproportionately more mice in the test group are expected to be in lowest lung damage categories and more mice in the control group with higher lung damage scores. The expected odds ratio in favour of the test intervention is 2. What sample size is required to detect $OR_{new} = 2$ with 95% confidence and 80% power?

Here the proportions and cumulative proportions for the test group must be estimated from the control group data and the planned odds ratio (Machin et al. 2018). First, calculate the expected cumulative proportion

Table 21.4: Mouse lung tissue histopathology: Calculating expected proportions in each histopathology category when the odds ratio for a new study is pre-specified.

Histopathology category i	Proportion of mice		New average $\bar{p}$	Cumulative proportion	
	Control, p_0	Expected test p_1		Control, $P_{0,i}$	Expected test $P_{1,i}$
0	0.35	0.519	0.435	0.35	0.519
1	0.25	0.231	0.241	0.60	0.750
2	0.30	0.197	0.249	0.90	0.947
3	0.10	0.053	0.076	1.00	1.000
Total	1.00	1.00			

$P = OR_{new} \cdot P_1/[1 - P_1 + (OR_{new} \cdot P_1)].$

Then

$P_1 = (2 \cdot 0.35)/[1 - 0.35 + (2 \cdot 0.35)] = 0.519,$

$P_2 = (2 \cdot 0.60)/[1 - 0.60 + (2 \cdot 0.60)] = 0.750,$

$P_3 = (2 \cdot 0.90)/[1 - 0.90 + (2 \cdot 0.90)] = 0.947,$

and $P_4 = 1$.

The expected proportions p_i for each category are back-calculated as the differences between cumulative proportions: $p_1 = 0.519$; $p_2 = (0.750 - 0.519) = 0.231$, and so on (Table 21.4). Then the new $\frac{3}{1 - \sum_{i=1}^{k} \bar{p}^3}$ is $3/(1 - 0.112) = 3.38$, and

$$N_k = 3.38 \cdot 4 \cdot \frac{(1.96 + 0.8416)^2}{(\ln 2)^2}$$

$= 220.8$, or approximately 221.

21.4 Paired or Matched Ordinal Data

Before-after and crossover studies have each subject serving as its own control so that data are paired. Matched case–control studies match subjects on pre-specified characteristics except for the condition of interest (Julious and Campbell 1998; Julious et al. 1999). These designs result in correlated outcomes.

In paired studies with binary outcomes (positive = 1, negative = 0), there are four possible outcomes for the first and second responses (Table 21.5).

To estimate sample size for paired data based on a pre-specified odds ratio, Julious et al. (1999) suggest first estimating the expected total number of discordant paired observations. The proportion of discordant pairs p_{disc} is $s + t$ and the pre-specified odds ratio OR is s/t. The expected number of discordant pairs n_{disc} for the pre-specified odds ratio is

$$n_{disc} = \frac{[z_{1-\alpha/2}(OR + 1) + 2 \cdot z_{1-\beta} \sqrt{OR}]^2}{(OR - 1)^2}$$

Table 21.5: In paired studies with binary outcomes (positive = 1, negative = 0), there are four possible outcomes for the first and second responses.

	First response	Second response	Sum of first and second responses
Both negative (agree)	0	0	r
First negative, second positive (discordant)	0	1	s
First positive, second negative (discordant)	1	0	t
Both positive (agree)	1	1	u
Total			$r + s + t + u$

Finally, the total number of paired observations N_{pair} is

$$N_{pair} = \frac{\left[z_{1-\alpha/2}(OR+1) + z_{1-\beta}\sqrt{p_{disc}(OR+1)^2 - (OR-1)^2}\right]^2}{p_{disc}(OR-1)^2}$$

If there is a single dominant category, Julious et al. (1999) suggest that preliminary sample size estimates can be obtained by dichotomising the data into the dominant group and one other by pooling observations from the other categories, then estimating sample size as above.

Example: Stress Reduction in Healthy Dogs

Mandese et al. (2021) reported that 27/44 dogs (61%) showed a major increase in stress levels relative to baseline following separation from owners during routine physical exams. There were 16/44 (36%) dogs that showed a reversal in stress levels between baseline and the subsequent assessment (discordance).

Suppose a new study was to be designed to detect a 50% reduction in risk of stress ($OR = 0.5$) with 80% power and 95% confidence. Then, the number of discordant paired observations required will be

$$n_{disc} = \frac{\left[1.96(0.5+1) + 2 \cdot 0.8416\sqrt{0.5}\right]^2}{(0.5-1)^2} \approx 68$$

From the previous data, it was expected that $s + t = 0.36$. The new $OR = 0.5$. Therefore, the approximate sample size is $68/0.36 \approx 189$

21.5 Sample Size for Observer Agreement Studies

Rater agreement studies are essential for validating scoring systems by quantifying variation caused by both the measurement system and the personnel performing the measurements. A valid scoring system will produce consistent results when different observers score the same animal or specimen independently or if the same rater scores the same animal or specimen multiple times. The goal is to assess both *reproducibility* (the similarity in scores made by multiple observers on the same specimen) and *repeatability* (the similarity in scores for the same observer over multiple assessments on the same specimen). The reliability of the measurements is interpreted in the context of *a priori* definition of the minimally acceptable difference between observers (Hallgren 2012). Detailed coverage of the design of reproducibility and repeatability (R & R) studies is beyond the scope of this chapter; however, detailed guidance can be found elsewhere (e.g. Montgomery and Runger 1993a, b; https://sixsigmastudyguide.com/repeatability-and-reproducibility-rr).

Rater agreement studies are a special case of repeated binary or categorical ordinal data (Fleiss and Cohen 1973; Fleiss 1981). In these studies, two or more observers assess the same specimens or items, often several times. Cohen's kappa coefficient (κ) is used to measure both inter-rater and intra-rater reliability for two raters only. Fleiss' κ is used for >2 raters (Fleiss 1971) but is not a generalisation of Cohen's κ. For estimating inter-rater reliability, the number of specimens to be assessed, N, can be estimated from the expected proportion of discordant ratings p_{disc} and Cohen's κ as follows:

$$\kappa = \frac{p_o - p_e}{1 - p_e}$$

where p_o is the observed agreement between raters, and p_e is the agreement expected by chance (Appendix 21.A). Then the number of specimens N to be assessed by at least two individual raters with precision δ, a pre-specified κ, and confidence interval $100(1 - \alpha)\%$ is

$$N = z_{1-\alpha/2}^2 \cdot \frac{4(1-\kappa)}{\delta^2}$$
$$\cdot \left[(1-\kappa)(1-2\kappa) + \frac{\kappa(2-\kappa)}{2 \cdot p_{disc}(1-p_{disc})}\right]$$

To estimate within-observer reproducibility, the number of specimens to be assessed by each rater is given by

$$N_{reprod} = z_{1-\alpha/2}^2 \cdot \frac{2 \cdot p'(1-p')[1-2p' \cdot (1-p')]}{(1-2 \cdot p') \cdot \delta^2}$$

where

$$p' = \frac{1 - \sqrt{1 - 2 \cdot p_{disc}}}{2}$$

Example: Rater Agreement

Two pathologists evaluated the same 50 histology specimens for evidence of cancerous lesions. The data are as follows (Table 21.6):

The observed agreement between two raters p_0 is calculated as the proportion of total ties, calculated as the sum of [(both scored negative) + (both scored positive)] divided by the total number of readings:

$$p_0 = (21 + 16)/50 = 37/50 = 0.74$$

The probability of agreement by chance p_e is the proportion of times each rater classified responses as either positive or negative:

$$p_e = [(4 + 16)/50] \cdot [(9 + 16)/50] + [(21 + 9)/50]$$
$$\cdot [(21 + 4)/50] = 0.20 + 0.3 = 0.5$$

Then $\kappa = (0.74 - 0.5)/(1 - 0.5) = 0.48$. Good agreement is indicated by κ close to 1. In this example, the agreement is not particularly good. The proportion of discordant observations p_{disc} is $13/50 = 0.26$.

Then the number of specimens to be assessed to capture a precision δ of 10% with 95% confidence is

$$N = 1.96^2 \cdot \frac{4\,(1-0.48)}{0.1^2} \cdot \left[(1-0.48)(1-2 \cdot 0.48) \right.$$
$$\left. + \frac{0.48(2-0.48)}{2 \cdot (0.26)(1-0.26)} \right] \approx 1532.$$

Sample sizes can be reduced substantially when raters are trained up to a consistent high standard before the study begins so that the number of discordant observations are minimised. For example, for good agreement $\kappa = 0.8$ and $p_{disc} = 0.1$, then

$$N = 1.96^2 \cdot \frac{4\,(1-0.8)}{0.1^2} \cdot \left[(1-0.8)(1-2 \cdot 0.8) \right.$$
$$\left. + \frac{0.8(2-0.8)}{2 \cdot (0.1)(1-0.1)} \right] \approx 373.$$

Table 21.6: Data for two pathologists evaluating the same 50 histology specimens.

	Rater 1	Rater 2	Number of specimens
Both negative (agree)	0	0	21
First negative, second positive (discordant)	0	1	9
First positive, second negative (discordant)	1	0	4
Both positive (agree)	1	1	16
Total			50

21.A Sample SAS Code For Calculating Cohen's κ from Raw Data

```
data test;
input rater1 rater2 count @@;
datalines;
. . . . . . . . . . .
;
run;
proc freq data=test;
tables rater1*rater2 /agree;
weight count;
test kappa;
exact kappa;
title "Cohen's Kappa Coefficients";
run;
```

The macro MKAPPA to calculate κ for more than two raters is described by Chen et al. (2012)

References

Agresti, A. (2013). *Categorical Data Analysis*, 3rd ed. New York: Wiley.

Campbell, M.J., Julious, S.A., and Altman, D.G. (1995). Estimating sample sizes for binary, ordered categorical, and continuous outcomes in two-group comparisons. *BMJ* 311: 1145–1148.

Chen, B., Zaebast, D., and Seel, L. (2012). A macro to calculate kappa statistics for categorizations by multiple raters. https://support.sas.com/resources/papers/proceedings/proceedings/sugi30/155-30.pdf (accessed 2022).

Fleiss, J.L. (1971). Measuring nominal scale agreement among many raters. *Psychological Bulletin* 76: 378–382.

Fleiss, J.L. (1981). *Statistical Methods for Rates and Proportions*, 2nd ed. New York: Wiley.

Fleiss, J.L. and Cohen, J. (1973). The equivalence of weighted kappa and the intraclass correlation coefficient as measures of reliability. *Educational and Psychological Measurement* 33 (3): 613–619. https://doi.org/10.1177/001316447303300309.

Hallgren, K.A. (2012). Computing inter-rater reliability for observational data: an overview and tutorial. *Tutorial in Quantitative Methods for Psychology* 8 (1): 23–34. https://doi.org/10.20982/tqmp.08.1.p023.

Julious, S.A. and Campbell, M.J. (1998). Sample size calculations for paired or matched ordinal data. *Statistics in Medicine* 17: 1635–1642.

Julious, S.A., Campbell, M.J., and Altman, D.G. (1999). Estimating sample sizes for continuous, binary, and ordinal outcomes in paired comparisons: practical hints. *Journal of Biopharmaceutical Statistics* 9 (2): 241–251.

Machin, D., Campbell, M.J., Tan, S.B., and Tan, S.H. (2018). *Sample Sizes for Clinical, Laboratory and Epidemiology Studies*, 4th ed. Wiley-Blackwell.

Mandese W, Griffin F. Reynolds P, Blew A, Deriberprey A, Estrada, A (2021). Stress in client-owned dogs related to clinical exam location: a randomised crossover trial. *Journal of Small Animal Practice*, 62(2):82–88. https://doi.org/10.1111/jsap.13248.

Montgomery, D.C. and Runger, G.C. (1993a). Gauge capability analysis and designed experiments. Part I: basic methods. *Quality Engineering* 6 (1): 115–135.

Montgomery, D.C. and Runger, G.C. (1993b). Gauge capability analysis and designed experiments. Part II: experimental design models and variance component estimation. *Quality Engineering* 6 (2): 289–305.

Whitehead, J. (1993). Sample size calculation for ordered categorical data. *Statistics in Medicine* 12: 2257–2271.

22 Dose-Response Studies

22.1 Introduction

Preclinical dose-response studies are often used for preliminary investigations of efficacy, safety, or toxicity (Salsberg 2006; Dmitrienko et al. 2007, Machin et al. 2018). Because they quantify both the magnitude and shape of responses, dose-response designs are more informative than the classic two-group or 'one-way' comparisons (Wong and Lachenbruch 1996).

The goal of a dose-response study is to compare the effect of one or more agents on a biological response relative to a control. Biological responses are measured along an ordered dose gradient. Specific objectives (Dmitrienko et al. 2007, Bretz et al. 2008) include establishing if a dose-response relationship exists at all (proof of concept or proof-of-activity); determining efficacy of a test compound relative to a standard or control; defining form and shape of dose-related trends; estimating optimal doses; and estimating the optimal therapeutic window (Box 22.1).

> **BOX 22.1**
> *Dose-Response Studies*
>
> Dose-response studies determine
>
> *Efficacy* = relative potency of the test intervention relative to a control
>
> Outcome measures:
>
> Shape of the dose-response relation
> Minimum effective dose
> Maximum dose
> Median effective dose ED50:
> Change in response per unit change in dose (slope)
> Shift.

Informative dose-response studies depend on the existence of a positive relationship or trend between dose and response, rather than statistically significant 'effects' that are not dose-related (Kodell

A Guide to Sample Size for Animal-based Studies, First Edition. Penny S. Reynolds.
© 2024 John Wiley & Sons Ltd. Published 2024 by John Wiley & Sons Ltd.

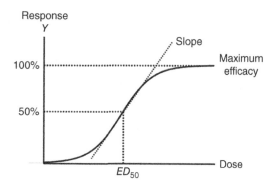

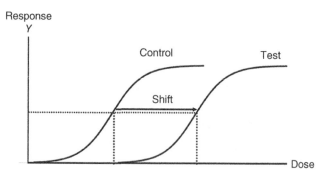

Figure 22.1: Key metrics describing dose-response curves. See text for details.

and Chen 1991). Therefore, before proceeding with a dose-response study, the researcher should first establish if responses to candidate agents are likely to differ from placebo, and second, if a dose-response trend is likely to exist. If a trend is possible, further information should be collected to determine the choice of dose, dose range, and type of control. Supporting evidence for decision to proceed with a formal dose-response trial is based on prior information or proof of concept studies (Laird et al. 2021).

The dose-response curve is quantified by four metrics (Figure 22.1):

Maximum efficacy: Ceiling, or maximum, response (position on the *y*-axis);

Potency: Position of the curve along the dose axis (*x*-axis). It measures the activity of the drug in terms of the dose or concentration of the drug required to produce a defined effect (*y*-axis). *Shift* is the displacement of the test agent curve along the *x*-axis relative to control;

Median effective dose ED_{50}: The dose that on average induces a response in 50% of subjects;

Slope: Change in response per unit change in dose.

Efficacy outcome measures must be clearly identified, quantitative, and measurable. These can be binary (e.g. death yes/no; tumour presence/absence), counts (e.g. cell counts), time to event (e.g. time to predetermined tumour size or humane endpoint), or continuous (e.g. tumour area, biomarker concentrations, fluorescence) (Salsberg 2006, Bretz et al. 2008). Baseline characteristics for each dose group are described by means, standard deviations, and group size n. Measures of effect are expressed in terms of mean or standardised mean differences, with the placebo or vehicle control group as the reference group.

Biologically relevant trends are rarely associated with 'statistically significant' differences between dose groups, and significance tests performed on each dose point separately are likely to be both misleading and invalid. Hypothesis testing should focus either on parameter estimates for curve metrics or area under the curve (AUC) rather than attempting multiple testing on each dose point (Altman 1991). Simple data plots and summary statistics will be most effective for initial determinations of magnitude and shape.

22.2 Sample Size Requirements

'Good' information for dose-response studies means precision of metric estimates (smallest possible standard error) and power for model discrimination and tests of efficacy (Wong and Lachenbruch 1996). Simulations suggest that the traditional design of four dose groups with five animals of each sex for each group will be underpowered to detect differences as large as 30% (Slob et al. 2005). The goal of a dose-response study is usually to demonstrate that a test intervention is 'better' than a control by some biologically relevant amount. Therefore, controls are usually negative (placebo, vehicle). Dose-response studies without a control cannot be used to conclude that a drug effect exists.

Sample size determinations will require the following information:

Number of treatment arms. The basic design is a parallel two-arm study, with intervention treatments and doses randomly allocated to animals. Sample size can be greatly reduced by crossover designs (Senn 2002), factorial designs (Dmitrienko et al. 2007), or Monte Carlo-based methods that simultaneously assess multiple heterogeneous tumour models (Ciecior et al. 2021).

Efficacy metric type: binary, ordinal, time to event, continuous. Much smaller sample sizes are needed when the outcome is continuous compared to categorical or time-to-event outcomes.

Dose range. The dose range must be wide enough to cover the expected response range and discriminate different models of trend (e.g. linear versus quadratic or nonlinear). Doses should be chosen to bracket the benchmark dose (usually zero) and the likely maximum dose (Slob et al. 2005). Choice of the maximum dose must be made with care. If the dose is too high, unexpected or severe adverse events related to safety and tolerance may occur, and if too low, efficacy effects will not be detectable (Pinheiro et al. 2006, Bretz et al. 2008).

Dose placement. Selection of dosage levels and how they are spaced are crucial design components for reliable descriptions of the dose-response relationship. At least five or six dose groups are recommended, especially when outcomes are highly variable (coefficient of variation of 18% or more; Slob et al. 2005). When the shape of the response is not known beforehand, equally spaced designs are more robust than optimal designs (e.g. D-optimal and c-optimal designs with dose levels determined by optimisation techniques; Holland-Letz and Kopp-Schneider 2015).

Expected response frequency. It is usually assumed that higher doses will result in more pronounced responses compared to lower doses (Dmitrienko et al. 2007).

Model choice. The shape expected for the response curve will determine the form of the model to be fitted to the data, and therefore the number of regression coefficients to be estimated (Bretz et al. 2008). Linear regression is useful for first-pass approximations of the overall dose-response relation, and for initial estimates of nonlinear parameter values. Splines have the advantage of being able to fit almost any curve

shape with only two parameters (Crippa and Orsini 2016). Interpreting appearance of curve shapes based on historical data must be performed with care, as artefact can be introduced by small sample sizes, lack of randomisation, confounders, large variability, and 'fishing' (analysing large number of endpoints) (Thayer et al. 2005). However, design features that increase precision, such as dose placement and number of dose levels, coupled with consistency in experimental procedures, are more important than model choice (Slob et al. 2005).

22.2.1 Translational Considerations

Dose-response sample size planning should consider three major factors affecting translation potential of the model: sex, allometric scaling, and application of 3Rs principles. These are described in more detail in Chapter 6.

Sex. National Institute of Health best-practice standards for animal-based research strongly encourage the inclusion of female animals (Clayton and Collins 2014 and consideration of sex as a biological variable (Miller et al. 2017). Regulatory guidelines may mandate testing of both sexes at each dose level (e.g. OECD Test Guidelines `https://www.oecd.org/`). Sample sizes can be minimised by careful choice of statistically–based experimental designs.

Allometric scaling. Care must be taken when determining dose ranges derived from interspecific or intraspecific allometric dose conversions. General scaling relationships may be useful for initial approximations. However, allometric predictions for pharmacokinetic relationships should be based on quality data and validated up-to-date methodologies (Blanchard and Smoliga 2015).

3Rs principles. Regulatory agencies promote the incorporation of reduction and refinement methods and processes into protocols, and increasingly, replacement with non-animal models, such as cell and tissue-based assays. For example, OECD guidelines for chemical testing in animals[1] provide practical

---

[1] `https://www.oecd.org/chemicalsafety/testing/oecdguidelinesforthetestingofchemicals.htm`

recommendations for application of 3Rs methods in acute and chronic toxicity assays. Mouse oncology models commonly use tumour burden (tumour size, number of tumours, tumour recurrence) to assess tumorigenicity, progression, and response to treatment. Because tumour burden is a critical humane endpoint, follow-up time and terminal tumour sizes must be predefined, and humane limits strictly adhered to (Nunamaker and Reynolds 2022).

22.3 Dose-Response Regression Models

The basic dose-response design is the parallel-line assay, modelled as a linear regression:

$$Y = \beta_0 + \beta_1 \cdot D + \beta_2 \cdot X$$

Here β_0 is the intercept, β_1 is the slope describing the change in Y with dose D, and β_2 is the coefficient for describing the difference between control ($X = 0$) and test treatment ($X = 1$) arms (Finney 1978; Figure 22.2).

When the response is binary, logit or probit transformations are used to linearise the data. Nonlinear regression models are used to estimate more complex relationships between dose and response. For continuous response data, the 3-parameter or 4-parameter nonlinear E_{max} models are commonly used. When dose concentrations span a very wide range or increase exponentially, analyses can

sometimes be simplified by logarithmic transformation to linearise the model. Residuals and fit statistics should be examined before analysis to determine if the chosen model provides the best fit to the data.

Model building can begin with ordinary least-squares linear regression on prior or exemplary data or anticipated values. Preliminary slope estimates can be obtained from the difference of the expected mean values for the highest and lowest dose, divided by the dose range:

$$\beta_{1,plan} = \frac{mean_{max} - mean_{min}}{dose_{max} - dose_{min}}$$

Preliminary estimates for minimum and maximum efficacy are obtained from the anticipated maximum and minimum values of the response variable. Preliminary estimates of the median effective dose ED_{50} can be obtained from the median value of planned doses covering the range of the anticipated therapeutic effect.

Hypothesis tests are one-tailed. This is because most dose-response studies are designed to assess differences in response to increasing doses, so there is only one pre-specified test direction.

22.4 Sample Size: Binary Response

Binary responses are modelled as proportions p for each dose group. Therefore the distribution of the response will be sigmoid, and the relation is linearised by the logit or probit transformation of proportions (Finney 1978, Kodell et al. 2010). The form of the regression is

$$Y = F^{-1}(P) = \beta_0 + \beta_1 \cdot \log_{10}(D) + \beta_2 X$$

where P is the proportion of subjects exposed to dose D that shows the response. For logit analysis, F is the logistic cumulative distribution function (cdf), where $F(x) = 1/(1 + e^{-x})$. For probit analysis, F is the normal (Gaussian) cdf.

Efficacy (ρ) is the ratio of the median dose for the test agent relative to control:

$$\rho = \frac{ED_{50}\ (Test)}{ED_{50}\ (Control)}$$

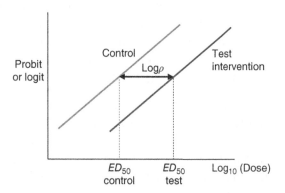

Figure 22.2: The probit regression model for quantifying efficacy between a test and control group. Efficacy is the horizontal difference between parallel test and control regression lines. The Y axis is the probit score for a binary response, and the X axis is the log-transformed doses.

(Finney 1978). The difference in efficacy ($\log_{10}\rho$) between test and control groups is the difference in the $\log_{10}$ transformed values. It is shown graphically as the horizontal distance between parallel test and control regression lines (Figure 22.2):

$$\log_{10}(\rho) = \log_{10}[ED_{50}\ (Test)] \\ - \log_{10}[ED_{50}\ (Control)]$$

The null hypothesis is $\log_{10}(\rho) = 0$ versus the alternative hypothesis H_A: $\log_{10}(\rho) > 0$.

For g dose groups, the sample size per dose group n, is

$$n \cong \frac{2\left(t_{df,1-\alpha} + t_{df,1-\beta}\right)^2 \cdot \left(1/\sum_{i=1}^{g} w_i\right)}{[b_1 \cdot \log_{10}(\rho)]^2}$$

where $b \cdot \log_{10}(\rho)$ is the effect size obtained from b (the estimate of the logit or probit slope based on prior data), ρ (the target efficacy), and w_i are regression weights. The regression weights w_i are obtained from the expected target response proportions P_i in each dose group.

For logit analysis, the logit transformation weights are

$$w_i = P_i(1 - P_i)$$

For probit regression, the probit weights are

$$w_i = \frac{\varphi[\Phi^{-1}(P_i)]^2}{P_i(1 - P_i)}$$

where φ is the normal probability density function (pdf), and Φ^{-1} is the cumulative density function (cdf) of the standard normal distribution (SAS code is given in Appendix 22.A.1).

To solve for power, the equation is re-arranged to obtain the critical t-value for $1 - \beta$ over a range of candidate sample sizes n_i

$$t_{df,1-\beta} = \sqrt{n \cdot \left(\frac{[b_1 \cdot \log_{10}(\rho)]^2 \cdot \sum_{i=1}^{g} w_i}{2}\right)} - t_{df,1-\alpha}$$

then power is calculated iteratively from the cumulative density function for each t value. The degrees of freedom df for the t-distribution depend on the number of dose groups, not the number of subjects. Therefore, $df = 2g - 3$. For example, if there are $g = 5$ dose groups, the degrees of freedom are $2(5) - 3 = 7$, and the critical t values for $\alpha = 0.05$ and $\beta = 0.2$ are $t_{7,0.95} = 1.895$, and $t_{7,0.80} = 0.896$,

respectively. The total sample size N is the product of the sample size per group, the number of groups in each treatment arm, and the number of treatment arms, so that $N = 2 \cdot g \cdot n_i$ (Kodell et al. 2010).

Example: LD_{50} Radiation Lethality Study in Mice

Kodell et al. (2010) reported the results from a 10-day radiation lethality study in mice. Eighty mice were randomly allocated to one of five radiation dose levels and one of two treatment groups (drug versus control) so that there were 8 mice in each of 10 groups. Probit regression was fitted to these data to obtain the common slope b and the estimated LD_{50} values for each treatment group (Appendix 22.A.2).

The probit slope is 28.51. The $\log_{10}LD_{50}$ (control) is 0.95117 and $\log_{10}LD_{50}$(drug) is 1.01814. Log efficacy $\log_{10}(\rho)$ is

$$\log_{10}(\rho) = \log_{10}[LD_{50}(Drug)] - \log_{10}[LD50(control)] \\ = 1.018 - 0.951 = 0.067$$

The estimate of efficacy is obtained from the anti-log of the difference:

$$\rho = \text{anti-log}_{10}(0.067) = 1.167$$

Therefore, relative efficacy is approximately 1.17.

Suppose a new study is planned to detect a range of relative efficacies of 1.05, 1.1, and 1.2 based on these data. What sample sizes are required to detect these relative efficacies with 95% confidence and power 80%?

1. Obtain the anticipated response rates for each dose group, and convert to proportions. In this example, prior data suggested that anticipated response rates will be approximately 5%, 27.5%, 50%, 72.5%, and 95% for each of the five dose groups ($p_i = 0.05$, 0.275, 0.50, 0.725, 0.95).

2. Calculate regression weights. Logit regression weights are calculated as $w_i = P_i(1 - P_i)$. Probit regression weights are calculated from the cdf and pdf functions (Appendix 22.A.1).

3. Calculate power over a range of candidate n (Appendix 22.A.3). For 95% confidence,

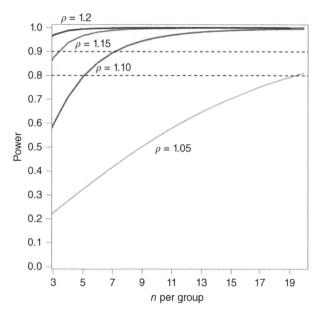

Figure 22.3: Power curves for a range of sample size per group n and efficacy ρ.

Source: Data from Kodell et al. (2010).

$\alpha = 0.05$, and the t value $t_{7,0.95}$ is 1.895. Based on the previous data, a common slope of 28.5 is assumed for the calculations. Power is calculated by deriving the cdf for each value of $t_{1-\beta}$, iterating over the entire range of candidate n and choosing the n that results in power >0.8. Total sample size is then calculated as $2(5) \cdot n$.

Results are shown in Figure 22.3. A sample size of $n = 8$ per group (or total $N = 80$) has power of only 46% to detect relative efficacy of 1.05 with $\alpha = 0.05$. Approximately 20 mice per group ($N = 200$) are required to detect efficacy this small with 95% confidence. In contrast, to detect a larger relative efficacy of $\rho = 1.2$, only two mice per group ($N = 20$) are required for power $>90\%$. However, it must be determined ahead of time if an efficacy of this magnitude is even biologically feasible before considering such small sample sizes for a study.

22.5 Sample Size: Continuous Response

22.5.1 Linear Dose-Response

When the response is continuous and normally distributed, and the relation between response and dose levels is linear, total sample size N can be approximated by:

$$N = g \cdot n = g \cdot \frac{\left(z_{1-\alpha} + z_{1-\beta}\right)^2}{D \cdot (b/s)^2}$$

(Machin et al. 2018). Here g is the number of dose levels, and D is the adjustment for the doses d_i and dose range:

$$D = \sum_{i=0}^{g-1} (d_i - \bar{d})^2$$

If the doses d_i are equally spaced, then D reduces to $g(g^2 - 1)/12$. The effect size b/s is estimated by the regression slope b obtained from the difference in expected mean response values at the maximum and minimum dose: $b = (\mu_{max} - \mu_{min})/(d_{max} - d_{min})$, and the anticipated standard deviation s.

Example: Dose-Response Effects of Vitamin D on Mouse Retinoblastoma

(Data from Dawson et al. 2003). To determine the effectiveness of a vitamin D analogue in inhibiting retinoblastoma, transgenic mice ($N = 175$) were fed a vitamin D- and calcium-restricted diet, then randomised to five dose groups of a vitamin D analogue: 0.0 (vehicle control), 0.1, 0.3, 0.5, or 1.0 µg. Outcomes were tumour area (µm) and serum calcium (mg/dL). The authors concluded there was no dose-dependent response. However, examination of the data suggests that the most dramatic tumour reductions occurred at doses between 0 and 0.1. It was observed that undesirable adverse events, such as hypercalcemia (calcium $>$ 10 mg/dL) and mortality, were minimised at very low doses.

Suppose a new dose-response study is planned for candidate doses $<$0.1 µg, with the objective of detecting a per cent reduction in tumour area of at least 65% with standard deviation of 20%, one-sided $\alpha = 0.05$, and power $1 - \beta = 0.8$. The candidate doses were 0.025, 0.05, and 0.1 µg against vehicle control of 0 µg. In the original study, the initial tumour area was approximately 90×10^3 µm³ for the control group. Then the expected tumour size at 0.1 µg is 31.5, the slope is $(90 - 31.5)/(0 - 0.1) = -585$, and $s = 0.2^*585 = 117$.

There will be $g = 4$ new dose groups. Doses are not equally spaced, so $D = \sum_{i=0}^{4-1} (d_i - \bar{d})^2$. The mean dose is $(0 + 0.025 + 0.05 + 0.1)/4 = 0.044$, and $D = (0.0-0.044)^2 + (0.025-0.044)^2 + (0.05-0.058)^2 + (0.1-0.058)^2 = 0.005$. Approximate sample size is

$$N \geq 4 \cdot \frac{(1.645 + 0.8416)^2}{0.005 \cdot (585/117)^2} \approx 198$$

or approximately 50 mice per dose level.

22.5.2 Nonlinear Dose-Response

Many dose-response patterns are nonlinear. Models are fitted by nonlinear regression, and parameter estimates and associated confidence intervals obtained for all relevant metrics.

The E_{max} model is a common choice for nonlinear dose-response curves (MacDougall 2006). The 3-parameter model describes a concave-down functional response:

$$Y_i = E_0 + \frac{E_{max} D}{ED_{50} + D}$$

The 4-parameter model describes a sigmoid functional response:

$$Y_i = E_0 + \frac{E_{max} D^h}{ED_{50}{}^h + D^h}$$

In both models, D is dose, E_0 is the baseline, or zero-dose effect when no drug is present (placebo or vehicle control), E_{max} is the ceiling or maximum dose effect, and ED_{50} is the dose which produces 50% of the maximum effect. The fourth parameter is h, the Hill or slope parameter. When $h = 1$ the model reduces to the 3-parameter form. Initial parameter estimates for fitting nonlinear models can be obtained from either prior or exemplary data (MacDougall 2006). Nonlinear models cannot produce R^2 values for coefficient estimates. Individual parameter estimates and confidence intervals are used to assess statistical significance of the regression coefficients.

The area under curve (AUC) is an alternative analysis method (Altman 1991), if initial nonlinear parameter estimates are unavailable, if the nonlinear model cannot be fitted to the raw data, or the dose range is too narrow to obtain sufficient precision of key metrics (i.e. the confidence intervals are too wide to be practical). Disadvantages of this method are that comparisons of different groups require the same values for dose concentrations, and AUC units of measurement are difficult to interpret.

Sample size can be approximated by using a prespecified set of linear contrasts to capture anticipated differences in curve shape and test hypotheses (Tukey et al. 1985, Stewart and Ruberg 2000, Chang and Chow 2006, Dmitrienko et al. 2007). Contrasts are a set of integers that code for expected differences between pairs of treatment means. Contrasts are constrained to sum to zero. They can be customised to compare specific effects rather than just the overall difference between null versus alternative hypotheses. Contrasts can be coded to compare the mean for a specified dose level to the overall mean or to make specific comparisons for user-defined pairs. For example, if it was desired to test the mean response for the placebo group against responses for three other dose groups, then the contrasts would be $(-3\ 1\ 1\ 1)$.

Minimum sample size depends on the correct specification of the model shape, and therefore choice of contrast coefficients (Chang and Chow 2006). Therefore, the major disadvantage of this method is that contrasts need to be determined *a priori*, and model misspecification will result in a loss of power.

The null hypothesis for evaluating means for k dose groups is:

$$H_0 : c_1\theta_1 = c_{12}\theta_2 = ... = c_k\theta_k$$

where c_i are contrasts for each expected change in response θ_i from the reference level. The value of the contrasts must sum to zero. The null hypothesis is $H_o : \sum_{i=0}^k c_i \cdot \theta_i = 0$, and the alternative hypothesis for effect is $\sum_{i=0}^k c_i \cdot \theta_i = \varepsilon$, where ε is the expected improvement in response for the ith dose group.

Sample size per dose group n is

$$n = \left[\frac{(z_{1-\alpha} + z_{1-\beta}) \cdot s}{\varepsilon} \right]^2 \cdot \sum_{i=0}^k \frac{c_i^2}{f_i}$$

where c_i are the expected values for the contrasts. The standard deviation s can be estimated from prior data. The sample size fraction f_i is a measure of the balance of sample sizes across dose groups. For example, if there are four dose groups, then a balanced design has $f_i = 1/4 = 0.25$ (Chang and Chow 2006).

Example: Dose-Response Effects of Vitamin D on Mouse Retinoblastoma: Contrasts Method

A new mouse retinoblastoma study was planned to test the hypothesis of a linear dose-response with a relative change of 15% in tumour areas of 90 and 30 μm. The relative change in tumour area at each dose level (expressed as a proportion relative to placebo) can be approximated as $\theta_0 = 0, \theta_1 = 0.15, \theta_2 = 0.30, \theta_3 = 0.65$, with standard deviation s of approximately 0.3.

To test the initial hypothesis 'Is there a response?' the contrasts are set up to compare the control group against the other three dose levels: $c_i = (-3, 1, 1, 1)$. Then

$$\varepsilon = (-3 \cdot 0) + (1 \cdot 0.15) + (1 \cdot 0.3) + (1 \cdot 0.65) = 1.1.$$

The contrasts are $c_i = (-3, -1, 1, 3)$. Then $\varepsilon = (-3) \cdot 0 + (-1) \cdot 0.15 + (1(0.3)) + (3) \cdot 0.65 = 2.1$. For one-sided $\alpha = 0.05$ and power of 80%, the approximate sample size is

$$n = \left[\frac{(1.645 + 0.842) \cdot (0.3)}{1.1} \right]^2 \cdot 4$$
$$\left[(-3)^2 + (1)^2 + (1)^2 + (1)^2 \right]$$
$$\approx 27, \text{ for a total sample size of } 27 \times 4 = 108.$$

The R package 'DoseFinding' provides a comprehensive set of tools for design, contrast generation, multiple contrast tests, and nonlinear model fits for dose-response models (Bornkamp et al. 2023; https://cran.r-project.org/web/packages/DoseFinding).

22.A SAS Code for Dose-Response Calculations

22.A.1 Calculating Logit and Probit Regression Weights

```
*input anticipated response for each dose
group as a proportion;
data prob;
input P @@;
datalines;
0.05 0.275 0.5 0.725 0.95
;
data prob;
set prob;

*calculate logit weights;
w_logit=P*(1-P);

*calculate probit weights;
norminv = quantile('normal', P);
f=pdf('NORMAL',norminv,0,1);
w_probit=f*f/w_logit;
run;
```

22.A.2 Determining Relative Efficacy in a Dose-Response Study. (Data from Kodell et al. 2010.)

```
/*TRT is the two treatment groups where C is
the control group and
T is the test drug treatment group */
/*N is number of mice, Response Y is number of
deaths */

data probit;
input trt $ dose N Y;
ldose=log10(dose); *log10 transformation for Y;
datalines;
C  7 8 0
C  8 8 1
C  9 8 3
C 10 8 8
C 11 8 8
T  7 8 0
T  8 8 0
T  9 8 0
T 10 8 4
T 11 8 5
;
run;

*The response Y is a proportion, therefore Y =
number responding divided by number per group;

*probit model;
proc probit log10 plot=predpplot;
class trt;
model Y/N=Dose trt / lackfit inversecl
itprint;
output out=P p=Prob std=std xbeta=xbeta;
run;

*logistic model: note that the distribution
function is 'logistic';
proc probit log10 plot=predpplot;
class trt;
```

```
model Y/N=Dose trt / d=logistic lackfit
inversecl itprint;
output out=L p=Prob std=std xbeta=xbeta;
run;
```

22.A.3 Solving for Power Using the Iterative Exact Method

```
%let nummax = 20;              *maximum sample size
n per group;
data prob;
do n = 2 to #max by 1; *set increment to
desired step size;
        alpha=0.05;            *confidence;
        rho=1.2;               *target efficacy;
        g=5;                   *number of dose
                                groups;

        DF=2*g-3;              *degrees of freedom;

        *calculate critical t value for alpha;
        t_alpha=quantile('T',1-alpha,DF);

        *summed probit weights wi;
        sum_p = 0.22394 + 0.55843 + 0.63662+
0.55843 + 0.22394;

        *calculate effect size;
ES = (28.51*log10(rho))**2;
sqrtes = sqrt(n*(ES*sum_p/2));

    *calculate t value for beta;
        tbeta=sqrtes-t_alpha;

    *calculate power;
        power=cdf('T',tbeta,DF);
    output;
 end;
run;

proc print;
    var n power;
run;
```

References

Altman, D.G. (1991). *Practical Statistics for Medical Research*. Chapman & Hall/CRC.

Blanchard, O.L., and Smoliga, J.M. (2015). Translating dosages from animal models to human clinical trials – revisiting body surface area scaling. *FASEB Journal*, 29(5):1629–34. https://doi.org/10.1096/fj.14–269043.

Bornkamp, B., Pinheiro, J., Bretz, F., and Sandig, L. (2023). DoseFinding: planning and analyzing dose finding experiments. https://cran.r-project.org/web/packages/DoseFinding/index.html (accessed 2023).

Bretz, F., Hsu, J., Pinheiro, J., Liu, Y. (2008). Dose finding – a challenge in statistics. *Biometrical Journal Biometrische Zeitschrift*, 50(4):480–504. https://doi.org/10.1002/bimj.200810438.

Chang, M. and Chow, S.-C. (2006). Chapter 14: Power and sample size for dose response studies. In: *Dose Finding in Drug Development*, Statistics for Biology and Health (ed. N. Ting), 220–241. New York: Springer https//doi.org/10.1007/0-387-33706-7_9.

Ciecior, W., Ebert, N., Borgeaud, N. et al. (2021). Sample-size calculation for preclinical dose-response experiments using heterogeneous tumour models. *Radiotherapy and Oncology* 158: 262–267. https://doi.org/10.1016/j.radonc.2021.02.032.

Clayton, J.A. and Collins, F.S. (2014). Policy: NIH to balance sex in cell and animal studies. *Nature* 509: 282–283.

Crippa A, Orsini N (2016). Dose-response meta-analysis of differences in means. *BMC Medical Research Methodology*, 16:91. https://doi.org/10.1186/s12874-016-0189-0.

Dawson, D.G., Gleiser, J., Zimbric, M.L. et al. (2003). Toxicity and dose-response studies of 1-alpha hydroxyvitamin D2 in LH-beta-tag transgenic mice. *Ophthalmology* 110 (4): 835–839. https://doi.org/10.1016/S0161-6420(02)01934-6.

Dmitrienko, A., Fritsch, H.J., and Ruberg, S.R. (2007). Chapter 11: Design and analysis of dose-ranging clinical studies. In: *Pharmaceutical Statistics Using SAS: A Practical Guide* (ed. A. Dmitrienko, C. Chuang-Stein, and R. 'Agostino), 273–312. Cary: SAS Institute.

Finney, D.J. (1978). *Statistical Method in Biological Assay*, 3e. New York: Macmillan.

Holland-Letz T, Kopp-Schneider A (2015). Optimal experimental designs for dose-response studies with continuous endpoints. *Archives of Toxicology*, 89 (11):2059–68. https://doi.org/10.1007/s00204-014-1335-2.

Kodell, R.L. and Chen, J.J. (1991). Characterization of dose-response relationships inferred by statistically significant trend tests. *Biometrics* 47: 139–146.

Kodell, R.L., Lensing, S.Y., Landes, R.D. et al. (2010). Determination of sample sizes for demonstrating efficacy of radiation countermeasures. *Biometrics* 66: 239–248. https://doi.org/10.1111/j.1541-0420.2009.01236.x.

Laird, G., Xu, L., Liu, M., Liu, J. (2021). Beyond exposure-response: a tutorial on statistical considerations in dose-ranging studies. *Clinical and Translational Science*, 14(4):1250–1258. https://doi.org/10.1111/cts.12998.

Macdougall, J. (2006). Analysis of dose–response studies – E_{max} model, chap 9. In: *Dose Finding in Drug Development*, Statistics for Biology and Health (ed. N. Ting), 127–145. New York: Springer https://doi.org/10.1007/0-387-33706-7_9.

Machin, D., Campbell, M.J., Tan, S.B., and Tan, S.H. (2018). *Sample Sizes for Clinical, Laboratory and Epidemiology Studies*, 4e. Wiley-Blackwell.

Miller, L.R., Marks, C., Becker, J.B., et al. (2017). Considering sex as a biological variable in preclinical research. *FASEB Journal*, 31(1):29–34. https://doi.org/10.1096/fj.201600781R.

Nunamaker, E.A. and Reynolds, P.S. (2022). 'Invisible actors' – how poor methodology reporting compromises mouse models of oncology: a cross-sectional survey. *PLoS ONE* 17 (10): e0274738. https://doi.org/10.1371/journal.pone.0274738.

Pinheiro, J.C., Bornkamp, B., and Bretz, F. (2006). Design and analysis of dose-finding studies combining multiple comparisons and modeling procedures. *Journal of Biopharmaceutical Statistics* 16: 639–656.

Salsberg, D. (2006). Dose finding based on preclinical studies, chap 2. In: *Dose Finding in Drug Development*, Statistics for Biology and Health (ed. N. Ting), 18–29. New York: Springer https://doi.org/10.1007/0-387-33706-7_9.

Senn, S. (2002). *Crossover Trials in Clinical Research*. Chichester: Wiley.

Slob, W., Moerbeek, M., Rauniomaa, E., and Piersma, A. H. (2005). A statistical evaluation of toxicity study designs for the estimation of the benchmark dose in continuous endpoints. *Toxicological Sciences* 84: 167–185. https://doi.org/10.1093/toxsci/kfi004.

Stewart, W.H. and Ruberg, S.J. (2000). Detecting dose response with contrasts. *Statistics in Medicine* 19 (7): 913–921. https://doi.org/10.1002/(sici)1097-0258(20000415)19:7<913::aid-sim397>3.0.co;2-2.

Thayer, K.A., Melnick, R., Burns, K., Davis, D., Huff, J. (2005). Fundamental flaws of hormesis for public health decisions. *Environmental Health Perspectives*, 113(10), 1271–1276. https://doi.org/10.1289/ehp.7811

Tukey, J.W., Ciminera, J.L., and Heyse, J.F. (1985). Testing the statistical certainty of a response to increasing doses of a drug. *Biometrics* 41: 295–301.

Wong, W.K., and Lachenbruch, P.A. (1996). Tutorial in biostatistics. Designing studies for dose response. *Statistics in Medicine*, 15(4):343–59. https://doi.org/10.1002/(SICI)1097-0258(19960229)15:4<343::AID-SIM163>3.0.CO;2-F.

Index

Please note that page references to Figures will be followed by the letter 'f', to Tables by the letter 't'.

A Guide to Sample Size for Animal-based Studies, First Edition. Penny S. Reynolds.
© 2024 John Wiley & Sons Ltd. Published 2024 by John Wiley & Sons Ltd.